Plant-Parasitic Nematodes of Coffee

Plant-Parasitic Nematodes of Coffee

Ricardo M. Souza
Editor

Universidade Estadual do Norte Fluminense Darcy Ribeiro, Brazil

Springer

Editor
Prof. Ricardo M. Souza
Universidade Estadual do Norte Fluminense Darcy Ribeiro,
CCTA/Lab. Entomologia e Fitopatologia
Av. Alberto Lamego, 2000
Campos dos Goytacazes (RJ), Brazil
ricmsouza@censanet.com.br

ISBN: 978-1-4020-8719-6 e-ISBN: 978-1-4020-8720-2

Library of Congress Control Number: 2008930208

Cover picture: Histological section of a coffee root showing *Meloidogyne exigua* females and eggs (from Göldi, 1887) (published with permission)

Printed on acid-free paper

9 8 7 6 5 4 3 2 1

springer.com

This book is dedicated to
the best of my world: Claudia, Lara and Anya
to my mother, Maria, with whom I share an endless joy in reading and writing.
and to James G. Baldwin, my former PhD supervisor, and the professors in the Dept. of Nematology at the University of California at Riverside, who decisively contributed to my scientific formation.

Preface

When I conceived this book, what I had in mind was what I did *not* know about coffee-parasitic nematodes (CPNs). Indeed, after reading many papers and several chapters in books, I felt far from having a comprehensive understanding of the subject. Not only would it be a daunting task to retrieve the numerous articles, reports, theses and dissertations on CPNs published since 1878, but it would also be impossible to learn, on my own, from all the enormous experience acquired by nematologists and coffee growers in so many countries.

Therefore, this book is dedicated to those with restless minds, who want to know more about CPNs and their importance in coffee production worldwide. This book has been diligently written by top scientists in their areas of expertise or country, and it has been meticulously edited to guarantee precision without compromising an enjoyable read. I learned a lot from this book . . .I'm sure you will too.

Finally, I'd like to thank Zuzana Bernhart from Springer, who believed in this project and decided to publish it; Susan Casement, who revised all chapters for grammatical correctness; and all the contributors, without whom this book would never have became a reality.

Campos dos Goytacazes, RJ, Brazil — Ricardo M. Souza

Contents

Contributors

Amoncho Adiko

Laboratory of Nematology, Centre National de Recherche Agronomique, Abidjan, The Ivory Coast, amoncho.adiko@yahoo.com

Carlos Alberto Rivillas

Centro Nacional de Investigaciones del Café, Chinchiná, Colombia, carlos.rivillas@cafedecolombia.com.

François Anthony

Institut de Recherche pour le Développement, UMR RPB, Montpellier, France, anthony@mpl.ird.fr.

Francisco Anzueto

Anacafe, Guatemala City, Guatemala, franciscoa@anacafe.org.

Benoît Bertrand

Centre de Coopération Internationale en Recherche Agronomique pour le Développement, UMR RPB, TA A-98/IRD, Montpellier, France, benoit.bertrand@cirad.fr

Ricardo Bressan-Smith

Universidade Estadual do Norte Fluminense Darcy Ribeiro, Plant Physiology Laboratory, bressan@uenf.br.

Vicente P. Campos

Universidade Federal de Lavras, Lavras, Brazil, vpcampos@ufla.br.

Regina M.D.G. Carneiro

Embrapa-Recursos Genéticos e Biotecnologia, Brasilia, Brazil, recar@cenargen.embrapa.br

Lynn K. Carta

The United States Department of Agriculture/ARS, Nematology Laboratory, Beltsville, USA, lynn.carta@ars.usda.gov

Elis T. Cofcewicz
Embrapa-Recursos Genéticos e Biotecnologia, Brasilia, Brazil, eliscof@adylnet.com.br.

Hernando Cortina
Centro Nacional de Investigaciones del Café, Chinchiná, Colombia, hernando.cortina@cafedecolombia.com.

M. Dhanam
Central Coffee Research Institute, Coffee Research Station, Karnataka, India, wahdhanjhs@rediffmail.com.

Luiz Carlos C.B. Ferraz
Escola Superior de Agricultura Luiz de Queiroz/USP, Piracicaba, Brazil, lccbferr@carpa.ciagri.usp.br.

Alvaro Gaitán
Centro Nacional de Investigaciones del Café, Chinchiná, Colombia, alvaro.gaitan@cafedecolombia.com.

Philippe G. Gnonhouri
Laboratory of Nematology, Centre National de Recherche Agronomique, Abidjan, The Ivory Coast, philippe.gnonhouri@cnra.ci

Zafar A. Handoo
The United States Department of Agriculture/ARS, Nematology Laboratory, Beltsville, USA, zafar.handoo@ars.usda.gov

Adan Hernández
Procafe, Santa Tecla, El Salvador, aherna01@procafe.com.sv

Mário M. Inomoto
Escola Superior de Agricultura Luiz de Queiroz/USP, Piracicaba, Brazil, mminomot@carpa.ciagri.usp.br

Mirian P. Maluf
Embrapa Café, Campinas, Brazil, maluf@iac.sp.gov.br.

Josephine M. Namaganda
National Agricultural Research Organisation, Agricultural Research Institute, Kampala, Uganda, jnamaganda@kari.go.ug.

Claudio Marcelo G. Oliveira
Instituto Biológico, Campinas, Brazil, marcelo@biologico.sp.gov.br

Denis O. Seudieu
International Coffee Organization, London, UK, seudieu@ico.org.

Juliana R.C. Silva
Universidade Federal de Lavras, Lavras, Brazil, jucampos2006@hotmail.com.

Andrea M. Skantar
The United States Department of Agriculture/ARS, Nematology Laboratory, Beltsville, USA, andrea.skantar@ars.usda.gov

Ricardo M. Souza
Universidade Estadual do Norte Fluminense Darcy Ribeiro, Entomology and Plant Pathology Laboratory, Campos dos Goytacazes, Brazil, ricmsouza@censanet.com.br

K. Sreedharan
Central Coffee Research Institute, Coffee Research Station, Karnataka, India, crsento@yahoo.com.

Loang K. Tran
Western Highlands Agro-forestry Scientific and Technology Institute, Buon Ma Thuot City, Vietnam, bauloang@dng.vnn.vn.

Henrique D. Vieira
Universidade Estadual do Norte Fluminense Darcy Ribeiro, Plant Science Laboratory, Campos dos Goytacazes, Brazil, henrique@uenf.br.

Luc Villain
Centre de Coopération Internationale en Recherche Agronomique pour le Développement TA A-98/IRD, Montpellier, France, luc.villain@ird.fr

Soekadar Wiryadiputra
Indonesian Coffee and Cocoa Research Institute, Jember, Indonesia, soekadar@yahoo.com

Introduction

In 1878 the French naturalist Clément Jobert reported a disease affecting coffee plantations in the then Province of Rio de Janeiro, Brazil. Although he identified the causal agent, it was not until 1887 that the Swiss naturalist Emil A. Göldi described *Meloidogyne exigua*, as part of an extensive publication on that disease (see Chapter 12).

Since then, coffee-parasitic nematodes (CPNs) have grown to become a serious problem for coffee cultivation in many regions of the world, in which the extent of their direct and indirect impacts is yet to be estimated. Indeed, since the nineteenth century coffee cultivation has provided the first economic momentum of many tropical regions or whole countries. In recent decades, although industrialization and agricultural diversification have reduced the role of coffee trading in national GDPs, coffee cultivation remains crucial for the economic and social stability of millions of people across the globe. Under these circumstances, from the presumed yield losses that occur in the vast regions where no nematologists work, to the well-reported widespread decimation of plantations in Brazil, CPNs ought to be one of the most important nematode groups worldwide.

Despite their importance, CPNs have never until now been the subject of an in-depth review, in which hundreds of reports, papers published in national and international journals, dissertations and theses are critically examined. Instead of an individual work, a review prepared by several contributors provides different perspectives on CPNs, enriched by different educational backgrounds and by a broad range of expertise and research experiences. Furthermore, the review should also be a window to the nematode problems faced by coffee growers from several countries, and to the research efforts of and results obtained by these countries' nematologists.

This exchange of information is all the more important as one considers the technical and language difficulties that are still hurdles to the traffic of ideas and materials between nematologists located in tropical countries. All sort of difficulties, including poor internet connections, a lack of resources for foreign travel, the labyrinth of research funding and bureaucracy have created the present situation: there is virtually no international collaboration between nematologists dedicated to CPNs. Even in Brazil, where these nematodes have been studied for decades and by a sizable group of nematologists, virtually no one is aware of the nematode problems faced by coffee growers in Africa or Asia, for example, nor are they aware of the work performed by nematologists there.

The first chapter of this book introduces coffee – the plant and its cultivation - to those not familiar with it, providing a background for understanding many aspects of CPNs, such as their biology, interaction with their hosts, epidemiology and management.

In chapter 2, nematologists who often work on specific aspects of CPNs are invited to visit the evolution of the world coffee industry since the early twentieth century, and to see how its different phases and crises have influenced coffee cultivation and trading, research funding and technical support for growers.

From chapter 3 through 10, basic and applied aspects of the most damaging nematodes to coffee, *Meloidogyne* spp. and *Pratylenchus* spp., are discussed by top specialists in their areas of expertise. Chapter 11 reviews the information available on the many other nematode genera and species that have been reported associated with or as parasitic to coffee.

From chapter 12 through 17, nematologists from several countries review the landmarks in nematological work on CPNs in their countries, and present their research efforts, results and prospects.

Part I
The Crop

Chapter 1
Coffee: The Plant and its Cultivation

Henrique D. Vieira

Abstract This chapter aims to introduce coffee (*Coffea* sp.) to those not familiar with it, as a platform for understanding the following chapters. Initially, a few interesting events in coffee history are outlined, followed by diagrams and color images that explain aspects of coffee botanics that are directly related to production. The most important *Coffea* species, for production or breeding, are described. Important features of coffee cultivation, such as soil preparation, seedling production, harvest and postharvesting processing, are explained. A comparative discussion is carried out on the most important technological aspects of this crop, such as full sun *vs* shaded cultivation systems, arabica *vs* robusta coffee production and low *vs* high technological input.

Keywords Coffee origin · coffee cultivation · *Coffea* diversity · coffee botanics · coffee world production

1.1 Introduction

The word 'coffee' is probably derived from the former Kingdom of Kaffa (today part of Ethiopia), where coffee (*Coffea* sp.) was first cultivated from around the fifth to the eighth century. From its legendary origin in the Ethiopian highlands, the beverage was introduced into the Arab world through Egypt and Yemen, where it became widely consumed since alcoholic drinks were not allowed. In the Yemen, coffee was being cultivated commercially around the fourteenth century. It was introduced into Europe through Venice, and despite complaints about the 'Muslin beverage', its consumption slowly spread through this continent, the Americas and Asia (Neves, 1974; Anonymous, 2004).

Despite efforts from the Arabs to control coffee cultivation – by prohibiting the export of unroasted beans and seedlings – in the early eighteenth century the Dutch

H.D. Vieira
Universidade Estadual do Norte Fluminense Darcy Ribeiro/CCTA/LFIT, Campos dos Goytacazes, Brazil
e-mail: henrique@uenf.br

started its cultivation in their Asian and South American colonies, as did the French in the Caribbean. Today, coffee is cultivated in dozens of tropical countries, supporting regional or national economies (see Chapter 2). Coffee consumption *per* capita has increased, driven by its property of increasing the alertness of those who drink it and by the pleasant ambience it fosters when it is drunk socially. Many reports exist on its benefits to health when consumed moderately (Ascherio et al., 2001; Van Dan and Feskens, 2002; Encarnação and Lima, 2003).

This chapter focuses on introducing coffee – the plant, its cultivation and postharvest processing – to those who are not familiar with it; hence, aspects of botanics, diversity and agronomic practices are outlined to provide a background to the chapters that follow. Text and images have, therefore, been combined in the hope that reading this will be as enjoyable as drinking a good cup of coffee.

1.2 Coffee Botanics

Depending on the species, coffee grows as a perennial shrub or tree, with an extensive root system concentrated on the 0–60 cm soil zone, although roots are found growing down to three meters deep (Fig. 1.1). The distribution of the root system may nonetheless be altered by factors such as water availability and soil structure (Rena et al., 1986; Rena and Guimarães, 2000).

Above ground, arabica coffee (*C. arabica* L.) typically presents one main trunk; 'suckers' may appear but they are usually pruned. Robusta coffee (*C. canephora* Pierre ex A. Froehner) is typically multi-trunk. In both species, orthotropic branches grow vertically from the trunk; from these, the plant emits more or less horizontal plagiotropic branches, on which blooming and production occur (Figs. 1.2; 1.3A). Through trimming and pruning, the plant's natural architecture may be altered (Wormer and Gituanja, 1970).

Most coffee species have persistent leaves, although defoliation may occur because of abiotic (such as drought) or biotic (such as disease) stresses. Such defoliation is inversely related to production, and may be responsible for yield losses of up to 20%. Leaves are continuously emitted, but climate pattern and occasional stressful weather conditions determine when new leaf flushes occur (Gindel, 1962; Barros and Maestri, 1972).

Hermaphrodite flowers are emitted in inflorescences on the axiles of plagiotropic branches (Figs. 1.3B; 1.4A). Therefore, any factor that compromises the development of these branches will affect production. In a given geographic area, all plants bloom synchronously (Fig. 1.4B). The number of times plants bloom *per* year depends on the region's latitude and rainfall pattern; for example, in southeast Brazil, where marked dry and rainy seasons occur, the plants bloom two to three times/year, while in equatorial, rainy Costa Rica the plants may bloom up to fifteen times/year (Alvin, 1960). This has major implications for harvesting and control of pests and diseases. For arabica coffee, one important aspect in the relationship between or biotic stresses (including nematodes) and productivity is the fact

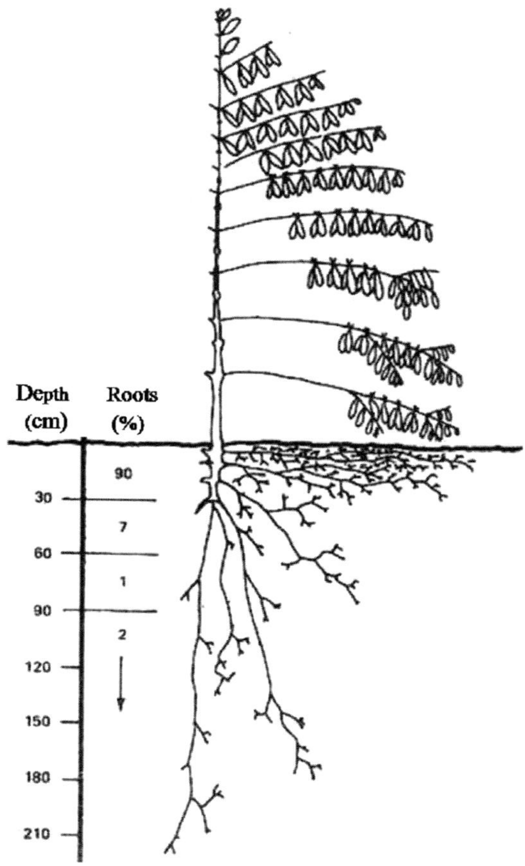

Fig. 1.1 Schematic representation of the root system of a cultivated coffee plant (from Rena, 1986, with permission)

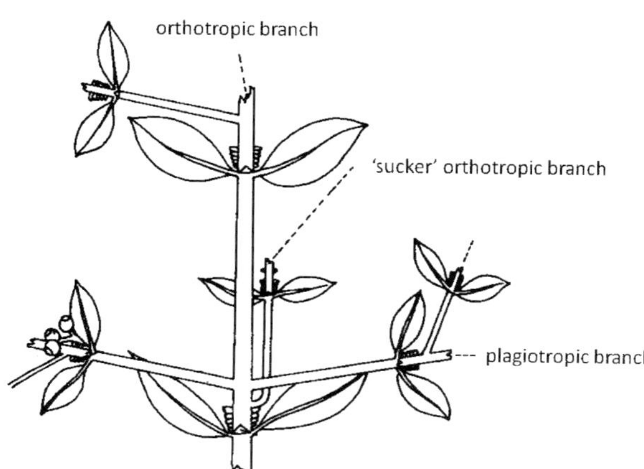

Fig. 1.2 Schematic representation of the aerial part of a cultivated coffee plant (adapted from Wormer and Gituanja, 1970, with permission)

A

B

Fig. 1.3 Coffee blooming and production. (**A**) on horizontal plagiotropic branches (Photo by H. Vieira). (**B**) anatomic details (from Köhler, 1887) (see color Plate 1, p. 315)

Fig. 1.4 Coffee blooming. (**A**) inflorescence on the axiles of a plagiotropic branch (Photo by F. Partelli, with permission). (**B**) synchronous blooming (Photo by H. Vieira) (see color Plate 2, p. 316)

that blooming occurs on the plagiotropic branches grown in the previous year. On robusta plants, blooming occurs on the branches grown in the current year (Dean, 1939; Moens, 1968).

In arabica coffee, ripe fruits ('berries') are red or yellow (Fig. 1.5A,B), with orange ones indicating cross pollination; in robusta plants, more hues occur. The format of the fruit, nearly round to oblong, varies with the *Coffea* species; the size of the fruit and of its endosperm (the 'bean') varies with the cultivar or variety planted and cultivation conditions. Usually two beans are produced/fruit (Fig. 1.3B). Most importantly, the bean contains proteins, caffeine, oils, sugars, dextrine, chlorogenic acid and several other substances that will determine the characteristics of the beverage; this will also be influenced by aspects of harvesting, processing and bean roasting (Rena et al., 1986).

Fig. 1.5 *Coffea* species. (**A, B**) *C. arabica*. (**C**) *C. dewevrei*. (**D**) *C. stenophylla* (Photos by H. Vieira) (see color Plate 3, p. 317)

1.3 Coffee Diversity

The genus *Coffea* belongs to the family Rubiaceae, being composed of 103 species (Davis et al., 2006). These are divided in the sections Eucoffea, Mascarocoffea, Argocoffea and Paracoffea; the first three originate from Africa and the latter from Asia. The section Eucoffea is the only one with economic and breeding relevance, for it includes arabica and robusta coffees as well as the species discussed below. In natural conditions, most *Coffea* species occur in tropical Africa, particularly in Madagascar and mainland surrounding countries. Some species occur in India. Part of *Coffea* sp. diversity has been preserved in germplasm banks, and a fraction of it has been screened for nematode resistance (see Chapter 9).

Apart from *C. arabica*, all species are diploids (2n = 22); the exception is probably a natural tetraploid hybrid (2n = 44), and it is autogamous, although about 10% of cross pollination occurs. *C. arabica* and *C. canephora* are virtually the only commercially cultivated species, with the former representing 70% of world production. Many cultivars, mutants and hybrids of arabica coffee are grown throughout the world or used in breeding programs (see Chapter 9); the same occurs with robusta (Carvalho, 1958; Medina Filho et al., 1984).

According to some authors, *C. congensis* A. Froehner may be one of the parentals that gave rise to *C. arabica*. That species and *C. liberica* W. Bull ex Hiern are cultivated in limited areas in Africa and Vietnam. *C. racemosa* Ruiz and Pav. is appreciated in Mozambique, being deciduous and remarkably resistant to high temperatures and drought; some plants are resistant to 'leaf miner'(*Leucoptera coffeella* Guerin-Mèneville and Perrottet). Because *C. dewevrei* De Wild. and T. Durand (Fig. 1.5C) produces poor beverage, it is not commercially cultivated; nonetheless, it is considered important for breeding programs due to its adaptability to poor soils and drought. Likewise, *C. eugenioides* S. Moore is not produced commercially, but it is maintained in germbanks as a repository of genes to be transferred to *C. arabica*. *C. stenophylla* G. Don (Fig. 1.5D) is of interest for its resistance to 'leaf miner' (Chevalier, 1947; Carrier, 1978; Bridson, 1982).

1.4 Coffee Cultivation

Special attention should be paid to agronomic and phytosanitary aspects of coffee seedlings, since the plantation is expected to have a life-span of at least 20 years. Seedlings are produced through seeds (in the case of the autogamous arabica coffee, Fig. 1.6A) or vegetative cloning from orthotropic branches (Fig. 1.6B), which is recommended for the allogamous robusta coffee to reduce variability in the plant stand. Alternatively, grafted seedlings may be produced (Fig. 1.6C,D,E), combining an arabica scion with a robusta rootstock, which may have been selected for nematode resistance (Matiello et al., 2005; Ferrão et al., 2007; see Chapter 9).

The necessary operations involved in establishing a plantation vary according to the previous use of the area, topography and availability of equipment and implements. In the full sun cultivation system (see below), the area is cleaned of

Fig. 1.6 Coffee seedling production and cultivation. (**A**) nursery. (**B**) seedlings vegetatively produced from orthotropic branches. (**C, D**) grafting of seedlings. (**E**) grafted seedling. (**F**) full sun cultivation (Photos by H. Vieira) (see color Plate 4, p. 318)

vegetation, the soil may be plowed, disced and receive fertilizers. In upland plantations, special care must be taken to establish the plantation along contour lines. In the shaded cultivation system, the original vegetation is maintained and its canopy is managed to allow suitable amounts of sunlight to reach coffee plants (Rena et al., 1986; Matiello et al., 2005).

The recommended plant density/hectare (ha) varies with the cultivar or variety planted, soil topography and fertility, climate and available labor. Generally speaking, higher densities reduce the productivity *per* plant and increase it in terms of area used; on the other hand, higher densities create a microclimate that is favorable to 'leaf rust' (caused by *Hemileia vastratrix* Berk and Br.) and the 'berry borer' (*Hypothenemus hampei* Ferrari); no relationship between plant density and infestation has been established for nematodes. In full sun, plant density varies from three to 10 thousand plants/ha. In the shade, plant density is even more variable. Currently, there is a tendency to plant at higher densities in a number of countries, such as Brazil, Colombia and Mexico.

With regard to exposure to sunlight, there exists a great divide in coffee cultivation. Virtually all plantations in Brazil are in full sun (Fig. 1.6F), which presents higher productivity/plant and area in comparison to the shaded system; it also allows mechanization and intercropping (Fig. 1.7A). This system has been introduced in countries where shaded plantations (Fig. 1.7B) is predominant, such as those in Central America, particularly Colombia. Full sun plantations are nevertheless exposed to higher risks of hydric stress; in regions of higher technological input, irrigation may be used (Fig. 1.8A).

Most coffee plantations in Central America, India, Vietnam and Indonesia are shaded (see Chapters 13–16). This system is more commonly adopted in regions of accentuated topography, low technological input or where coffee is just one of several crops cultivated by smallholders. It has the advantage of causing less environmental disturbance and providing protection from soil erosion (Rena et al., 1986).

The coffee industry has yet another divide: arabica and robusta coffees. The former is better adapted to higher altitudes and milder climate; it has higher market value and provides a better beverage. However, the most commonly grown cultivars and varieties are susceptible to leaf rust and root-knot nematodes (*Meloidogyne* sp.). In comparison, robusta coffee is better adapted to hydric deficit; it is resistant to 'leaf miner' and 'leaf rust', but more susceptible to mites, 'berry borer' and *Colletotrichum* spp. It is more often used to produce instant coffee, or it is mixed with arabica coffee to produce 'blends' (Anonymous, 1985; Matiello et al., 2005).

As regards the production system, throughout the world coffee is cultivated under a variety of agronomic practices and input levels. For example, the plant architecture may be left unmanaged, or the grower may trim or prune the plants routinely or when he is trying to recover a plantation that has suffered abiotic or biotic stresses. Robusta coffee plants are more often trimmed than arabica ones so as to manage the former's multi-trunk habit, and to facilitate harvesting. For example, in India robusta plants are continuously trimmed to keep them short and easy to harvest (Fig. 1.8B). In Vietnam, plants are trimmed so that plagiotropic branches are emitted from the

A

B

Fig. 1.7 Coffee cultivation. (A) full sun plantation intercropped with beans (Photo by F. Partelli). (B) shaded plantation (Photo by K. Sreedharan, with permission) (see color Plate 5, p. 319)

plant's top; upon production, these branches incline downward, giving the plant the aspect of an open umbrella (Jansen, 2005).

As regards technological input, coffee plantations may be managed entirely without fertilization, irrigation or pest and disease control. In most regions, such inputs vary according to the traditions of coffee cultivation, the grower's financial resources and the prospects of profit from upcoming harvests; naturally, the growers' profits are greatly influenced by the world coffee market (see Chapter 2). In some areas in Brazil, plantations receive a high technological input, which includes routine fertilization, proper control of pests and diseases, and irrigation. Alternatively, 'organic' coffee, which receives low agrochemical-input, is being increasingly produced in Brazil and other countries, despite technical difficulties, high cost of certification and labor and reduced productivity. Mexico remains the largest producer of 'organic' coffee.

Fig. 1.8 Coffee cultivation and harvest. (**A**) plantation being irrigated (Photo by D. Barbosa, with permission). (**B**) harvesting of robusta coffee (Photo by K. Sreedharan, with permission) (see color Plate 6, p. 320)

1.5 Coffee Harvesting and Processing

Harvesting is the most important operation in coffee cultivation. When done by hand, it employs 50% of the man-hours required by this crop, and it represents 25–35% of the production cost. It also has a strong influence on the quality of the beverage obtained. The harvest season varies with the region's climate, rainfall and the cultivar or variety grown. For example, in Brazil most plantations are harvested from June through September (the dry season); occasionally, harvesting may take place from March through May, or in November and December.

Ideally, only ripe coffee berries should be harvested because they provide the best beverage. Nonetheless, in most production systems practical constraints lead the growers to conduct a less selective harvest, which includes unripe and overripe berries. These should not represent more than 20% of the production if a high quality beverage is to be produced. The grower should also pay attention to dirt, debris, insect-bored or defective berries which compromise product classification and the grower's revenue.

In Brazil, 90% of the plantations are harvested manually; the berries are stripped from the plant branches (Fig. 1.9A) and fall on the ground, into baskets or on fabric or plastic strips laid under the plant (Fig. 1.9B,C). Letting the berries fall on the ground is not recommendable because dirt, debris, moldy and rotten berries end up being collected as well.

In many countries, harvesting is a nearly continuous operation because the plants bloom several times a year, which results in marked inconsistencies in the ripeness of berries collected; in these cases, stripping the trees results in a high percentage of unripe berries mixed with ripe ones. In such cases, growers selectively pick ripe berries only. Although this system requires much labor, the product reaches a better market price, and its consistent quality results in a top-quality beverage.

In Brazil, mechanical harvesting (Fig. 1.10A,B) has been increasingly used because it is so difficult to hire, manage and pay the large labor force required for manual harvesting; operational costs may drop by 40%. Mechanical harvesting is more suitable for medium to large plantations in areas with slopes of up to 20% incline (Matiello et al., 2005).

Upon harvesting, the berries undergo either dry or wet processing. In the former, debris and some of the damaged berries are eliminated through flotation in washing channels. Right after this, the berries are spread out on terraces and turned several times a day until they have dried evenly under the sun (Fig. 1.10C). Depending on weather conditions, this process may take weeks to complete, during which time mold and bacteria must not develop on the berries. Alternatively, drying machines may be used to quicken this process.

In the wet processing method, debris and part of the damaged berries are eliminated in washing channels. The berry's outer layer and part of its pulp is mechanically removed; the remaining pulp is usually removed by fermentation and washing. Therefore, in this method, the coffee beans, not the berries, are sun dried.

After being sold by the growers, the beans undergo further processing, which is generally conducted by industry: hulling, polishing, cleaning, sorting by size, density or color, grading, roasting and grinding, which results in top-quality coffee beans (Fig. 1.10D,E,F).

1.6 Coffee Production Worldwide

About 60 tropical and subtropical countries (Fig. 1.11) produce coffee extensively, with 21 of these producing over one million 60 kg-bags/year; the top 15 producers are listed in Table 1.1. By continent, about 60% of the coffee produced comes from the Americas, 24% from Asia, 14.5% from Africa and 1.5% from Oceania (Matiello et al., 2005; Anonymous 2008b).

As regards types of coffee grown, arabica coffee is largely predominant in the Americas, although Brazil has reached the mark of seven to nine million bags/year of robusta coffee. In Africa, 60% is robusta coffee, which is also predominant in Asia.

A

B

C

Fig. 1.9 Coffee harvest. (**A**) strip harvesting (Photo by F. Partelli, with permission). (**B, C**) harvested coffee in basket and fabric strip, respectively (from Anonymous, 1985, with permission) (see color Plate 7, p. 321)

Fig. 1.10 Coffee harvesting and processing. (**A, B**) mechanical harvesting (from Anonymous, 1985, with permission). (**C**) coffee berries being sun dried. (**D, E, F**) damaged, high grade and roasted coffee beans, respectively (Photos by H. Vieira) (see color Plate 8, p. 322)

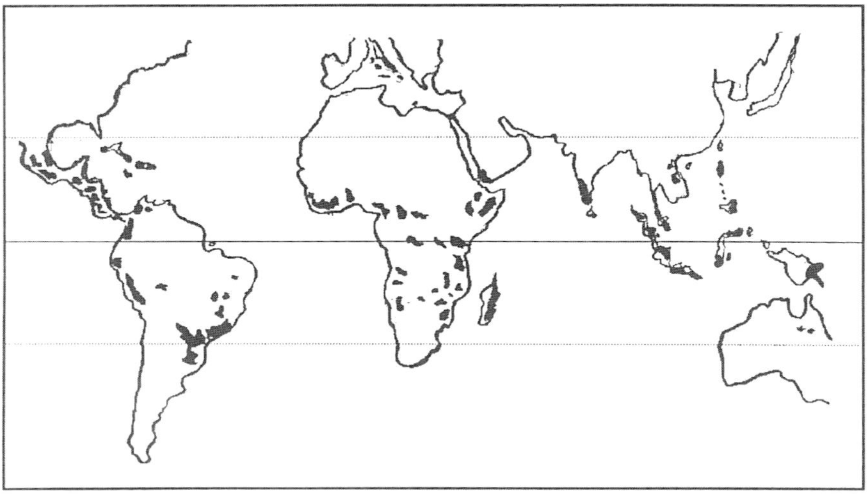

Fig. 1.11 World inter-tropical coffee-growing region (adapted from Matiello et al., 2005, with permission)

Table 1.1 Ranking of the top 15 coffee-growing countries according to 2006/2007 data, and their proportion of arabica and robusta production

Countries	Arabica coffee (%)	Robusta coffee (%)	Production (2006/2007) (× 1,000 60-kg bags)[a]
Brazil	65	35	38,000
Vietnam	10	90	13,200
Colombia	100	0	11,000
Indonesia	10	90	6,600
India	40	60	4,800
Mexico	97	3	4,100
Ethiopia	100	0	4,000
Guatemala	90	10	3,700
The Ivory Coast	0	100	2,900
Uganda	10	90	2,900
Peru	100	0	2,900
Honduras	100	0	2,900
Costa Rica	100	0	2,000
El Salvador	100	0	1,500
Nicaragua	100	0	1,400

[a]Anonymous (2008b).

In the last 30 years, world coffee production has increased at the rate of about one million bags/year, from 65–70 million in the early 1970s to 110 to 115 million nowadays. It is forecast that production will soon reach 120 million bags. This rise in production has not been matched by demand, which has caused a downward trend in international coffee prices for nearly a decade; this has had major consequences for the whole industry (Matiello et al., 2005; Anonymous, 2008a; see Chapter 2).

References

Alvin PT (1960) Physiology of growth and flowering in coffee. Turrialba 2:57–62.

Anonymous (1985) Cultura do café no Brasil: manual de recomendações. MIC/IBC/GERCA, Rio de Janeiro.

Anonymous (2004) Café no Mundo. In: Café. Consórcio Brasileiro de Pesquisa e Desenvolvimento/Café, Brasilia.

Anonymous (2008a) Consumo de café torrado, moido e solúvel no Brasil. http://www.cicbr.org.br/pensa/tela1.php?ic=2&is=3&c=demanda&ip=36. Visited on March 25th 2008.

Anonymous (2008b) Primeiro levantamento de café 2007/2008. http://www.conab.gov.br/conabweb/download/ safra/3boletim cafe.pdf. Visited on Feb 15th 2008.

Ascherio A, Zhang SM, Hermán NA et al (2001) Prospective study of caffeine consumption and risk of Parkinson's disease in men and women. Ann Neurol 50:56–63.

Barros RS, Maestri M (1972) Periodicidade do crescimento em café. Rev Ceres 19:424–448.

Bridson DM (1982) Studies in *Coffea* and *Psilanthus*. Kew Bull 36:817–859.

Carrier A (1978) La structure genetique des caféiers spontanes de la region Malgashe (Mascarocoffea). Leur relations avec les caféiers d'origine africaine (Eucoffea). Mem ORSTOM 97:1–223.

Carvalho A (1958) Advances in coffee production technology. Recent advances in our knowledge of coffee trees. 2. Genetics. Coffee Tea Ind 81:30–36.

Chevalier A (1947) Lés caféiers du globe. III Systhematique des caféiers et fauxcaféiers, maladies et insects nuisibles. Paul Lechevalier, Paris.

Davis AP, Govaerts R, Bridson DM et al (2006) An annotated taxonomic conspectus of the genus Coffea (Rubiaceae). Bot J Linnean Soc 152:465–512.

Dean LA (1939) Relationship between rainfall and coffee yields in the Kona district Hawaii. J Agric Res 59:217–222.

Encarnação RO, Lima DR (2003) Café e Saúde Humana. EMBRAPA/CAFÉ, série documentos, Brasilia.

Ferrão RG, Fonseca AFA, Bragança SM et al (2007) Café Conilon. INCAPER, Vitória.

Gindel L (1962) Ecological behavior of the coffee plant under semi-arid conditions. Turrialba 4:49–63.

Jansen AE (2005) Plant Protection in Coffee. Recommendations for the Common Code for the Coffee Community. Deutshe Gesellschaft Für Technische Zusammenarbeit (GTZ) GmbH, Eschborn.

Köhler FE (1887) Medizinal Pflazen. Verlag von Fr. Eugen Köhler, Gera-Untermhaus.

Matiello JB, Santinato R, Garcia A et al (2005) Cultura do café no Brasil. Novo manual de recomendações. MAPA/SARC/PROCAFÉ – SPAE/DECAF, Rio de Janeiro.

Medina Filho HP, Carvalho A, Sondahl MRI et al (1984) Coffee breeding and related evolutionary aspects. Plant Breed Rev 2:157–193.

Moens P (1968) Investigaciones morfológicas, ecológicas y fisiológicas sobre cafetos. Turrialba 18:209–233.

Neves C (1974) A estória do café. Instituto Brasileiro do Café, Rio de Janeiro.

Rena AB, Guimarães PTG (2000) Sistema radicular do cafeeiro: estrutura, distribuição, atividade e fatores que influenciam. EPAMIG, série documentos, Belo Horizonte.

Rena AB, Malavolta E, Rocha M et al (1986) Cultura do cafeeiro: fatores que afetam a produtividade. POTAFOS, Piracicaba.

Van Dan RM, Feskens, EJ (2002) Coffee consumption and risk of type 2 diabetes mellitus. Lancet 390:1477–1478.

Wormer TM, Gituanja J (1970) Seasonal patterns of growth and development of arabica coffee in Kenya. II. Flower initiation and differentiation in coffee East of the Rift Valley. Kenya Coffee 35:270–277.

Chapter 2
The Coffee Industry: History and Future Perspectives

Denis O. Seudieu

Abstract This chapter focuses on changes which have characterized the world coffee industry since its development as a marketable commodity, and the impact of these changes on coffee research. Three main periods have been identified through these changes. The first one is the free market, with Brazil dominating it until the early 1950s; this was followed by the period of controlled market within the frame of international cooperation between exporting and consuming countries (1960s through 1980s); the third period is the current free market situation within the framework of international cooperation, which started in mid-1989. During the first period, efforts to increase yields were undertaken through scientific research supported mainly by Governments. The public sector in Brazil and Colombia was the major driver of research and development in the coffee industry. In the second period, also known as the post-war period, the increased investment in agricultural research encouraged the development of new techniques for intensive production and better management of nematodes, pests and diseases. To address price fluctuations, governments set up price regulation mechanisms through international cooperation, creating the International Coffee Organization to manage it. Governments and their parastatals were driving coffee industry in producing countries and specialized assistance was available to farmers; in many countries research institutions benefited from substantial funding. The current period is characterized by the return to a free market, with the government withdrawing from the coffee industry. In many countries this new environment has weakened research institutions and extension services, since the private sector has not been prepared to replace the government in providing core services.

Keywords Coffee market changes · coffee research · coffee regulation · coffee production

---

D.O. Seudieu
International Coffee Organization, London, United Kingdom
e-mail: seudieu@ico.org

R.M. Souza (ed.), *Plant-Parasitic Nematodes of Coffee*,

2.1 Introduction

The economic performance and development prospects of many developing countries are largely dependent on commodity exports. The heavy dependence of these countries on a few commodities exposes them to adverse economic impacts, sometimes with harmful consequences for growth and poverty reduction. As a labour-intensive crop, coffee (*Coffea* sp.) is one of the main generators of employment in producing countries, and it plays a vital role in their social structure and development. Over 60 countries in Latin America, Africa and Asia/Oceania produce coffee, providing a livelihood for some 25 million coffee farming families (Clarence-Smith and Topik, 2003).

Since mid-1998 the world coffee market has experienced a significant imbalance in supply and demand resulting in a sharp fall in prices. This situation has led to a serious deterioration in the living conditions of a large number of coffee growers who depend on coffee for most of their income. The purpose of this article is to review historical changes in the world coffee industry and analyze prospects for the future. Three main periods characterize the development of the world coffee industry: the period of the free market, which lasted until the early 1950s, in which one producing country dominated the market; the period of controlled market with international cooperation between exporting and consuming countries (1960s through 1980s); and the current period of the free market within the framework of international cooperation, which started in mid-1989.

2.2 The Era of the Free Coffee Market

Until the early 1950s, coffee was cultivated in a limited number of countries, with average world production being less than 40 million bags. This crop had been a major export from Latin America, shaping both the economy and the natural landscape of the region. Latin American countries dominated coffee production and exports, with Brazil as the main actor. Producing countries, notably Brazil and Colombia, were backed by banks to regulate the supply of coffee. In other words, the governments of these countries intervened to maximize export revenues or to act as last–resort buyers in times of surplus production. For example, Brazil tried many times to buoy up world coffee prices by holding back stocks, having launched its last unilateral price support scheme in 1953/1954. Roasting and retailing sectors were still relatively fragmented.

To meet the needs of a growing number of coffee consumers, efforts to increase production or yields were made through scientific research supported by governments, private sector and international research bodies. The public sector in Brazil and Colombia was at the forefront of research and development efforts to expand coffee production, which started in Brazil with the creation of the Imperial Agronomic Station of Campinas in 1887, followed by several research institutes created later in the 1920s (Beintema et al., 2001). In Colombia, research activities were introduced by the National Federation of Coffee Growers with the creation of the National Coffee Research Center (abbreviated Cenicafe in Spanish) in 1938 (Anonymous, 1998; Beintema et al., 2000). Research institutions were created in other countries as well, such as the Mysore Coffee Experimental Station in India in 1925 and the Institut Français du Café et du Cacao in France (1958). It may be noted that this Institute conducted research on coffee and cocoa in many African countries, including the Ivory Coast and Cameroon (Priovolos, 1981).

2.3 Era of the Controlled Coffee Market

Profound political, economic, social and technological developments in the 1950s and 1960s combined to bring about a new era for the coffee industry. As far as technological development is concerned, before the 1950s efforts to increase coffee yields were based mainly on traditional methods using selection techniques. During the post-war period, sophisticated breeding techniques were introduced as well as intensive production methods to increase yields. Agricultural research expenditure increased in the 1960s through 1980s. Similarly, insect pest problems were addressed by a wide range of pesticides as well as new techniques to manage the problems of nematodes and diseases. Various control measures against pests and diseases in coffee such as 'leaf rust' (caused by *Hemileia vastatrix* Berk et Br.), 'black rot' (*Pellicularia koleroga* Cook), 'red root' [*Ganoderma philippi* (Bres and P. Henn.) Bres.] and 'coffee wilt' (*Fusarium xylarioides* Steyaert) were adopted. It is important to note that research was also boosted by overseas research institutions in the former colonizing countries. Most national research institutions in producing countries continue to be supported by their counterparts in former colonial powers such as France and the United Kingdom. With their assistance, the management of major diseases and pests has been facilitated in many producing countries.

During the post-war period, the colonial powers like France, the United Kingdom, and Belgium encouraged their overseas territories to increase coffee production with the dual purpose of creating alternative sources of supply within their currency zones and strengthening their economies by developing coffee as a key cash crop. New trade links were formed and, with the end of the colonial era, Africa emerged as a major supplier for the European market in competition with Latin America, which had previously dominated coffee exports.

This new era may be called the era of processed coffee. During this time we witnessed the emergence of soluble coffee and vacuum packaging, the advent of the supermarket and high-powered selling of brands. The clash between traditional methods of trading, roasting and retailing and the new techniques of mass marketing, backed by national advertising campaigns, was intense. The roasting industry became concentrated in fewer hands and big multinational companies emerged. As major purchasers, these companies came to exert a powerful influence on world trade.

The world coffee market has been traditionally subject to substantial short-term fluctuations for a number of reasons, including economic situations, supply variations, climatic shocks, low elasticity of supply and demand relating to prices or revenues, and a time lag in supply response to major price movements. Just after the Korean War, prices rose to unprecedented heights. But in the second half of the 1950s and early 1960s, they fell drastically due to overproduction. Faced with the collapse of prices, producing countries sought to defend their economies through joint defence strategies. This gave rise in the 1950s and 1960s to moves to regulate the coffee market through international agreements. The origins of the International Coffee Organization (ICO) reflect these circumstances of the 1960s. The switch from a free market to a controlled economy had the support of many importing countries. It fitted in with the then current United Nations thinking; the colonial powers wanted to help their former territories emerging as newly independent nations and the United States, traditionally a bastion of free trade, had embarked on a policy of Western hemisphere solidarity (Lucier, 1988).

International coffee Agreements marked a major departure from traditional trading methods. The power players on the world scene were now Governments and their parastatals. During this period, technical assistance was available to farmers as subsidies from Governments contributed to funding research and extension services. This era was characterized by major research funding in many producing countries, and it was favourable to the social stability of coffee producing regions, resulting in good and stable profitability for coffee growers. This system has benefited many farmers in almost all coffee producing countries, although some negative experiences were recorded where cumbersome public administration affected many farmers.

It appears that a stable political and economic environment is fundamentally important to the successful development of a country's coffee sector. Many countries that have faced years of civil war or political upheaval have experienced a serious decline in their coffee industry. Angola, the Democratic Republic of Congo and the Ivory Coast illustrate the negative impact of political instability.

On the exporting side, this era elevated to new importance the parastatal Coffee Boards and 'Caisses de stabilisation'. On the importing side, the big roasters became key advisers. The work of the ICO from 1963 to 1989 in general covered the periods of relatively low but stable prices in which the economic clauses of the Agreements, involving a system of export quotas, were fully operative for market regulatory purposes, as well as the periods of high prices in which the free market prevailed.

During this period the ICO diversification fund helped to reduce individual countries' dependency on coffee and created an awareness of the need to plan production within the context of national economic policy. There was also cooperation in marketing coffee through operations financed by the ICO Promotion Fund in Japan, Europe and the United States, as well as the information programs on coffee and health.

2.4 Era of the Free Market Within the International Cooperative Framework

The regulatory role of the ICO began to decline in the late 1980s, after more than two decades of success, because setting prices by means of non-market control mechanisms (quotas) gave rise to overproduction, and developing countries naturally tried to maximize their revenue by increasing production, regardless of the fact that demand did not necessarily match supply. Uncontrolled production, together with the ascendancy of the liberal view that perhaps the market itself could best adjust prices in the medium and long term, contributed to the ICO losing some of its ability to intervene in the market.

Since 1989, we have moved back to a free market situation as the world coffee market has undergone far-reaching changes affecting production, consumption and trade. Coffee producers are much more exposed to market forces than in the past. However, it would be a mistake to think that the clock has been turned back to a pre-quota age. The transformations that took place in the 1960s through 1980s still have a big impact on the current situation. The international political, social and economic ambience in which we now live is very different from that of 40 years ago. The pattern of production and consumption has changed. Asia has emerged as one of the major producing regions, with Vietnam becoming the world's second largest producer and exporter. Some producers such as Brazil have improved the structure of their coffee industry to become more efficient with a reasonably low production cost. Increasingly, more emphasis is being given to quality, and the 'gourmet' sector of the market has become a growth area.

In many other producing countries, the internal marketing system and the whole coffee industry have been liberalized. In most countries, commodity sectors that were previously insulated from world market price developments (market competition) have become part of the world commodity economy, exposing the market participants to new, unfamiliar and significant competitive pressure, and to price risks that were previously absorbed by government entities. The withdrawal of governments and the fragmentation of the marketing systems have led credit systems to collapse in many countries, mainly in Africa, negatively affecting productivity and forcing farmers to sell their product directly after the harvest, thus exposing them to the vagaries of seasonal price behavior. In many cases, extension systems were weakened and the budget for research institutions substantially reduced due to the withdrawal of governments from the sector. In the Ivory Coast, prior to liberalization, the state body controlled input supply to farmers for export crops. Fertilisers, pesticides, and seeds were supplied to farmers, frequently free of charge. The farmer paid for these services through deductions made by the board or cooperatives from the farm's gate price. Credit requirements were therefore limited.

At the same time as liberalization in coffee producing countries, there has been a growing concentration amongst trade houses, roasters and the distribution networks, consumers have become more price conscious, and the markets have become more competitive and volatile. The large roasters increasingly focus on what they perceive to be their core business, the roasting and selling of coffee. They expect their green coffee suppliers to take care of everything else, and their counterparts in the marketing chain are expected to offer fully integrated services ranging from the procurement of the right quality green coffee to its 'just-in-time' delivery at roasting plants. The consumer, the final drinker of the beverage, is becoming more demanding. Moreover, consumers are no longer looking for a standardized coffee beverage as they expect pleasure, excitement and a variety of flavours. It is noticeable that standardization has resulted in a fall in consumption in traditional markets in Europe and North America. However, the growth of specialty coffee, which is based on differentiation, has helped to improve the situation. Competition from other beverages such as soft drinks, fruit juices and tea is fierce. Most industries have moved towards greater integration of their operations. This is a consequence of the advances of modern technology, the speed of communication and the enormous costs of development.

It is in this new environment that another coffee crisis began in mid-1998, with world prices subject to a sustained downward trend, reaching catastrophic levels not experienced by the coffee industry in exporting countries for more than 30 years. The annual average of the ICO composite indicator price, which was 133.91 US cents/lb (~453 g) in 1997, recorded a level of 45.60 cents/lb in the year 2001 and 47.74 cents in 2002, before rising slightly to 51.91 cents in 2003 and 62.15 cents in 2004. It is important to note that a substantial improvement was recorded in 2005, with an average price of 89.96 cents/lb during the first 10 months (January through October). The value of exports by exporting countries during 2004 was estimated at US$ 6.88 billion for a transaction involving 90.7 million 60 kg bags, compared to US$ 12.8 billion for total exports of 80.26 million bags in 1997.

It may be noted that adverse consequences of the crisis include in many cases social, environmental and economic effects. The impact of the coffee price crisis on poverty, which lasted nearly five years, has been well documented (Anonymous, 2003; 2004). Evidence provided by coffee producing countries to the ICO is compelling. In many countries, reductions in the cash income of farmers mean less money for basic items such as health and education. In the latter item, girls are particularly at risk of being kept from school. In El Salvador, the United Nations' World Food Programme has had to distribute emergency rations to 10 thousand coffee-growing families. There have been widespread increases in unemployment. Moreover, the crisis has led in many areas to the abandonment of farms, population movement to urban areas and illegal migration. Problems of low prices have also increased incentives to plant narcotic drugs.

This coffee crisis therefore constitutes a clear stumbling block to sustainable development in the affected areas and countries. Figure 2.1 indicates this unprecedented decline in international coffee prices.

At the level of market fundamentals, many changes have occurred during this new era. As indicated above, Vietnam has emerged as the second largest producer while Brazil, with relatively low production costs, has increased its production capacity to an average of over 45 million bags *per* year. World coffee production for the crop year 2004/2005 was estimated at 115 million bags and world consumption

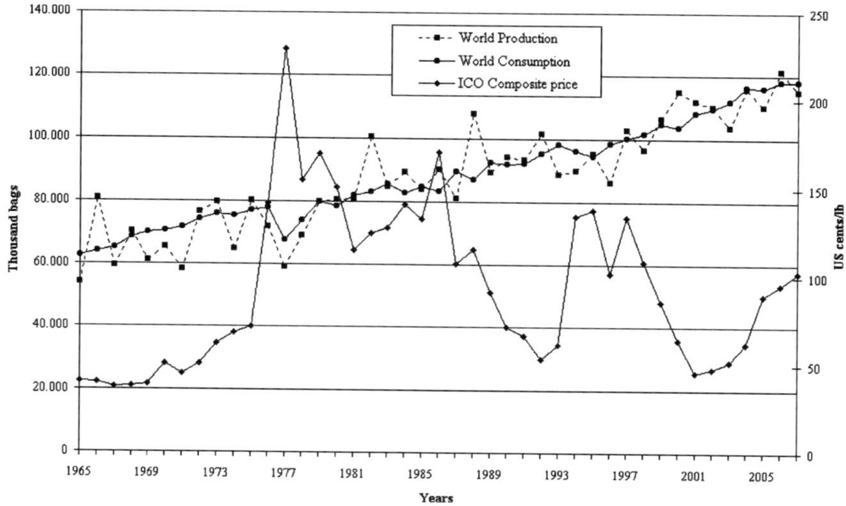

Fig. 2.1 Coffee production, consumption and ICO compositor price since 1965

at 115 million bags for the calendar year 2004, thanks to an increase in domestic consumption in exporting countries. Consumption in many importing countries seems to have reached saturation point, while significant potential in exporting countries still needs to be tapped.

2.5 Concluding Remarks

In this new environment, what should the role of multilateral organizations be in an increasingly less interventionist world? The scope for potential action is large, and the ICO has in fact redefined its role in the prevailing free market conditions, establishing an international development strategy for coffee as a framework for its future work (Anonymous, 2004). One of the key areas covered by the ICO is promoting coffee development projects, including those designed to combat coffee diseases and strengthen research and extension services in exporting countries. Extension and research are vital functions that affect the performance of the coffee sector. Indeed, the contribution of research in scientific and technical areas as well as in economic, health, social and environment issues is an integral part of the ICO development strategy.

A global research network has been set up to gather scientific information, to harmonize research programs, and to avoid duplication and waste of resources. Although it has not yet started to operate, this global research network should, within a few years, provide a database linking ICO members indirectly to several years of research project work.

With market liberalization in many coffee producing countries, a variety of structures have emerged to provide extension services, including coffee-specific research and extension funded by the industry, and privatization of research and extension services contracted to private firms. It is therefore appropriate to look into the various ways to deliver research and extension services to farmers, assess their costs and effectiveness, with the aim of improving the provision of these services to farmers. Some producing countries are experiencing this new partnership between research institutes and the private sector. For instance, in the Ivory Coast the Centre National de Recherche Agronomique has been conducting research activities in partnership with the private sector. The Government of India provides adequate grants for research, extension and training programs implemented exclusively for coffee by the Coffee Board of India. Similarly, the Brazilian Agricultural Research Corporation (abbreviated Embrapa in Portuguese) and Colombia's Cenicafe employ resources derived from the private sector for research, extension and training. Although contributions from the Government continue to dominate, Colombia has diversified the sources of funding of its agricultural research through special research programs in partnership with various national and international organizations.

In conclusion, although market fundamentals have appeared supportive to prices over recent months, efforts must continue to assure a balance between supply and demand. Indeed, to maintain a sustainable coffee economy, it is important to ensure that increases in supply are matched by corresponding growth in demand. In market conditions such as those prevailing since mid-1998, where supply has consistently exceeded demand, leading to a crisis of low prices, it is particularly important that actions are taken to increase consumption by improving quality and through promotional and educational projects. It is clear from what has been said above that the ICO continues to have an important role to play in improving the coffee sector. The Organization has always adopted a dynamic approach to its work, adapting to changing circumstances and ensuring that it continues to address the problems facing the coffee community through international cooperation.

References

Anonymous (1998). Cenicafé – serving colombian coffee growers since 1938. Federación Nacional de Cafeteros de Colombia, Bogota

Anonymous (2003). Impact of the coffee crisis on poverty in producing countries. international coffee organization ICC 89-5 Rev.1, London

Anonymous (2004). Development strategy for coffee. International coffee organization EB-3768/01 Rev.3, London

Beintema NM, Avila, AFD, Pardey PG (2001). Agricultural R&D in Brazil: policy, investments, and institutional profile. IFPRI, EMBRAPA and FONTAGRO, Washington, DC

Beintema NM, Romano L, Pardey, PG (2000) Agricultural R&D in Colombia: policy, investments, and institutional profile. IFPRI and FONTAGRO, Washington, DC

Clarence-Smith WG, Topik S (2003). The global coffee economy in Africa, Asia, and Latin America, 1500–1989. Cambridge University Press, Cambridge

Lucier RL (1988). The international political economy of coffee: from Juan Valdez to Yank's Diner. Praeger Publishers, New York

Priovolos T (1981). Coffee and the Ivory Coast. Lexington Books, Toronto

Part II
The Root-Lesion Nematode, *Pratylenchus* spp.

Chapter 3
Taxonomy, Morphology and Phylogenetics of Coffee-Associated Root-Lesion Nematodes, *Pratylenchus* spp.

Zafar A. Handoo, Lynn K. Carta and Andrea M. Skantar

Abstract This review includes a synthesis of information on eight species of root-lesion nematodes (*Pratylenchus* spp.) that parasitize coffee or inhabit its rhizosphere. It includes a table of important morphological characters, a diagnostic key, photographs of anterior ends and tails of specimens from the USDA nematode collection, and a phylogenetic tree based on ribosomal DNA with drawings of scanning electron microscopic face-patterns. Information sources are evaluated and future research needs are outlined.

Keywords Diagnostic key · phylogeny · taxonomy · phylogenetic tree · evolution

3.1 Introduction

Root-lesion nematodes (*Pratylenchus* spp.) are among the most common and damaging to coffee (*Coffea* sp.) aside from root-knot nematodes and a few other genera. The genus *Pratylenchus* is comprised of 97 valid species of worldwide distribution and economic importance, which parasitize a wide variety of plant species. Members of this genus are called root-lesion nematodes because they produce lesions on feeder roots and occasionally on other underground plant parts as a result of their feeding. They are sometimes referred to as meadow nematodes due to their frequent occurrence in that environment.

The first described root-lesion nematode was *Tylenchus pratensis* De Man (de Man, 1880), which was redescribed and illustrated by De Man (1884). The genus name *Pratylenchus* was established by Filipjev in 1936, with *P. pratensis* (de Man) Filipjev as the type species. Sher and Allen (1953) first put the taxonomy of the genus on a basis familiar to modern taxonomists. Loof (1960; 1978; 1991) reviewed in detail the anatomy, morphology, distribution, systematics, variability and identification of the genus, and presented a key to its species. Key and comprehensive compendia including histories of the morphological work performed

Z.A. Handoo
United States Department of Agriculture/ARS, Nematology Laboratory, Beltsville, USA
e-mail: zafar.handoo@ars.usda.gov

by various authors have been given by Handoo and Golden (1989) and Frederick and Tarjan (1989).

The uncertain state of taxonomy within the genus *Pratylenchus* is well illustrated by the widely diverging synonymies that have been proposed, from 46 taxa by Loof (1991) compared to 12 more taxa noted by Ebsary (1991). Even more divergence exists in the increasing number of species recognized as valid within this genus. Fortuner (1984) recognized 58 valid species, while Ryss (1988) recognized 45; Frederick and Tarjan (1989), Cafe Filho and Huang (1989) and Handoo and Golden (1989) listed 49, 57 and 63 species, respectively. Loof (1991) and Ebsary (1991) recognized 46 and 58 species, respectively, while Siddiqi (2000) did so for 89.

An excellent review on nematodes reported to occur on coffee (Campos et al., 1990) have included the following five *Pratylenchus* species: *P. brachyurus* (Godfrey) Filipjev and Schuurmans Stekhoven, *P. coffeae* (Zimmermann) Filipjev and Schuurmans Stekhoven, *P. goodeyi* Sher and Allen, *P. loosi* Loof and *P. pratensis* (de Man) Filipjev. *P. vulnus* Allen and Jensen and *P. zeae* Graham have recently been added to this list (Campos and Villain, 2005). *P. panamaensis* Siddiqi, Dabur and Bajaj, to which *P. gutierrezi* Golden, López and Vilchez has been synonymized (Siddiqi, 2000), also parasitizes coffee (Siddiqi et al., 1991; Golden et al., 1992).

The common difficulty of identifying to species many coffee-associated *Pratylenchus* populations (Campos and Villain, 2005) logically suggests that improved, more accessible diagnostic methods, which are detailed later in this chapter, will uncover new species parasitic to this crop. From this updated group of eight species, *P. coffeae* is of notable quarantine importance worldwide. In the United States, the state of Florida has adopted internal phytosanitary measures against this nematode (Inserra et al., 2005b). Besides parasitizing coffee, *P. brachyurus* is a pathogen of peanut and soybean (Corbett, 1976; Schmitt and Barker, 1981). *P. loosi* is a major pest of tea (Seinhorst, 1977; Gnanapragasam and Mohotti, 2005), and so is *P. goodeyi* for bananas (Machon and Hunt, 1985). *P. penetrans* (Cobb) Filipjev and Schuurmans Stekhoven affects potatoes, woody perennials, soybean and cereals (Mai et al., 1977; Schmitt and Barker, 1981), while *P. zeae* is primarily a pest of cereals (Loof, 1991). However, the status of coffee as host to *P. zeae* deserves further study (Kubo et al., 2004).

As a contribution to improve the taxonomy of coffee-associated *Pratylenchus* species, this review offers an identification key with light and scanning electron microscopic (SEM) images and a molecular phylogenetic tree. This chapter also discusses the literature and future research possibilities.

3.2 Taxonomy

Order Tylenchida Thorne, 1949
Suborder Tylenchina Chitwood, 1950
Superfamily Tylenchoidea Orley, 1880
Family Pratylenchidae Thorne, 1949

Subfamily Pratylenchinae Thorne, 1949
Genus *Pratylenchus* Filipjev, 1936

Emended diagnosis (after Siddiqi, 2000): Pratylenchinae. Stout, cylindroid nematodes less than 1.0 mm long. No marked sexual dimorphism in anterior region. Lateral fields each with four to six incisures, occasionally with oblique median markings or striae. Incisures of adult females frequently absent owing to stretching of the cuticle. Deirids absent. Phasmids usually near middle of the tail or located one-third of tail length or more behind the anus. Cephalic region low, flattened anteriorly or rarely rounded, continuous with body contour; sclerotization massive. Labial disc inconspicuous, in SEM dumb-bell-shaped, with six labial pits around a minute oral aperture; amphidial apertures pore-like, near labial disc, indistinct. Lip region bearing two to four annules (one to three striae) set off by a narrowing of the head. Stylet strong, 20 μm or less long, with round, anteriorly flat or indented basal knobs. Median bulb oval to round, very muscular. Basal bulb extending back over intestine, usually in a lateroventral position. Three prominent esophageal nuclei. Esophageal lumen and intestine joined by an obscure muscular valve. Excretory pore prominent, about opposite to the nerve ring. Hemizonid slightly anterior to excretory pore. Position of the vulva from the nematode anterior end in relation to the body length (V%) usually at 70–80%. Pseudo monoprodelphic, with only the anterior ovary functional. Postvulval uterine sac present, with or without rudiments of posterior ovary. Spermatheca large, rounded or sometimes oval to square, usually axial. Female tail subcylindrical to conoid, usually about two to three anal body widths; terminus smooth or annulated. Males known in most species. Bursa enclosing tail terminus. Spicules slightly arcuate with subterminal pore on dorsal side. Gubernaculum simple, trough like, male tail pointed. Caudal alae enveloping tail.

For this review, specimens of the coffee-associated root-lesion nematodes, *P. coffeae, P. loosi, P. brachyurus, P. panamaensis, P. pratensis, P. goodeyi, P. vulnus* and *P. zeae* have been examined from the USDA Nematode Collection at Beltsville, Maryland (USA). These species had been previously mounted in glycerin, and the examinations have been made with a compound light microscope. Morphometric data have been obtained with an eyepiece micrometer, and the measurements have been made in micrometers (μm) unless otherwise stated. The morphometric data for the most important diagnostic characters have been updated and organized according to the compendium format adopted by Handoo and Golden (1989). Photomicrographs of female's cephalic region ('head') and tails have been made with a 35-mm camera. Original descriptions and any subsequent redescriptions or other related data have also been used to assess species.

3.2.1 Identification Characters, Techniques Used and Problems for Species Identifications

De Man's morphometric ratios are essential for diagnosis of *Pratylenchus* species, with V% being the most reliable (Siddiqi, 1997). Other discrete characters commonly used to distinguish them are body length, head shape and number of annules, length of stylet, shape of stylet knobs, structure of lateral field, presence/absence and shape of spermatheca, length and structure of posterior uterine branch, shape of female tail terminus, presence or absence of males and shape and length of spicule and gubernaculum. Loof (1991) discussed these characters in detail, with particular emphasis on the intraspecific variation that limits their reliability. Table 3.1 contains updated morphometric data for the most important diagnostic characters of the coffee-associated *Pratylenchus* species. Figures 3.1 through 3.4 contain photomicrographs of female's heads and tails.

Identification of *Pratylenchus* species is hampered by the similarity among species in some cases, and by the significant intraspecific variability of both morphological and morphometric diagnostic characters in other cases. Several authors have described morphological variations within species that make it difficult to separate species accurately using traditional microscopy. Variable features include tail shape, the number of annules in the ventral part of tail, the lateral field throughout the body (Corbett and Clark, 1983), and the presence of one supplementary lip annule in some specimens (Baujard et al., 1990). This situation has prompted researchers to discover alternate methods and features for more accurate identification. Although not considered routine, SEM is a technique sometimes used for morphological analysis (see for example Sher and Bell, 1975; Corbett and Clark, 1983; Trett and Perry, 1985; Baujard et al., 1990; López and Salazar, 1990; Sakwe and Geraert, 1994; Inserra et al., 1998; 2005a; Duncan et al., 1999; Hernández et al., 2001; Carta et al., 2001; 2002). Fortunately, for taxonomic purposes, SEM has shown the stability and reliability of several nematode surface features, with the lip and face region, lateral field, and tail receiving the most attention (see for example Anderson and Townshend, 1980; Inserra et al., 2005a). In an SEM study of the surface features of nine *Pratylenchus* species the lip region has been demonstrated to be a particularly good taxonomic character; species have been separated into three groups according to the pattern of the first lip annule and the oral disc (Hernández et al., 2001).

3.2.2 The Coffee-Associated Root-Lesion Nematodes, **Pratylenchus** *spp.*

The eight species listed below have been reported from the roots of coffee (Campos et al., 1990). However, a recently described species from northern Europe, *P. brzeskii* Karssen, Waeyenberge and Moens, is morphologically similar to *P. coffeae* and *P. loosi* (Karssen et al., 2000), species both known to parasitize coffee in more southern climates and difficult to distinguish by morphology (Pourjame et al., 1999). To our knowledge, *P. brzeskii* has not been examined for possible parasitism on coffee. This closely related species is another indication that *P. coffeae* and relatives represent a densely populated species complex (Campos and Villain, 2005). Uncertainty about the identity of some amphimitic *Pratylenchus* populations has been recorded from Brazil (Siciliano-Wilcken et al. 2002a,b; Silva and Inomoto,

Table 3.1 Compendium of eight *Pratylenchus* species known to parasitize coffee. Values represent range, average

Species	Body length (mm)	Stylet (μm)	No. lip annules	a[a]	b	c	V%	No. tail annules	Tail shape[b]	Tail terminus	Spicule length (μm)	Guber-naculum length (μm)	Spermatheca	Males
P. brachyurus	0.39–0.75	17–22	2	15–29	5–10	13–28	82–89	15–21	SCYL,	SMO	16–17	5–6	Empty	Very rare
	0.59	20.7		23	8	25	85.6	18.5	TRC,					
									SHM,					
									SMO					
P. coffeae	0.46–0.70	15–18	2	21–30	5–8	17–27	76–82	17–24	HEM-BLP,	IND,	15–18	4.2–6	Large, broadly oval to round	Common
	0.53	16.5		25	6.5	22	80	20		TRC, BR,	16.5			
									SMO	SMO				
P. panamaensis (= *P.gutierrezi*)	0.43–0.55	16–18	2	15–24.9	3.5–4.5	17–25	74–84	18–23	SCYL	BLR,	15.5–21	3.2–4.4	Oval to round	Common
	0.5	16.8		19.8	3.9	19.9	80	21		ANN	16.8	3.7		
P. goodeyi	0.40–0.68	14–17	4	24–37	5.5–7.3	14–18	73–75	21	SC	TP,	15.5–21	5–6	Large, oblong	Common
	0.52	15.7		32	6.5	16.5	74			SMO				
P. loosi	0.48–0.64	14–18	2	28–36	5.7–7.1	18–25	79–85	27–34	NAR, SA	NAR,	16–20	4–7	Oval	Common
	0.57	16.5		32	6.4	21	82.5			SMO	17.5	5.5		

Table 3.1 (continued)

Species	Body length (mm)	Stylet (μm)	No. lip annules	a[a]	b	c	V%	No. tail annules	Tail shape[b]	Tail terminus	Spicule length (μm)	Gubernaculum length (μm)	Spermatheca	Males
P. pratensis	0.40–0.63	12–16	3	21.8–30.3	5.5–7.6	14–27	76–80	20–28	SCYL	RND, Ob, ASY,	17–19	6–7	Oval	Present
	0.52	14.5		24	6	23.5	78	25.5		ANN				
P. vulnus	0.46–0.91	16–18	3–4	26.6–39.2	5.3–7.7	14–28	78–84	20–34	TAP	BLP, NAR	14–20	4–6	Oval, oblong	Common
	0.75	16.5		33.5	6.5	23	81				17	5		
P. zeae	0.36–0.58	15–17	3	25–30	5–8	17–21	66–76	16–25	TAP	NAR, SA	14–15	4–5	Round	Unknown
	0.43	16		27	6.5	18.5	72	21			14.5	4.5		

[a] Morphometric ratios a, b, c, V%: a = de Man's ratio of body length/widest body width; b = de Man's ratio of body length/esophagus length (lip to pharyngeal-intestinal valve); c = de Man's ratio of body length/tail length; V% = distance of lip to vulva/body length.

[b] Tail shapes: NAR = narrowly rounded; SA = subacute; SC = sharply conical; SCYL = subcylindrical; TAP = tapering. Shape of tail terminus: ASY = asymmetrical; BLP = bluntly pointed; BLR = bluntly rounded; BR = broadly rounded; HEM = hemispherical; IND = indented; Ob = oblique; RND = round; SHM = subhemispherical; STP = small terminal peg; SA = subacute; TP = terminal projection; TRC = truncate. Tail tip annulation: SMO = smooth; ANN = annulated. (Some images are available in Frederick and Tarjan, 1989).

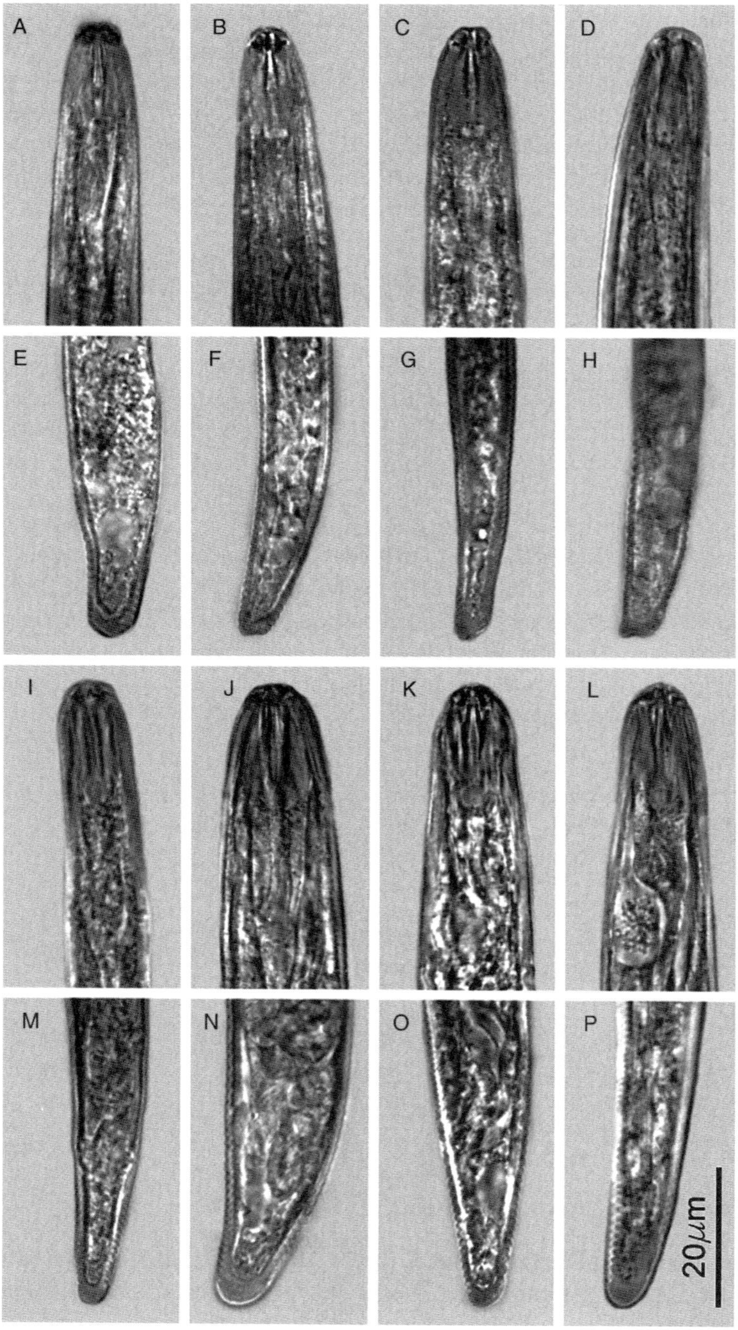

Fig. 3.1 Photomicrographs of female heads and tails, showing variations in tail shape. (**A–H**) *P. coffeae*, (**I–P**) *P. brachyurus*. (Photos by Z.A. Handoo)

2002), Guatemala (Villain et al., 1998; Villain, 2000), El Salvador and Costa Rica (Herve, 1997; Duncan et al., 1999) based on host range and genetic information.

Drawings of *Pratylenchus* species other than *P. goodeyi* and *P. pratensis* (see below) may be found at the USDA website (http://ars.usda.gov/Main/docs.htm?docid = 9866) and at the University of Nebraska websites http://nematode.unl.edu/pratkey7.htm#pratkey7, http://nematode.unl.edu/pracoff.htm, http://nematode.unl.edu/ploos.htm and http://nematode.unl.edu/prapse.htm.

3.2.2.1 *Pratylenchus coffeae* (Zimmermann, 1898) Filipjev and Schuurmans Stekhoven, 1941

The taxonomy of *P. coffeae* (Fig. 3.1 A–H) has been the subject of numerous studies (Sher and Allen, 1953; Loof, 1960; 1978; 1991; Roman and Hirschmann, 1969; Siddiqi, 1972; Rashid and Khan, 1978; Bajaj and Bhatti, 1984; Inserra et al., 1996; 1998; 2001; Mizukubo, 1992; Duncan et al., 1999; Ryss, 2002a; Van Den Berg et al., 2005).

This species is the most widespread and damaging on coffee. It occurs in the Dominican Republic, El Salvador, Guatemala, Puerto Rico, Costa Rica, Brazil, India, Southeast Asia, Barbados, Martinique, Tanzania, Madagascar, Indochina, Java, Indonesia and Venezuela. On other hosts this species is found throughout the tropics and in many subtropical regions. Specific locations include Japan, Australia, South Africa, Brazil, Oman (Campos and Villain, 2005) and southern parts of the United States (Norton et al., 1984).

3.2.2.2 *Pratylenchus brachyurus* (Godfrey, 1929) Filipjev and Schuurmans Stekhoven, 1941

The taxonomy of *P. brachyurus* (Fig. 3.1 I–P) has been advanced by authors in Europe and the Americas (Sher and Allen, 1953; Loof, 1960; 1978; 1991; Roman and Hirschmann, 1969; Corbett, 1976; Corbett and Clark, 1983; López and Salazar, 1990; Hernández et al., 2001; Ryss, 2002a).

In South America, this was one of the first root-lesion nematodes known on coffee (Campos and Villain, 2005). It is found primarily in the tropics and subtropics, and specifically in Australia, Brazil, Peru, USA, Turkey, West Africa, South Africa and Japan. In some areas of Brazil it may be more widespread than *P. coffeae* (Campos and Villain, 2005).

3.2.2.3 *Pratylenchus loosi* Loof, 1960

Taxonomic and morphological studies of *P. loosi* (Fig. 3.2 A–H) have been published in various review papers (Seinhorst, 1977; Loof, 1978; 1991; Inserra et al., 1996; 2001).

This species has been reported on coffee in Sri Lanka (Hutchinson, 1963 cited by Whitehead, 1968). On other hosts its geographic distribution includes Sri Lanka, India, Japan (Seinhorst, 1977; Campos and Villain, 2005), Korea (Park et al., 2002),

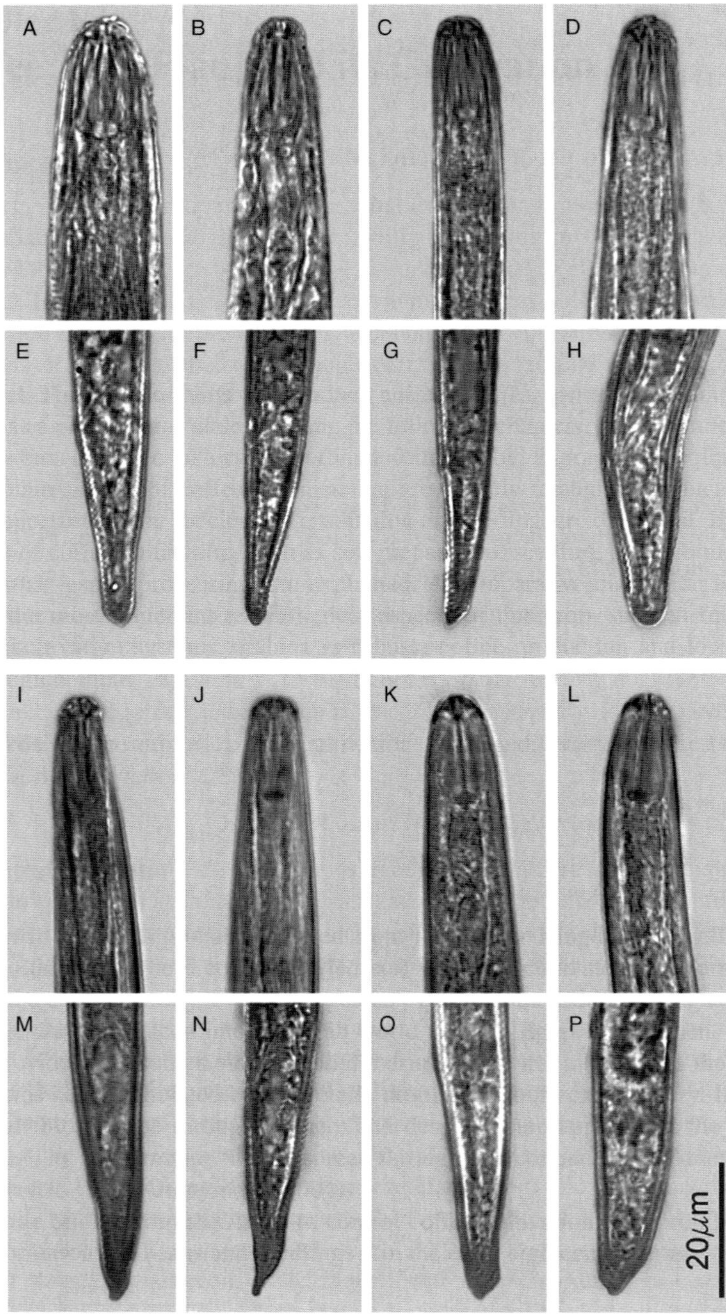

Fig. 3.2 Photomicrographs of female heads and tails, showing variations in tail shape. (**A–H**) *P. loosi*, (**I–P**) *P. goodeyi*. (Photos by Z. A. Handoo)

American Samoa (Brooks, 2004), Guadeloupe (Van Den Berg and Quénéhervé, 2000) and Iran (Hajieghrari et al., 2005). In the United States, it occurs in Florida, Louisiana and Kansas (Inserra et al., 1996; 2001; Norton et al., 1984; Powers, 2008).

3.2.2.4 *Pratylenchus goodeyi* Sher and Allen, 1953

The taxonomy and morphological variation of *P. goodeyi* (Fig. 3.2 I–P) have been described in various reviews (Loof, 1960; 1978; 1991; Corbett and Clark, 1983; Machon and Hunt, 1985).

This species has been reported on coffee in Tanzania (Bridge, 1984). On other hosts, its geographic distribution includes East Africa, Canary Islands, Kenya, Tanzania, England, Russia and the USA (Norton et al., 1984; Machon and Hunt, 1985). Diagnostic drawings of *P. goodeyi* may be viewed at plpnemweb.ucdavis.edu/Nemaplex/images/G105S12.gif.

3.2.2.5 *Pratylenchus panamaensis* Siddiqi, Dabur and Bajaj, 1991 [syn. *Pratylenchus gutierrezi* (Golden, López and Vilchez, 1991) Siddiqi, 2000]

The morphological variation in *P. panamaensis* (Fig. 3.3 A–H) has been reported as a new species (*P. gutierrezi*) and characterized by Duncan et al. (1999) and Inserra et al. (1998). This species has been found parasitizing coffee in Panama (Siddiqi et al., 1991), the central plateau of Costa Rica (Golden et al., 1992), Guatemala (Inserra et al., 1998) and Oman (in USDA Nematode Collection, entry #1546).

3.2.2.6 *Pratylenchus pratensis* (de Man, 1880) Filipjev, 1936

Studies on the taxonomy and morphological variation of *P. pratensis* (Fig. 3.3 I–P) have been advanced by diverse authors (Sher and Allen, 1953; Loof, 1960; 1974; 1978; Seinhorst, 1968; Roman and Hirschmann, 1969; Frederick and Tarjan, 1989; Ryss, 2002a). Diagnostic drawings of *P. pratensis* can be found at plpnemweb.ucdavis.edu/nemaplex/images/G105S45.gif.

According to Whitehead (1968), Somasekhar (1959) had reported this species on coffee in south India. *P. pratensis* has been mistaken for *P. crenatus* Loof, *P. penetrans, P. brachyurus, P. coffeae* and possibly *P. loosi* (Loof, 1960; 1974). The coffee-parasitic status of *P. pratensis* is uncertain also because nematode identification could not be confirmed by voucher slides, nor was the original coffee population examined with molecular methods. However the occurrence of the related *P. vulnus* on coffee (Monteiro et al., 2001) gives some plausibility to Somasekhar's report.

The geographic distribution of *P. pratensis* on various crops includes Europe, South Africa and India (Loof, 1974). A consensus of opinion suggests that this species does not occur in the Americas (Norton et al., 1984), but it is fairly common in Europe (Peña et al., 2007). However, the morphologically similar species *P. pratensisobrinus* Bernard occurs in Alaska (Bernard, 1984) and *P. pseudopratensis* Seinhorst in Konza Prairie, eastern Kansas (USA) (Powers, 2008). This

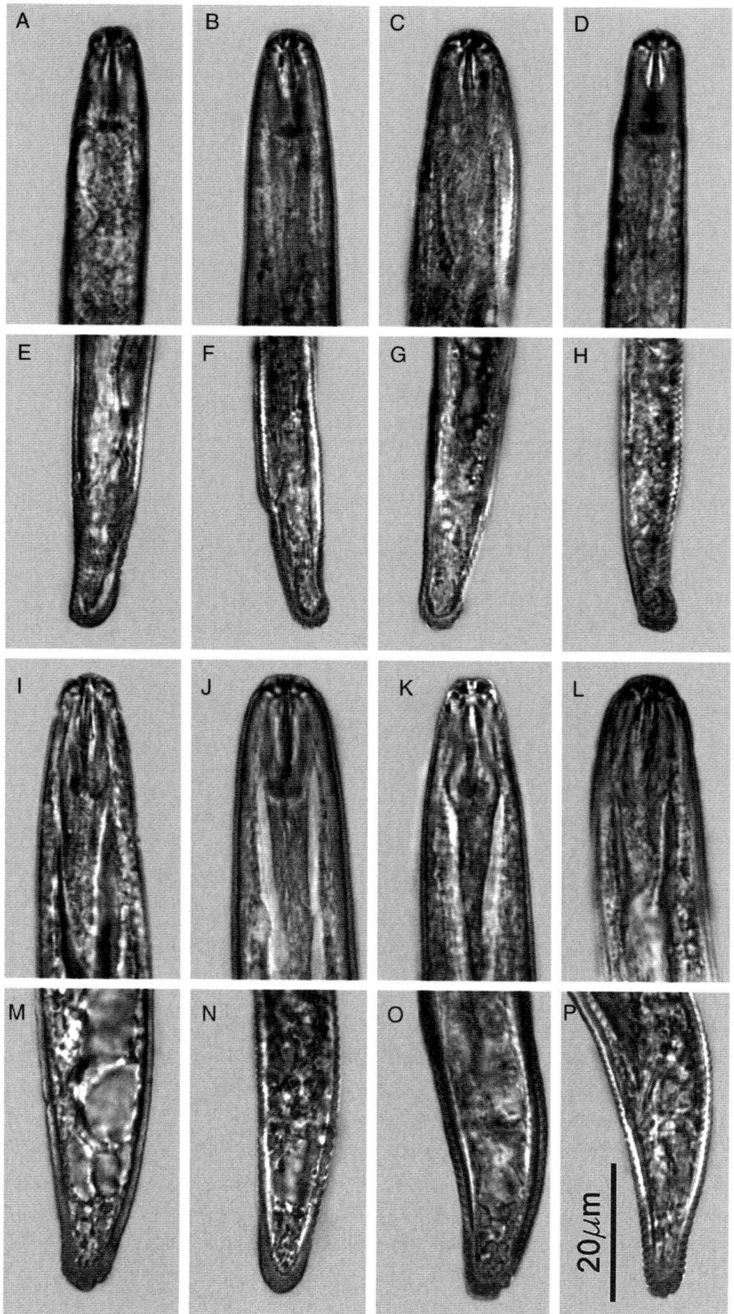

Fig. 3.3 Photomicrographs of female heads and tails, showing variations in tail shape. (**A–H**) *P. panamaensis*, (**I–P**) *P. pratensis*. (Photos by Z. A. Handoo)

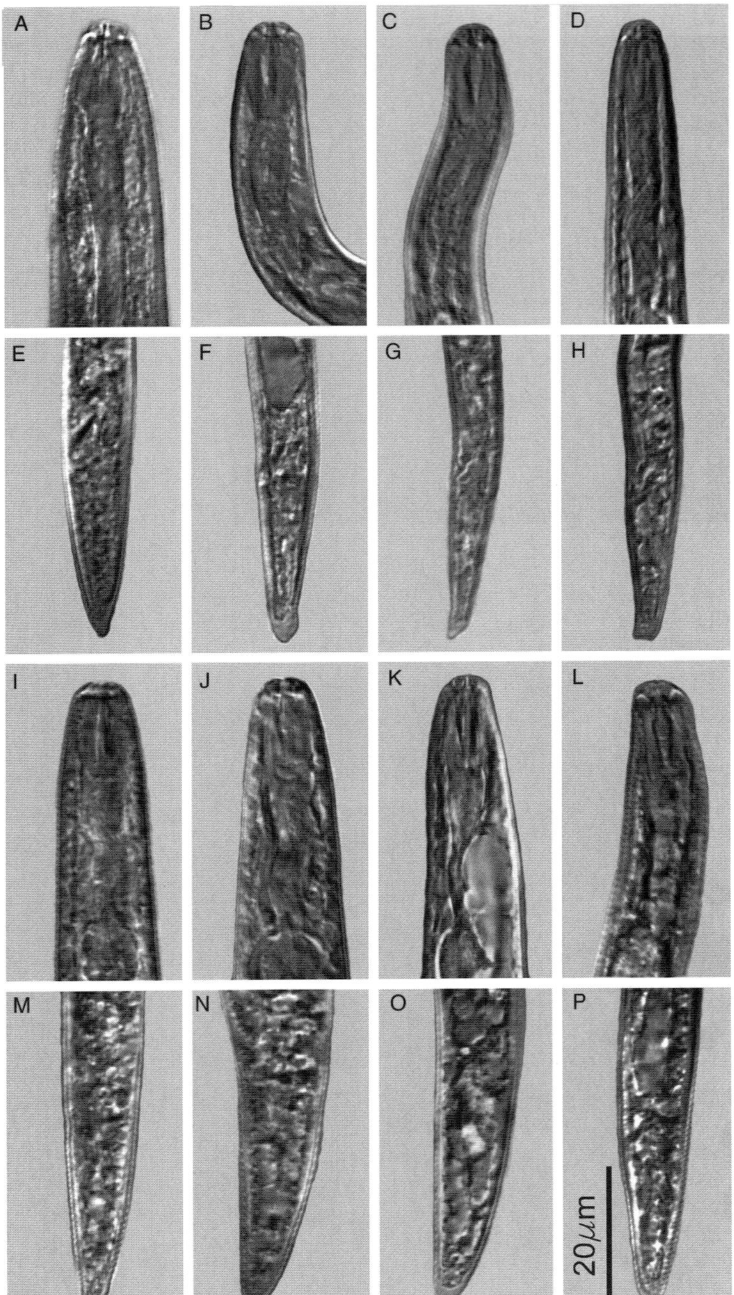

Fig. 3.4 Photomicrographs of female heads and tails, showing variations in tail shape. (**A–H**) *P. vulnus*, (**I–P**): *P. zeae*. (Photos by Z. A. Handoo)

issue will not be satisfactorily resolved until molecular sequences are available for *P. pratensis*-like taxa.

3.2.2.7 *Pratylenchus vulnus* Allen and Jensen, 1951

The taxonomy and morphological variations in *P. vulnus* populations (Fig. 3.4 A–H) have been described in several reviews (Corbett, 1974; Doucet et al., 1996; 1998; 2001; Gao et al., 1999). This species has been recently discovered on coffee in Brazil (Monteiro et al., 2001). It has also been found on other crops in Southern Europe, Russia, Egypt, South Africa, India, Japan, China, Philippines, New Zealand, USA, Mexico, Cuba and Argentina (Corbett, 1974; Gao et al., 1999; Lax et al., 2004).

3.2.2.8 *Pratylenchus zeae* Graham, 1951

The taxonomy and morphological variation of *P. zeae* (Fig. 3.4 I–P) have been described in various papers in the more than 50 years since it was described (Roman and Hirschmann, 1969; Fortuner, 1976; Olowe and Corbett, 1983, 1984a,b; Troccoli et al., 1996; Inserra et al., 2005a).

This species occurs on coffee in Brazil (Ferraz, 1980; Campos, 2002) and Colombia (in USDA Nematode Collection, entry #4686). It is also distributed on other crops worldwide in USA, Cuba, Trinidad, Venezuela, Brazil, throughout Africa, Madagascar, Egypt, Iraq, India, Japan, Australia (Fortuner, 1976) and in Indonesia (in USDA Nematode Collection, entry #2356).

3.2.3 *Key to Coffee-Associated* Pratylenchus *Species*

1 Lip region composed of 2 annules (rarely three) 2
1a Lip region composed of 3–4 annules 5

2(1) Tail terminus smooth 3
2a Tail terminus annulated or indented . . . . ***P. panamaensis*** (= ***P. gutierrezi***)

3(2) Lip region low, with outer margins angular; stylet typically 19–22 μm long (in Loof (1960) range is 17–22, but 17 very rare) with massive rounded knobs; V% = 82–89; tail subcylindrical, with truncate to subhemispherical or broadly rounded terminus; males rare. . . . . ***P. brachyurus***
3a Lip region high, roundly convex; stylet less than 18 μm long, V% = 76–85; tail narrowly rounded to subacute with hemispherical to bluntly pointed, truncate smooth terminus 4

4(3a) V% = 78(76–82); a = 25(21–30); tail terminus truncate or broadly rounded, occasionally indented . . . . ***P. coffeae***
4a V% = 82(79–85); a = 32(28–36); tail terminus narrowly rounded to subacute . . . . ***P. loosi***

5(1a) Tail terminus smooth 6
5a Tail terminus annulated; stylet 12–16 μm long; lip region with three annules; V% = 76–80; tail subcylindrical with obtuse to rounded asymmetrical annulated terminus . . . . ***P. pratensis***

6(5) V% = 66–76. 7
6a V% = 78 – 84; body slender, a = 25–40; spermatheca oval, oblong filled with sperm, posterior uterine sac long, tail terminus bluntly pointed to narrowly rounded; males common . . . . ***P. vulnus***

7(6) Lip region low with 3 annules; stylet 15–17 μm long with broad, anteriorly flattened knobs, tail terminus narrowly rounded to subacute; males extremely rare . . . . ***P. zeae***
7a Lip region high with 4 annules; stylet 16–17 μm long with rounded flattened knobs; tail sharply conical with dorsal tail contour characteristically sinuate anterior to terminus; tail terminus with a small terminal peg; males common . . . . ***P. goodeyi***

3.3 Phylogenetic Trees and Molecular Characterization

Based on morphological data, Ryss (2002a,b) have presented multi-entry and mono-entry keys and diagnostic relationships within *Pratylenchus* sp., along with proposals for phylogeny and evolution of this genus. Also, a morphological tree for a somewhat different set of taxa has been constructed using cladistic methods (Carta et al., 2002). Not surprisingly, these morphological frameworks are often inconsistent with some molecular phylogenetic trees inferred from 28S rDNA sequences (Al-Banna et al., 1997; Duncan et al., 1999; Carta et al., 2001; De Luca et al., 2004). This is partly due to a more limited and different set of species used in the molecular studies, and also due to selection of outgroups, which can have a major impact on branching order (Carta et al., 2001).

The first phylogenetic study of some *Pratylenchus* species with *Radopholus* sp., *Hirschmanniella* sp. and *Nacobbus* sp. demonstrated a polyphyletic tree using the D3 segment of rDNA (Al-Banna et al., 1997). A second study on the *P. coffeae* species complex subdivided and defined many populations into genetic units using both D2 and D3 rDNA regions (Duncan et al., 1999). A study conducted using more *Pratylenchus* species and different outgroups has restored *Pratylenchus* monophyly (Carta et al., 2001). An analysis of sequences of multiple individuals of one or more populations of *P. thornei* Sher and Allen, *P. neglectus* (Rensch) Filipjev and Schuurmans Stekhoven, *P. mediterraneus* Corbett, *P. pinguicaudatus* Corbet and *P. vulnus* has demonstrated high variability among individuals of *P. neglectus* (De Luca et al., 2004). A different assemblage of taxa using variously coded morphological characters has been used to construct trees (Ryss, 2002b; Carta et al., 2002) with different topologies from molecular trees; these differences cannot simply be attributed to differences in species composition.

While the number of sequences and taxa used for testing hypotheses of relationships within the genus *Pratylenchus* has grown through the last decade, it is clear that molecular trees will continue to require expansion, clarification and eventual integration with morphological data.

Molecular methods are often essential to confirm species identity, as with the discovery of *P. jaehni* Inserra, Duncan, Troccoli, Dunn, Santos, Kaplan and Vovlas, which has been revealed from a 28S rDNA phylogeny (Duncan et al., 1999; Inserra et al., 2001). While straightforward PCR-RFLP diagnostics are available for some coffee-parasitic *Pratylenchus* species (Pourjame et al., 1999), many such tests have not been validated with multiple populations or related species. Obtaining the DNA controls necessary for this standardization may also present a challenge, as some species may be difficult to obtain or require labor-intensive culture methods to maintain. Nevertheless, when characterizing potentially new or economically important populations, the generation of gene sequences for comparison with those in GenBank® is highly recommended.

To construct a phylogenetic tree of the coffee-associated *Pratylenchus* spp., 28S and 18S rDNA sequences have been obtained either from GenBank® or our own unpublished data. For the 28S rDNA D2-D3 region these include the following species and GenBank® accession numbers: *P. panamaensis* (= syn. *P. gutierrezi*) isolate K1, AF170440; *P. loosi* isolate N1, AF170437; *P. coffeae* isolate M1, AF170435; *P. zeae*, AF303950; unpublished sequence for peanut-parasitic *P. brachyurus* NL8 isolate from Florida; *Radopholus similis* (Cobb) Thorne, outgroup D3, U47558. ClustalW alignments (Thompson et al., 1994) have been made for sequences of 28S D2-D3 rDNA from the five *Pratylenchus* species from coffee listed previously plus two species from other hosts (*P. hexincisus* Taylor and Jenkins, AF303949 and *P. pseudocoffeae* Mizukubo, AF170444) plus two outgroups: *Meloidogyne exigua* Göldi, AF435804 and *Hirschmanniella pomponiensis* Abdel-Rahman and Maggenti, DQ077795.

Sequences for 18S rDNA include: *P. brachyurus*, AY279545; *P. goodeyi*, AJ966498; *P. pratensis*, AY284611; *P. vulnus*, AY286311 and *R. similis*, outgroup, AJ966502. A separate ClustalW alignment has been made for 18S rDNA of these five species from coffee, plus seven from other hosts: *P. crenatus*, AY284610; *P.* cf. *flakkensis* Seinhorst, DQ080595 (species unconfirmed); *P. hexincisus*, AY919242; *P. neglectus*, AY279544; *P. penetrans*, AY286308; *P. scribneri* Steiner in Sherbakoff and Stanley, AY286309; *P. thornei*, AJ966499, plus one outgroup (*R. similis*, AJ966502).

Based on the branch order in the two corresponding rDNA trees with overlapping taxa (not shown), a single synthetic composite has been constructed using PAUP* version 4.0b10 (Swofford, 1998), with parenthetical NEXUS tree format as an unresolved ladder-like topology in TreeView ver. 1.6.6 (Page, 1996). The resulting tree has been decorated with face views drawn from SEM images of nematodes obtained from published literature: *R. similis* (Sher and Bell, 1975), *R. neosimilis* Sauer (Sauer, 1985), *P. zeae* (Baujard et al., 1990; López and Salazar, 1990), *P. goodeyi* (Corbett and Clark, 1983; Hernández et al., 2001), *P. vulnus* (Corbett and Clark, 1983; Sauer, 1985; Hernández et al., 2001), *P. pratensis* (Corbett and Clark, 1983), *P. brachyurus* (Corbett and Clark, 1983; Baujard et al., 1990; López and Salazar,

1990), *P. gutierrezi* (Golden et al., 1992; Inserra et al., 1998; Duncan et al., 1999), *P. loosi* (Corbett and Clark, 1983; Baujard et al., 1990; Duncan et al., 1999; Pourjame et al., 1999; Inserra et al., 2001) and *P. coffeae* (Corbett and Clark, 1983; Inserra et al., 1998; Duncan et al., 1999; Inserra et al., 2001).

The schematic phylogenetic tree of coffee-associated nematodes, including drawings based upon SEM face views, is shown in Fig. 3.5. Compared to a previous

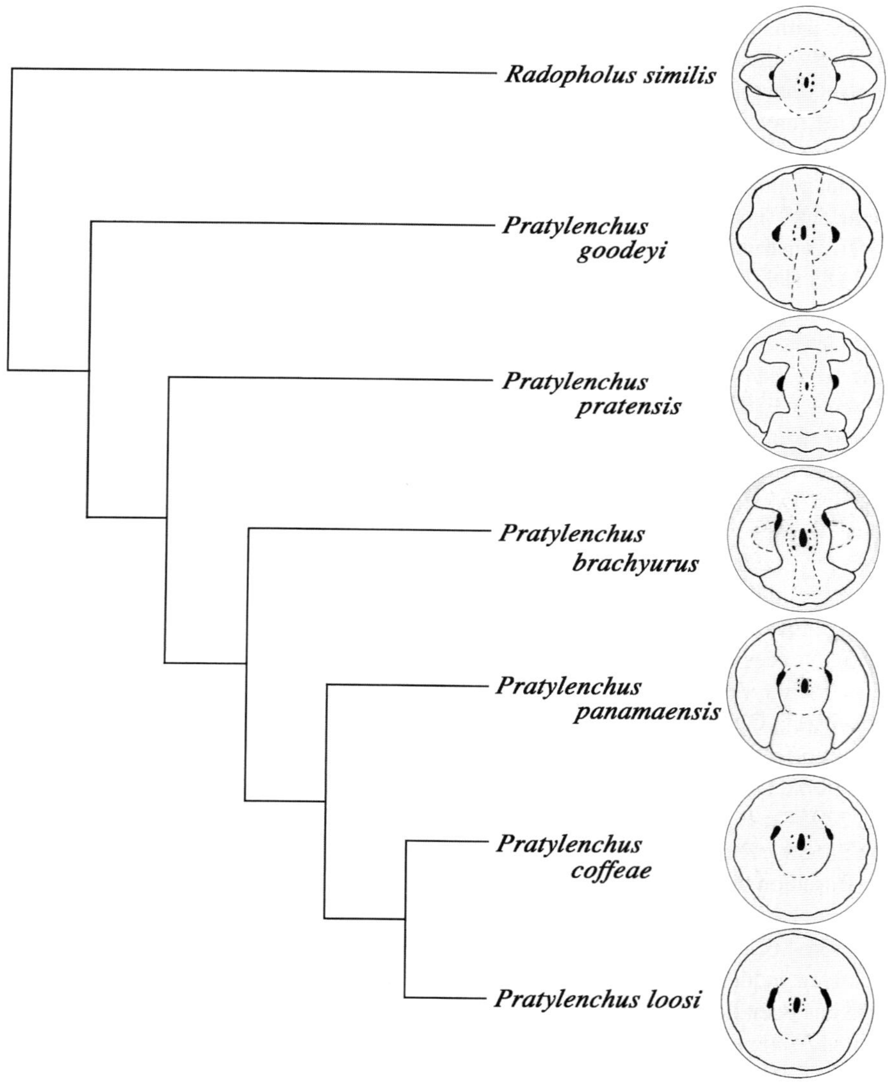

Fig. 3.5 Synthetic composite tree of six root-lesion nematode species derived from 28S rDNA and 18S rDNA trees with overlapping taxa, based on branch order and constructed with PAUP and TreeView. Scanning electron microscopic face views were drawn from the literature

molecular tree based solely on the D3 region of the 28S rDNA (Carta et al., 2001), the phylogenetic position of *P. brachyurus* inferred from the composite tree is more distant from *P. coffeae* and relatives than before. This updated position for *P. brachyurus* is more in line with the phylogenetic tree position based on morphology (Ryss, 2002b). In addition, *P. pratensis* and *P. coffeae* also appear highly divergent in the synthetic tree, unlike their position within the same clade in the morphological tree (Ryss, 2002b). The topology of the tree in Fig. 3.5 is congruent with the one shown in the most recent molecular phylogeny of this group (Inserra et al., 2007).

3.4 Concluding Remarks

Integrated studies on the morphological variation of *Pratylenchus* populations combined with molecular sequencing should result in improved methods of species delimitation. Of particular concern is the presence in the literature of nematode SEM face images that are sometimes variable and of poor quality, in part due to the use of formalin-fixed and dried specimens. This problem could be solved through more widespread application of low temperature-SEM (LT-SEM), a technique that reveals morphological features undistorted by chemicals and drying under pressure (Carta et al., 2003). Rapid cryo-fixation has revealed distinguishing features in root-lesion nematode faces even to the subspecies level (Carta et al., 2002). Determination of the number of lip annules is another serious problem, especially when few specimens are available for examination. This situation may improve through the use of new microscopic technology, such as the modular, high-resolution CytoViva® condenser (CytoViva Inc., Auburn, USA), with a cardioid annular ring that can achieve more than twice the resolution of standard circular condensers (Vainrub et al., 2006).

Increases in speed and capability and decreases in cost should lead to more frequent use of DNA sequencing by diagnostic labs for routine or selective species verification. Rapid new pyrosequencing technology, which generates short fragments (Shendure et al., 2004), may drive the development of rapid new diagnostics which are based upon short DNA fragments from multiple molecular markers.

Comparative pathogenicity studies have not been conducted for most coffee-parasitic root-lesion nematodes (Campos and Villain, 2005). Such studies would be highly desirable to assist in pest management decisions after a nematode species has been identified in a field. Systematic comparisons among species parasitizing either *C. arabica* L. (a commodity representing about 75% of world coffee exports, mostly in South and Central America) or the easier grown *C. canephora* Pierre ex Froehner (about 25% of exports, mostly grown in Africa and Asia) (Anonymous, 1986; Campos and Villain, 2005), would be especially valuable.

A concerted international research effort to centralize collection, preservation and molecular analysis of specimens, with satellite locations to perform morphology and pathogenicity studies, could greatly advance effective crop management of coffee-parasitic *Pratylenchus* species.

Acknowledgments The authors thank Donna Ellington, Maria Hult and Sharon Ochs for technical assistance. Mention of trade names or commercial products in this publication is solely for the purpose of providing specific information and does not imply recommendation or endorsement by the U. S. Department of Agriculture.

References

Al-Banna L, Williamson V, Gardner SL (1997) Phylogenetic analysis of nematodes of the genus *Pratylenchus* using nuclear 26S rDNA. Mol Phylogenet Evol 7:94–102

Anderson RV, Townshend L (1980) Variation of the first head annule in Canadian populations of *Pratylenchus penetrans* (Nematoda: Pratylenchidae) from three host plants. Can J Zool 58:1336–1340

Anonymous (1986) Cultura do cafe no Brasil. Instituto Brasileiro do Cafe, Rio de Janeiro

Bajaj HK, Bhatti DS (1984) New and known species of *Pratylenchus* Filipjev, 1936 (Nematoda: Pratylenchidae) from Haryana, India, with remarks on intraspecific variations. J Nematol 16:360–367

Baujard P, Mounport D, Martiny B (1990) Etude au microscope electronique a balayage de quatre espèces du genre *Pratylenchus* Filip'ev, 1936 (Nemata: Pratylenchidae). Rev Nematol 13:203–210

Bernard EC (1984) Hoplolaimoidea (Nematoda: Tylenchida) from the Aleutian islands with descriptions of four new species. J Nematol 16:194–203

Bridge J (1984) Coffee nematode survey of Tanzania. Report on the visit to examine plant-parasitic nematodes of coffee in Tanzania. Commonwealth Institute of Parasitology, St. Albans

Brooks FE (2004) Plant-parasitic nematodes of banana in American Samoa. Nematropica 34:65–72

Cafe Filho AC, Huang CS (1989) Description of *Pratylenchus pseudofallax* n. sp. with a key to species of the genus *Pratylenchus* Filipjev, 1936 (Nematoda: Pratylenchidae). Rev Nematol 12:7–15

Campos VP (2002) Coffee nematode survey in Minas Gerais State, Brazil. Research report of the grant by PNP&D/cafe. EMBRAPA, Brasilia

Campos VP, Sivapalan P, Gnanapragasam NC (1990) Nematode parasites of coffee, cocoa and tea. In: Luc M, Sikora RA, Bridge J (eds) Plant-parasitic nematodes in subtropical and tropical agriculture. CAB International, Wallingford

Campos VP, Villain L (2005) Nematode parasites of coffee and cocoa. In: Luc M, Sikora RA, Bridge J (eds) Plant parasitic nematodes in subtropical and tropical agriculture, 2nd edition. CAB International, Wallingford

Carta LK, Handoo ZA, Skantar AM et al (2002) A redescription of *Pratylenchus teres* Khan and Singh, 1974 (Nemata: Pratylenchidae), with the description of a new subspecies from South Africa, and a phylogenetic analysis of related species. Afr Plant Prot 8:13–24

Carta LK, Skantar AM, Handoo ZA (2001) Molecular, morphological and thermal characters of 19 *Pratylenchus* spp. and relatives using the D3 segment of the nuclear LSU rRNA gene. Nematropica 31:195–209

Carta LK, Wergin WP, Erbe EF et al (2003) A comparison of low temperature and ambient temperature SEM for viewing nematode faces. J Nematol 35:78–81

Corbett DCM (1974) *Pratylenchus vulnus*. CIH descriptions of plant-parasitic nematodes, 3, # 37, Commonwealth Institute of Helminthology, St. Albans

Corbett DCM (1976) *Pratylenchus brachyurus*. CIH descriptions of plant-parasitic nematodes, 6, # 89, Commonwealth Institute of Helminthology, St. Albans

Corbett DCM, Clark SA (1983) Surface features in the taxonomy of *Pratylenchus* species. Rev Nematol 6:85–98

De Luca F, Fanelli E, Di Vito M et al (2004) Comparison of the sequences of the D3 expansion of the 26S ribosomal genes reveals different degrees of heterogeneity in different populations and species of *Pratylenchus* from the Mediterranean region. Eur J Plant Pathol 110:949–957

De Man JG (1880) Die einheimischen, frei in der reinen Erde und im süssen Wasser lebenden Nematoden monographisch bearbeitet. Tijdschrift der Nederlandse Dierk. Vereenig. 5:1–104

De Man JG (1884) Nematodes der Niederlandischen Fauna. EJ Brill, Leiden

Doucet M, Pinochet J, Di Rienzo JA (1996) Comparative analysis of morphological and morphometrical characters in six isolates of *Pratylenchus vulnus* Allen and Jensen, 1951 (Nemata: Tylenchida). Fundam Appl Nematol 19:79–84

Doucet ME, Lax P, Pinochet J (1998) Variability of some external characters in *Pratylenchus vulnus* Allen and Jensen, 1951 (Nematoda: Tylenchida). Fundam Appl Nematol 21:205–206

Doucet MP, Lax P, Di Rienzo JA et al (2001) Temperature-induced morphometrical variability in an isolate of *Pratylenchus vulnus* Allen and Jensen, 1951 (Nematoda: Tylenchida) Nematology 3:1–8

Duncan LW, Inserra RN, Thomas WK et al (1999) Molecular and morphological analysis of isolates of *Pratylenchus coffeae* and closely related species. Nematropica 29:61–80

Ebsary BA (1991) Catalog of the order Tylenchida (Nematoda). Research Branch, Agriculture Canada, Ottawa

Ferraz S (1980) Reconhecimento das especies de fitonematóides presente nos solos do estado de Minas Gerais. Experientia 26:255–328

Filipjev IN (1936) On the classification of the Tylenchinae. Proceedings Helminthol Soc Wash 3:80–82

Fortuner R (1976) *Pratylenchus zeae*. CIH descriptions of plant-parasitic nematodes, 6, # 77, Commonwealth Institute of Helminthology, St. Albans

Fortuner R (1984) List and status of the genera and families of plant-parasitic nematodes. Helminth Abstr, ser B 53:87–133

Frederick JJ, Tarjan AC (1989) A compendium of the genus *Pratylenchus* Filipjev, 1936 (Nemata: Pratylenchidae). Rev Nematol 12:243–256

Gao X, Cheng H, Fang C (1999) New diagnostic characters for *Pratylenchus vulnus*. Nematol Mediter 27:9–13

Gnanapragasam NC, Mohotti KM (2005) Nematode parasites of tea. In: Luc M, Sikora RA, Bridge J (eds) Plant parasitic nematodes in subtropical and tropical agriculture, 2nd edition. CAB International, Wallingford

Golden AM, Lopez RC, Vilchez RH (1992) Description of *Pratylenchus gutierrezi* n. sp. (Nematoda:Pratylenchidae) from coffee in Costa Rica. J Nematol 24:298–304

Hajieghrari B, Mohammadi M, Kheiri A et al (2005) A study about geographical distribution of root-lesion nematode (*Pratylenchus loosi*, Loof 1960) in tea gardens at Guilan Province, Iran. Commun Agr Appl Biol Sci 70:889–892

Handoo ZA, Golden AM (1989) A key and compendium to the species of the genus *Pratylenchus* Filipjev, 1936 (lesion nematodes). J Nematol 21:202–218

Hernández M, Jordana R, Goldaracena A et al (2001) SEM observations of nine species of the genus Pratylenchus Filipjev, 1936 (Nematoda: Pratylenchidae). J Nematode Morphol System 3:165–174

Herve G (1997) Caracterisation biologique et moleculaire de populations de nematodes phytoparasites dans les plantations de cafeiers d'Amerique Centrale. Ecole Nationale Superieure Agronomique de Rennes, D.E.A. dissertation, Rennes

Inserra RN, Duncan LW, Dunn D et al (1998) *Pratylenchus pseudocoffeae* from Florida and its relationship with *P. gutierrezi* and *P. coffeae*. Nematologica 44:683–712

Inserra RN, Duncan LW, Dunn D et al (2005a) *Pratylenchus jordanensis* a junior synonym of *P. zeae*. Nematropica 35:161–170

Inserra RN, Duncan LW, Troccoli A et al (2001) *Pratylenchus jaehni* sp. n. from citrus in Brazil and its relationship with *P. coffeae* and *P. loosi* (Nematoda: Pratylenchidae). Nematology 3:653–665

Inserra RN, Duncan LW, Vovlas N et al (1996) *Pratylenchus loosi* from pasture grasses in Central Florida. Nematologica 42:159–172

Inserra RN, Stanley JD, O'Bannon JH et al (2005b) Nematode quarantine and certification programmes implemented in Florida. Nematol Mediterr 33:113–123

Inserra RN, Troccoli A, Gozel U et al (2007) *Pratylenchus hippeastri* n. sp. (Nematoda: Pratylenchidae) from amaryllis in Florida with notes on *P. scribneri* and *P. hexincisus*. Nematology 9:25–42

Karssen G, Waeyenberge L, Moens M (2000) *Pratylenchus* sp. nov. (Nematoda: Pratylenchidae), a root-lesion nematode from European coastal dunes. Ann Zoolog 50:255–261

Kubo RK, Oliveira CMG, Antedomênico SR et al (2004) Occurrence of *Pratylenchus* in coffee plantations of São Paulo State. Nematol Bras 28:159–166

Lax P, Doucet M, Di Rienzo JA et al (2004) Inter-population variability in *Pratylenchus vulnus* Allen and Jensen, 1951 (Nematoda: Tylenchida). Nematology 6:257–260

Loof PAA (1960) Taxonomic studies on the genus *Pratylenchus* (Nematoda). Tijdschr Plantenziekten 66:29–90

Loof PAA (1974) *Pratylenchus pratensis*. CIH Descriptions of plant-parasitic nematodes, 4, # 52, Commonwealth Institute of Helminthology, St. Albans

Loof PAA (1978) The genus *Pratylenchus* Filipjev, 1936 (Nematoda: Pratylenchidae): A review of its anatomy, morphology, distribution, systematics and identification. Swedish Universities Agricultural Science Research Information Centre, Uppsala

Loof PAA (1991) The family Pratylenchidae Thorne, 1949. In: Nickle WR (ed) Manual of Agricultural Nematology. M. Dekker, New York

López R, Salazar L (1990) Morphology of some *Pratylenchus* spp. (Nemata: Pratylenchidae) found in Costa Rica, as seen with the scanning electron microscope. Agron Costarric 14:189–195

Machon JE, Hunt DJ (1985) *Pratylenchus goodeyi*. CIH Descriptions of plant-parasitic nematodes, 8, # 120, Commonwealth Institute of Helminthology, St. Albans

Mai WF, Bloom JR, Chen TA (1977) Biology and ecology of the plant-parasitic nematode *Pratylenchus penetrans*. Pennsylvania State University, Bulletin 815, University Park

Mizukubo T (1992) Morphological and statistical differentiation of *Pratylenchus coffeae* complex in Japan (Nematoda: Pratylenchidae). Appl Entom Zool 27:213–224

Monteiro AR, Antedomênico SR, Ferraz LCCB et al (2001) Primeira ocorrência de *Pratylenchus vulnus* Allen and Jensen, 1951 em cafeeiro. Proceedings XXIII Congr Bras Nematol:88

Norton DC, Donald P, Kimpinski J et al (1984) Distribution of plant parasitic nematode species in North America. Society of Nematologists, Lakeland

Olowe T, Corbett DCM (1983) Morphology and morphometrics of *Pratylenchus brachyurus* and *P. zeae*. I. Effect of fixative and processing. Indian J Nematol 13:141–154

Olowe T, Corbett DCM (1984a) Morphology and morphometrics of *Pratylenchus brachyurus* and *P. zeae*. II. Influence of environmental factors. Indian J Nematol 14:6–17

Olowe T, Corbett DCM (1984b) Morphology and morphometrics of *Pratylenchus brachyurus* and *P. zeae*. III. Influence of geographical location. Indian J Nematol 14:30–35

Page RDM (1996) Treeview: an application to display phylogenetic trees on personal computers. Comput Appl Biosci 12:357–358

Park B-Y, Choi D-R, Lee J-K et al (2002) An unrecorded species of *Pratylenchus loosi* Loof (Tylenchida: Pratylenchidae) from tea in Korea. Korean J Appl Entomol 41:299–303

Peña E, Karssen G, Moens M (2007) Distribution and diversity of root-lesion nematodes (*Pratylenchus* spp.) associated with *Ammophila arenaria* in coastal dunes of Western Europe. Nematology 9:881–901

Pourjame E, Waeyenberge L, Moens M et al (1999) Morphological, morphometrical and molecular study of *Pratylenchus coffeae* and *P. loosi* (Nematoda: Pratylenchidae). Mededelingen – Faculteit Landbouwkundige en Toegepaste Biologische Wetenschappen, Universiteit Gent 64:391–401

Powers T (2008) The *Pratylenchus* project. http://nematode.unl.edu/pratyspp.htm, visited on Feb 22nd 2008

Rashid A, Khan AM (1978) Morphometric studies on *Pratylenchus coffeae* with description of *Pratylenchus typicus* Rashid, 1974. Indian J Nematol 6:63–72

Roman J, Hirschmann H (1969) Morphology and morphometrics of six species of *Pratylenchus*. J Nematol 1:363–386

Ryss AY (1988) Parasitic root nematodes of the family Pratylenchidae (Tylenchida) of the world fauna. (In Russian) Nauka, Leningrad

Ryss AY (2002a) Genus *Pratylenchus* Filipjev: multientry and monoentry keys and diagnostic relationships (Nematoda:Tylenchida:Pratylenchidae). Zoosystemat Ross 10:241–255

Ryss AY (2002b) Phylogeny and evolution of the genus *Pratylenchus* according to morphological data (Nematoda:Tylenchida). Zoosystemat Ross 10:257–273

Sakwe PN, Geraert E (1994) Species of the genus *Pratylenchus* Filipjev, 1936 (Nematoda: Tylenchida) from Cameroon. Fundam Appl Nematol 17:161–173

Sauer MR (1985) A scanning electron microscope study of plant and soil nematodes. CSIRO, Northfield

Schmitt DP, Barker KR (1981) Damage and reproductive potentials of *Pratylenchus brachyurus* and *P. penetrans* on soybean. J Nematol 13:327–332

Seinhorst JW (1968) Three new *Pratylenchus* species with a discussion of the structure of the cephalic framework and of the spermatheca in this genus. Nematologica 14:497–510

Seinhorst JW (1977) *Pratylenchus loosi*. CIH Descriptions of plant-parasitic nematodes, 7, # 100, Commonwealth Institute of Helminthology, St. Albans

Shendure J, Mitra RD, Varma C et al (2004) Advanced sequencing technologies: methods and goals. Nature Rev Genet 35:335–344

Sher SA, Allen MW (1953) Revision of the genus *Pratylenchus* (Nematoda: Tylenchidae). Univ California public Zool 57:441–470

Sher SA, Bell AH (1975) Scanning electron micrographs of the anterior region of some species of Tylenchoidea (Tylenchida: Nematoda). J Nematol 7:69–83

Siciliano-Wilcken SR, Inomoto M, Ferraz LC et al (2002a) Morphometry of *Pratylenchus* populations from coffee, banana, ornamental plant and citrus in Brazil. Proceedings IV Int Cong Nematol:356

Siciliano-Wilcken SR, Inomoto M, Ferraz LC et al (2002b) RAPD of *Pratylenchus* populations from coffee, banana, ornamental plant and citrus in Brazil. Proceedings IV Int Cong Nematol:163

Siddiqi MR (1972) *Pratylenchus coffeae*. CIH Description of plant-parasitic nematodes, 1, # 6, Commonwealth Institute of Helminthology, St. Albans

Siddiqi MR (1997) Techniques and methodologies for nematode disease diagnosis and nematode identification. FAO Plant Production and Protection Paper # 144, Rome

Siddiqi MR (2000) Tylenchida Parasites of Plants and Insects, 2nd edition. CAB International, Wallingford

Siddiqi MR, Dabur KR, Bajaj HK (1991) Descriptions of three new species of *Pratylenchus* Filipjev, 1936 (Nematoda: Pratylenchidae). Nematol Mediterr 19:1–7

Silva RA, Inomoto MM (2002) Host-range characterization of two *Pratylenchus coffeae* isolates from Brasil. J Nematol 34:135–139

Swofford DL (1998) PAUP – Phylogenetic Analysis Using Parsimony (and other methods), version 4.0. Sinauer Associates, Sunderland

Thompson JD, Higgins DG, Gibson TJ (1994) CLUSTAL W: improving the sensitivity of progressive multiple sequence alignment through sequence weighting, positions-specific gap penalties and weight matrix choice. Nucleic Acids Res 22:4673–4680

Trett MW, Perry RN (1985) Functional and evolutionary implications of the anterior sensory anatomy of species of root-lesion nematode (genus *Pratylenchus*). Rev Nematol 8:341–355

Troccoli A, Costilla MA, Lamberti F (1996) Note Morfo-biometriche su *Pratylenchus zeae* Graham, 1951. Nematol Mediterr 24:43–47

Vainrub A, Pustovyy O, Vodyanoy V (2006) Resolution of 90 nm ($\lambda/5$) in an optical transmission microscope with an annular condenser. Optic Lett 31:2855–2857

Van Den Berg E, Quénéhervé P (2000) *Hirschmanniella caribbeana* sp. n. and new records of *Pratylenchus* spp. (Pratylenchidae: Nematoda) from Guadeloupe, French West Indies. Nematology 2:179–190

Van Den Berg E, Queneherve P, Tiedi LR (2005) Six known plant-feeding nematodes from Guadeloupe, Martinige and French Guiana (Nemata: Tylenchina). J Nematode Morphol Systemat 7:109–129

Villain L (2000) Caracterisation et bioecologie du complexe parasitaire du genre Pratylenchus (Nemata: Pratylenchidae) present sur cafeiers (Coffea spp.) au Guatemala. Ecole Nationale Superieure Agronomique de Rennes, PhD thesis, Rennes

Villain L, Baujard P, Molina A et al (1998) Morphological and biological characterization of three *Pratylenchus* spp. populations parasitizing coffee trees in Guatemala. Nematologica 44:600–601

Whitehead AG (1968) Nematodea. In: Le Pelley RH (ed) Pests of Coffee. Longmans, Green and Co. Ltd., London

Chapter 4
Coffee-Associated *Pratylenchus* spp. – Ecology and Interactions with Plants

Mário M. Inomoto and Claudio Marcelo G. Oliveira

Abstract This chapter focuses on the basic biology of coffee-parasitic *Pratylenchus* spp., and on their interaction with coffee plants at the cellular, tissue and physiological levels. The parthenogenic species *P. brachyurus* and the amphimitic *P. coffeae* are well adapted to tropical climates, being prevalent in Indian and Central American coffee plantations, while the former is prevalent in Brazil. Soil temperatures lower than 10°C and higher than 32°C, and soil moisture content below 2% are unfavorable to the survival of these species. Their survival in fallowing soil is less than four months, although they survive for at least nine months in decaying roots; alternate hosts are also important for these species' epidemiology. It seems that edaphic conditions do not play a role in the distribution of *Pratylenchus* spp. on coffee. Both *Pratylenchus* species cause extensive damage in coffee root tissues, particularly in *Coffea arabica*; consequently, water and nutrient uptakes, photosynthesis and downward transport of sucrose are reduced; these processes originate the symptoms observed in parasitized coffee plants: stunting, severe chlorosis and leaf shedding.

Keywords Biology · histopathology · root-lesion nematodes · survival · symptoms

4.1 Introduction

Seven *Pratylenchus* species are known to be parasitic to coffee (*Coffea* sp.): *P. brachyurus* (Godfrey) Filipjev and Schuurmans Stekhoven, *P. coffeae* (Zimmerman) Filipjev and Schuurmans Stekhoven, *P. goodeyi* Sher and Allen, *P. gutierrezi* Golden, López and Vilchez, *P. loosi* Loof, *P. panamaensis* Siddiqi, Dadur and Barjas, *P. pratensis* (de Man) Filipjev, and *P. vulnus* Allen and Jensen. Another species, *P. zeae* Graham, has only been found in soil samples associated with gramineous weeds in coffee plantations (Schenck and Schmitt, 1992; Kubo et al., 2004).

M.M. Inomoto
Escola Superior de Agricultura Luiz de Queiroz/USP, Piracicaba, Brazil
e-mail: mminomot@carpa.ciagri.usp.br

R.M. Souza (ed.), *Plant-Parasitic Nematodes of Coffee*,

Because only *P. brachyurus* and *P. coffeae* have a wide distribution, mainly in tropical countries, these are the most studied species in their biological, ecological and control aspects. This chapter reviews some features of the interactions between *Pratylenchus* spp. and coffee. Whenever useful, information on interactions of *Pratylenchus* spp. with other plant species was also brought to light.

4.2 Life Cycle

All *Pratylenchus* spp., the so-called root-lesion nematodes, are endoparasitic and migratory nematodes. Males are rare in those species that reproduce by mitotic parthenogenesis, such as *P. brachyurus*, or abundant in amphimitic ones, such as *P. coffeae*. In general, the *Pratylenchus* life cycle is similar to that of other plant-parasitic nematodes, comprising eggs, four juvenile stages (J1 through J4), and adults.

Eggs are laid singly in the roots or in the soil. Although it is difficult to determine the total number of eggs laid by *Pratylenchus* females, the available data indicate that they lay few eggs. According to Graham (1951), each *P. brachyurus* female lays four to eight eggs *per* day, over 11 days feeding on maize roots growing in a moist chamber, under controlled temperature (26.7–29.4°C). No such information is available for coffee plants.

The first moult takes place inside the egg. In the first study on the biology of *P. coffeae*, Zimmermann (1898) observed that juveniles hatch in 6–8 days when the eggs are incubated in water at 28–30°C. Conversely, Lordello (1986) reported that in coffee seedlings, the J2 of *P. coffeae* are first observed 14 days after the eggs have been laid, but comparison with Zimmermann's data is limited since Lordello did not indicate the experimental temperature, and neither author mentioned the embryonic stage at which they began their observations. In coffee roots, Lordello (1986) observed the J1 stage on the eighth day after the eggs had been laid; J2, J3 and J4 on the 14th, 21st and 28th day, respectively; and the adults on the 29th to 32nd day. In potato, one generation of *P. coffeae* was completed in 27 days at 25–30°C (Gotoh, 1964, cited by Siddiqi, 1972), and the highest reproduction rate of *P. coffeae* in *Citrus jambhiri* Lush. (rough lemon) was obtained at 29.5°C, with optimal temperature for reproduction ranging from 26 to 32°C (Radewald et al., 1971).

No data is available about the life cycle of *P. brachyurus* on coffee. However, Graham (1951) estimated that, in laboratory, eggs of this species hatch in 15–20 days at 23.8–26.7°C. The author also reported that one generation of *P. brachyurus* is completed in 35–40 days on maize roots growing in a moist chamber at 23.8–29.4°C. According to Olowe and Corbett (1976), the motility of *P. brachyurus* in sand is not affected by temperatures ranging from 15 to 35°C, but the nematode remains inactive at temperatures below 10°C and above 35°C. The temperature affects more decisively the reproduction of *P. brachyurus*, which is inhibited in maize roots growing at 5, 10 and 15°C, and is enhanced from 20 to 30°C. The nematode population's increase peaks at 30°C, but decreases at 35°C. These results are considered consistent with the wide distribution of *P. brachyurus* in tropical countries.

On the other hand, the tea root-lesion nematode, *P. loosi*, has lower optimum temperature for development, ranging from 15.6 to 21.1°C (Sivapalan, 1972), and a longer life cycle of 45 to 48 days (Gadd and Loos, 1941). *P. loosi* is the most economically important plant-parasitic nematode of tea in Asia, and it has also been reported in association with coffee in Sri Lanka, the former Ceylon (Hutchinson, 1963, cited by Whitehead, 1969).

Both juveniles and adults of *Pratylenchus* sp. are able to enter the host roots. According to Rosana Bessi (personal communication), *P. coffeae* penetrates the roots of *C. arabica* L. (arabica coffee) mainly at the root tip (Fig. 4.1A,B), while Kumar (1982) reported penetration at the piliferous region. This focused nematode penetration could explain the destruction of the tap root in arabica coffee parasitized by *Pratylenchus* spp., as opposed to the minor root damage suffered by *C. canephora* Pierre ex A. Froehner (robusta coffee), in which the nematodes do not focus their penetration on any root region (Kumar, 1982).

Kumar (1982) also reported that arabica coffee roots seemed easier for *P. coffeae* to penetrate, in comparison to robusta coffee. In the former, around 10% of the nematodes effectively penetrated the roots within four to five days of the inoculation, while only 3% of the nematodes penetrated robusta roots within six to eight days of

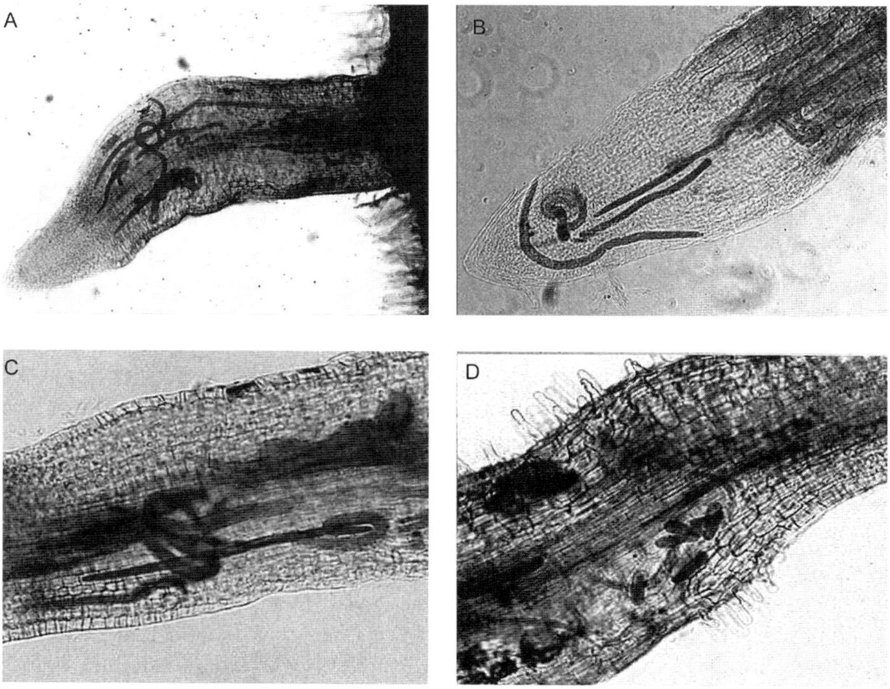

Fig. 4.1 *Pratylenchus coffeae* in the roots of *Coffea arabica* 'Catuai'. (**A**) massive nematode penetration at the root tip one day after inoculation (DAI). (**B**) nematodes at the root tip two DAI. (**C**) migrating and coiled-resting nematodes in root cortex four DAI. (**D**) Eggs laid in the cortex 20 DAI (Photos by Rosana Bessi, with permission)

the inoculation. In contrast, R. Bessi (personal communication) observed a massive penetration of *P. coffeae* just one day after inoculation of arabica coffee (Fig. 4.1A). These conflicting reports might stem from behavioral differences among *P. coffeae* populations or differences in the experimental conditions. Bessi also reported that after root penetration, the nematodes alternated among periods of migration through the cortex, resting in a coiled position inside the cells, and feeding on the cell contents (Fig. 4.1C), as Zunke (1990) had reported. Eggs were also observed in the cortical tissue (Fig. 4.1D).

In a recent study, Inomoto et al. (1998) observed that arabica coffee seedlings allowed a reproduction rate of just 0.9 for *P. brachyurus* 350 days after inoculation, while for *P. coffeae* the rate was 14 for the same experimental period. A lower fecundity of *P. brachyurus* females could explain this difference, although the possibility of a longer life cycle of *P. brachyurus* should be further investigated.

The genetic diversity in *Coffea* sp. should explain the high reproduction rate of *P. coffeae* in some genotypes of robusta coffee (as in 'IAC 4804' and 'IAC 4810'), and in the arabica coffee 'Mundo Novo', as opposed to the low reproduction observed in the robusta coffee 'IAC 4764' and 'IAC 4765' (Tomazini et al., 2005). As for *P. brachyurus*, Oliveira et al. (1999) found a consistently low reproduction rate in *C. canephora, C. salvatrix* Swynn. and Phil., and *C. congensis* A. Froehner, as well as in the interspecific hybrids Icatu and Sarchimor.

4.3 Survival

According to Feldmesser et al. (1960), two of the most important survival adaptations of plant-parasitic nematodes are the lack of host specificity (enabling completion of the life cycle on a variety of hosts), and the ability to undergo dormancy when unfavourable conditions, such as the absence of hosts, prevail. Possessing at least one of these adaptations enables the nematode to be a widespread, persisting parasite.

There has been only one study on the survival of *Pratylenchus* in coffee plantations (see below), with more information being available from other crops. In an apple orchard in Australia, Colbran (1954) removed all root pieces from the soil with the aid of a four-mesh sieve, and by using a biological assay he detected *P. coffeae* surviving in the soil for up to seven months in the absence of host plants. In Florida (USA), soil samples infested with *P. coffeae* were collected from a rough lemon orchard, and kept in the laboratory at different temperatures. The nematode survived for up to four months in moist soil kept near the field capacity at 10°C, but did not survive at temperatures above 38°C (Radewald et al., 1971). In South Africa, Koen (1967) collected *P. brachyurus*-infested soil from potato fields, and kept it at four temperatures (5, 8, 20 and 27°C) in the laboratory. After 20 weeks, the nematode had survived in all soil samples, but in lower numbers in those maintained at 5°C than in the samples maintained at 20 and 27°C. Also, more nematodes survived in the soil samples that had been kept wet (12% moisture, w/w) in comparison to those left to dry (5% moisture after 20 weeks).

Although these experiments were carried out under different regimes of temperature and soil moisture, they illustrate that *P. brachyurus* and *P. coffeae* are able to survive in the soil for at least four months in the absence of a host plant. Also, the nematode's survival seems to be shortened by extreme temperatures and low soil humidity.

A major component that enhances the survival of *Pratylenchus* is the presence of host root debris in the soil, which harbours and protects the nematodes. In a greenhouse experiment with soil temperature ranging from 18 to 29°C, Charchar and Huang (1991) reported that *P. brachyurus* survived for more than 3 months in soil mixed with root debris of *Melinis minutiflora*. In a laboratory trial, Feldmesser et al. (1960) reported that *P. brachyurus* remained infective after surviving for 21 months in soil mixed with debris of citrus roots. In South Africa, Koen (1967) observed that during the four-month-long winter, the *P. brachyurus* population in the soil dropped by 84% as the soil humidity decreased from 19% to 2% w/w. In the root debris, the nematode population dropped by only 39%, representing 66% of the total nematode population (soil + root debris) by the end of the winter. The survival of *P. brachyurus* in dried debris exposed to high temperatures was evidenced by Feldmesser and Rebois (1965).

These results clearly indicate a better survival of *P. brachyurus* in the host plant debris, where it remains protected from unfavourable temperature and desiccation. It is worth mentioning that none of the authors cited above made reference to which life stage(s) were involved in the *P. brachyurus* survival.

According to Kumar (1984a), *P. coffeae* persisted in the soil for up to nine months after infected coffee plants had been removed, but leaving the root system intact in the soil. In contrast, by removing the root debris and revolving the soil monthly caused the nematode population to drop to undetectable levels in just four months. No such information is available for *P. brachyurus*.

In coffee plantations, *P. brachyurus* and *P. coffeae* can survive by parasitizing weeds, previously cultivated crops or intercrops. Stradioto et al. (1983) reported that after maize harvest, maize-parasitizing *P. brachyurus* reproduced in the gramineous *Brachiaria* sp. and *Paspalum notatum* Fluegge during the 90-day off-season period. Lordello and Mello Filho (1969a) suggested that Pangola grass (*Digitaria eriantha* Steud. subsp. *pentzii*, formerly *D. decumbens*) could be a suitable host for *P. brachyurus*. Positive hosts for *P. brachyurus* include the gramineous *Melinis minutiflora* P. Beauv., *Hyparrhenia rufa* (Nees) Stapf., *B. purpurascens* (Henr. Blumea), *Chloris gayana* Kunth, *Cynodon dactylon* (L.) Persoon, *Panicum purpurascens* (Raddi) Henrard, sugarcane, and Sudan grass, as well as avocado, cassava, citrus, cotton, cowpea, yam (*Dioscorea* sp.), *Eucalyptus* spp., French bean, peach, peanut, pear millet, pineapple, *Pinus palustris* Mill, potato, rice, rubber tree, soybean, and tobacco (Lordello and Mello Filho, 1969b; Lordello, 1972; Corbett, 1976; Charchar and Huang, 1991). Although *P. brachyurus* reproduce poorly on coffee plants, high nematode populations can be found in plantations due to its reproduction in intercropped *Brachiaria decumbens* Stapf., which is often used as cover crop by coffee growers (Kubo et al., 2000).

P. coffeae, in turn, reproduces on several plant species besides coffee, such as on the gramineous *C. dactylon* and *Setaria verticilata* (L.) P.Beauv., on the trees albasia, rubber, mahogany, *Cinchona succirubra* Pav. ex Klotzsch, *Juglans regia* L., *Leucaena glauca* (Moench) Benth., and *Cassia tora* L., on the ornamental *Amaranthus lividus* L., snapdragon (*Antirrhinum majus* L.), camellia, caladium, oxalis, *Chrysanthemum* spp., dahlia, leopard plant (*Ligularia kaempferi* (DC.) Siebold. and Zucc., and marigold, on the algae *Nitella* sp., and on the aquatic plant *Potamogeton* sp., as well as on apple, bamboo, banana, potato, plum, red clover, strawberry, sweet potato, cocoa, grapevine, citrus, *Musa textilis* Née, lucerne, and tomato (Siddiqi, 1972; Kumar, 1984b; Mani et al., 1997).

These and many other plant species could render a crop rotation or fallowing against *P. coffeae* non-effective. For example, submitting a soil naturally infested with *P. coffeae* to a nine-month cultivation with weeds (*Eleusine indica* (L.) Gaertn., *Digitaria adscendens* (Kunth) Henrard and *Rumex acetosella* L.) or tomato resulted in high nematode reproduction in all plant species, specially *R. acetosella*, while the *P. coffeae* population was reduced to nearly undetectable levels in the fallow plots (Colbran, 1954).

4.4 Dispersion

According to Lordello and Mello Filho (1969a), *P. brachyurus* was disseminated throughout Brazil in the roots of Pangola grass cuttings, since this forage does not produce viable seeds for sowing. However, the production of coffee seedlings in *Pratylenchus*-infested soil is believed to be the most important way of disseminating the root-lesion nematodes. This is evident from the survey carried out by Reis (1965), who observed a high incidence of *Pratylenchus* sp. in coffee seedlings collected from several nurseries in the State of São Paulo, Brazil. These dissemination paths resulted in *P. brachyurus* being the most frequent root-lesion nematode in coffee plantations in the States of São Paulo and Minas Gerais (Souza et al., 1999; Kubo et al., 2004). The ability of *P. brachyurus* to survive in dried root debris, as discussed above, certainly contributes to this nematode's cosmopolitan distribution.

In India and Central America, *P. coffeae* is the most abundant root-lesion nematode associated with coffee. In India, *P. coffeae* is disseminated mostly by seedlings produced in nematode-infested soil. The soil is drawn directly from infested plantations or from areas with natural vegetation (Kumar, 1984b).

Since *P. brachyurus* and *P. coffeae* are polyphagous species, one might expect its introduction into new agricultural areas by way of seedlings, cuttings or tubers of several host plants. As these areas are turned into coffee plantations, severe problems with root-lesion nematodes may arise. For example, Moura et al. (2002) reported *P. coffeae* damaging a coffee plantation established in an area previously cultivated with yam (*Dioscorea cayennensis* Lam.). As the production and sale of tree seedlings is not regulated by adequate legislation concerning *Pratylenchus* spp. (Rosangela A. Silva, personal communication), one might also expect widespread introduction of root-lesion nematodes into tree farming areas, later turned into coffee plantations.

Therefore, every plant seedling should be produced free of plant-parasitic nematodes, and regulatory measures should be proposed and enforced to avoid dissemination of nematodes through seedlings of coffee or other plant species (Monteiro, 1981).

4.5 Edaphic and Climatic Conditions as Related to the Incidence of *Pratylenchus* spp. in Coffee Plantations

Some edaphic and climatic parameters have been reported to affect *Pratylenchus* populations, such as soil temperature, structure, pH, and humidity, as well as the environmental temperature. These reports suggest that both root-lesion nematodes and coffee plants have similar edaphic requirements, which probably do not play a significant role in the geographic distribution of *P. brachyurus* and *P. coffeae*, nor in its damage to coffee. Interactions with other soil nematodes or microorganisms seem to affect the incidence and density of *Pratylenchus* sp. in the soil. However, as previously mentioned, the main factor contributing to the localized incidence of root-lesion nematodes in coffee plantations is likely to be the efficiency of the dispersal agents. Furthermore, the great genetic variability and host preference amongst *P. coffeae* populations should also contribute to its localized incidence (Duncan et al., 1999; Silva and Inomoto, 2002; Wilcken et al., 2002).

According to Endo (1959), *P. brachyurus* reproduces better in strawberry and cotton grown in sandy loam soil than in clay loam, loam or sandy ones, indicating that soil texture does affect nematode activity. In laboratory, Olowe and Corbett (1976) reported that *P. brachyurus* moved better through sand particles sized from 0.375 to 0.750 mm, in comparison to smaller particles (0.096–0.300 mm). On the other hand, no correlation was found between *P. brachyurus* population and particle sizes in cotton fields in Brazil (Asmus, 2004).

Under laboratory conditions, pH does influence the viability of *P. brachyurus* juveniles. Nematodes incubated for one week in water acidified with HCl (pH 1, 3, 5 or 7) presented different rates of survival. At pH 1 and 3 the survival rates were 0 and 39.2%, while at pH 5 and 7 95% of the nematodes survived, with no statistical difference from the control (tap water at pH 7.3) (Koen, 1967). Considering that the optimal range of soil pH for coffee development is between 5.0 and 6.5 (Küpper, 1981), one should expect that under field condition *P. brachyurus* should not be affected by soil pH. Indeed, Cadet and Thioulouse (1998) observed that *P. brachyurus* was not influenced by the soil's physical and chemical characteristics, including pH, in tomato fields.

The limited data on the effects of environmental and soil temperatures on *Pratylenchus* sp. seem inconclusive, judging by the likely interference from other soil factors, such as texture and microorganisms. In coffee plantations in India, environmental temperature fluctuates very little over the year, and probably has no influence on *P. coffeae*. Nonetheless, higher nematode populations are observed in the monsoon months (July, August and September), corresponding to the period of increased rainfall and root activity (Kumar, 1984a). A contrary trend was observed in Guatemala, where vigorous growing of coffee roots and higher *P. coffeae* populations occur during dry months (December, January, and February) and at the beginning of the rainy season (June and July) (Villain et al., 1999). A sharp decrease in the nematode population, at the end of the rainy season, is believed to be associated with coffee root decay by secondary pathogens, which are favoured by the very moist soil.

In South Africa, potato field infestations with *P. brachyurus* are more severe during dry, hot summers (Koen, 1967), with the highest nematode population densities found in the upper 30 cm of soil during the summer, and between 20 and 40 cm during the winter. The author attributed the nematode migration to the winter's dessication of the upper soil layer.

Regarding the effects of other nematodes over *Pratylenchus* sp., Herve et al. (2005) observed competition between *P. coffeae* and *Meloidogyne exigua* Göldi, which was expressed by a strong negative correlation between root populations of these nematode species. Fourteen year-old arabica coffee trees harbouring high numbers of *P. coffeae* had low numbers of *M. exigua* on their roots, and vice-versa. In Costa Rica, the use of *M. exigua*-resistant genotypes of arabica coffee was linked to a significant build-up of *Pratylenchus* spp. populations (Villain et al., 1999), suggesting that coffee breeding programs should focus on both root-knot and root-lesion nematodes.

A greenhouse experiment demonstrated that coffee plants cultivated in low-phosphorus soil and infected with arbuscular mycorrhizal fungi (AMF) harboured more *P. coffeae* than uninfected plants, probably because the AMF enhanced the plant's uptake of phosphorus and root growth, with benefits to the nematodes. Such an effect was more evident when the AMF were inoculated four months prior to the nematode, in comparison to simultaneous inoculations. Prior inoculation with AMF also enhanced the plant's tolerance to *P. coffeae*, while simultaneous inoculation inhibited root colonization by the AMF, probably due to the destruction of root cortical cells by the nematodes (Vaast et al., 1998a).

4.6 Histopathology and Symptomatology

Kumar (1982) studied the histopathology of coffee roots infected with *P. coffeae*. After rupturing the epidermis, the nematodes invade the roots of both intolerant and tolerant (*sensu* Trudgill, 1991) coffees, respectively arabica and robusta. The nematodes migrate through epidermal and cortical cells by breaking down cell walls, and causing an enlargement of the cells just adjacent to the nematode's path, thus inducing a slight swelling of the infected root. In both coffee species, a depletion of starch was noticed in the cortical cells. A similar histopathology was observed in chickpea roots infected with *P. thornei* Sher and Allen (Castillo et al., 1998), and in soybean roots infected with *P. alleni* Ferris and *P. scribneri* Steiner in Sherbakoff and Stanley (Acosta and Malek, 1981). Kumar also reported that root lesions are formed by longitudinal migration of *P. coffeae* along the cortex. Generally, the nematodes migrate upwards, but eventually they do it towards the meristematic region. A dark brown substance, described by Kumar as 'wound-gum', is produced just adjacent to the injured region. This process is similar in both arabica and robusta coffees, although it is delayed in the latter (24 *vs*. 31 days after initial nematode penetration). Upon *P. coffeae* reproduction, the second generation may invade cortical and pericycle tissues. In arabica coffee the pericycle cells proliferate, resulting in a hyperplastic tissue which may girdle the stele, a process that does not happen in robusta coffee.

At the ultrastructural level, Townshend et al. (1989) showed that in alfalfa roots the cortical parenchyma cells penetrated and fed upon by *P. penetrans* (Cobb) Chitwood and Oteifa were generally devoid of cytoplasmatic content. Changes were also observed in the cells adjacent to those penetrated by the nematodes, including cortical parenchyma, endodermis, pericycle and vascular cells. Proximal cells had increased tannin deposits, degenerated mitochondria, increased numbers of ribosome and no internal membranous structure.

Several authors have characterized the symptoms caused by *Pratylenchus* spp. in coffee plants, under controlled conditions (Salas and Echandi, 1961; Inomoto et al., 1998; Kubo et al., 2003). Generally, *P. coffeae*-parasitized plants are stunted and exhibit pronounced leaf chlorosis and root shedding. In seedlings, the main root can be destroyed and lose the ability to sustain the shoot. Rootlets may exhibit altered colour, from dark brown to black, except those emerging from the most proximal part of the main root, probably because this region is not infected by the nematodes. Symptoms caused by *P. brachyurus* are similar but less severe than those described above (Fig. 4.2), perhaps because of its low reproductive rate on coffee, as discussed above.

Villain et al. (1999) observed different levels of pathogenicity to arabica coffee among populations of *Pratylenchus* sp. collected from three different coffee plantations in Guatemala. The authors considered the intraespecific diversity as the cause of such variability.

Kubo et al. (2003) described the symptoms caused by *P. coffeae* in arabica coffee seedlings at the stage of one or two pairs of leaves (Fig. 4.3). Similar symptomatology can also be observed in older plants, with six or seven pairs of leaves, of both arabica and robusta coffees (Tomazini et al., 2005). Furthermore, Inomoto et al. (2004) compared the aggressiveness of *P. coffeae* and *M. incognita* (Kofoid and White) Chitwood and concluded that both species cause a similar decay on coffee roots, resulting in poor shoot development.

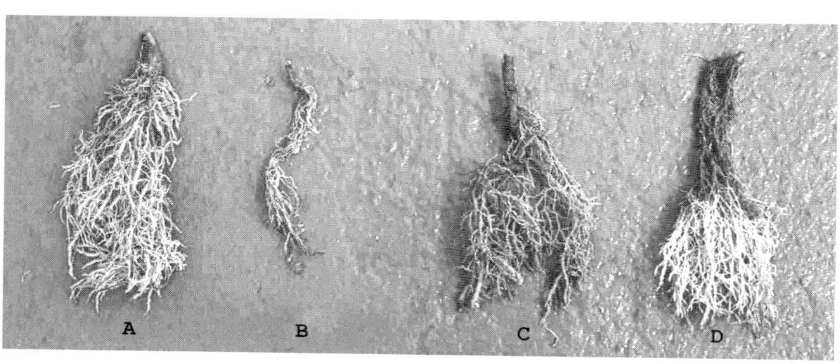

Fig. 4.2 (**A**) healthy coffee roots. (**B, C**) coffee roots infected by *Pratylenchus coffeae*. (**D**) infected by *P. brachyurus* (Photo by Mario Inomoto and Claudio Oliveira)

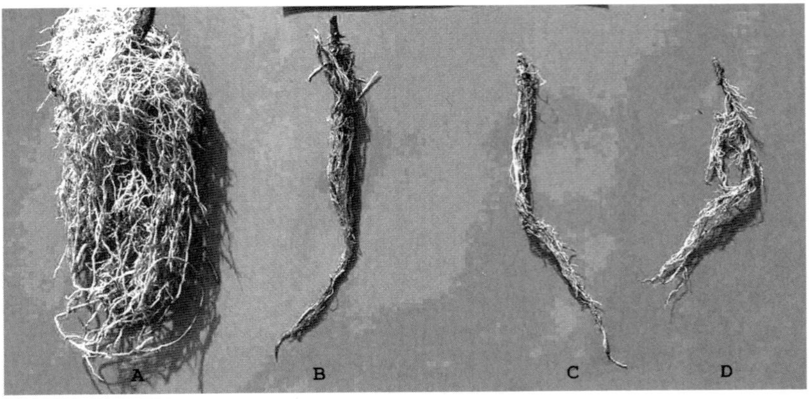

Fig. 4.3 (**A**) healthy arabica coffee roots, and symptoms caused by *Pratylenchus coffeae* at the inoculum level of 333 (**B**), one thousand (**C**), and three thousand (**D**) eggs and juveniles (Photo by Mario Inomoto and Claudio Oliveira)

Under field conditions, young coffee plants are very sensitive to *Pratylenchus* sp., and the symptoms are similar to those described above under greenhouse conditions (Kumar and Samuel, 1990). In some areas, heavily infected plants may develop a corky region at the base of the trunk (Schieber, 1968; Kumar 1984b). In Brazil, Lordello (1972) and Monteiro and Lordello (1974) reported that young coffee trees infected with *P. brachyurus* or *P. coffeae* were stunted and presented thin stems, nutrient deficiencies, and poor root system, and even plant death was observed. These symptoms have also been associated with *P. coffeae* in other coffee growing regions, such as Central America, East Africa and Asia (Salas and Echandi, 1961; Guiran, 1971; Kumar and Samuel, 1990). In Brazil, the present authors verified that mature plants parasitized by *P. coffeae* exhibited severe symptoms after short pruning, probably because newly grown roots were so damaged by the nematodes that the plant was not capable of producing healthy shoots, which is congruent with the symptoms described in India by Kumar and Samuel (1990).

As for *P. brachyurus*, field symptoms are more pronounced when *Brachiaria decumbens* or other good hosts are used as cover crop in coffee plantations (Kubo et al., 2000).

4.7 Physiology of the Parasitized Plant and its Relation to Yield Loss

It has been demonstrated that *Pratylenchus* sp. may cause extensive damage to coffee roots, resulting in reduced water and nutrient uptake. For example, *P. coffeae*-parasitized plants presented a significant reduction in ammonium and nitrate uptake, probably because the root integrity and function were affected by the nematode invasion. Indeed, abundant brown lesions were observed on the coffee roots as a result of death of cortical cells during nematode feeding and migration (Vaast et al., 1998b). These authors also noticed a lower concentration of nitrogen, phosphorus, potassium, calcium, magnesium and zinc in the leaves of coffee plants two months after the nematode inoculation, in comparison to uninoculated ones. This is congruent with the symptoms observed in coffee plantations parasitized by root-lesion nematodes.

Furthermore, these nematodes may also affect other aspects of the plant's physiology, such as photosynthesis and carbon partitioning. According to McClure (1977), the alteration in the physiology of carbohydrates in *Meloidogyne*-infected plants could be related to source–sink interactions, with the infected roots representing the sink. Depending on the strength of the sink (related to the number of nematode feeding sites in the roots), a high energy demand might induce an increase in the sucrose content of the leaves, via photosynthesis and starch hydrolysis, with subsequent transport to the roots.

Since *Pratylenchus* sp. are migratory nematodes that do not form feeding sites, the carbohydrate alterations in the host plant should not be related to a metabolic sink, but rather to the extensive root lesions caused by the nematodes. While examining carbon fixation and partitioning in *P. coffeae*-parasitized arabica coffee seedlings, Mazzafera et al. (2004) observed that a decrease in labelled sucrose in the roots was associated with an increase in the leaves. In addition, there was an increase in soluble sugar in the leaves, explained by the starch hydrolysis associated with a higher respiration rate. Both phenomena, a reduced transport of sucrose from the leaves and a higher respiration rate, could be a consequence of root damage by *P. coffeae*. Indeed, the physiological alterations were more pronounced in plants inoculated with eight thousand nematodes, which exhibited more root lesions, than in those plants inoculated with one thousand nematodes.

Interestingly, Mazzafera et al. (2004) also observed that the leaf chlorophyll content and the $^{14}CO_2$ fixation decreased in coffee seedlings parasitized by *P. coffeae*. Therefore, the authors hypothesized that the destruction of the root system by *P. coffeae* was readily felt by the leaves, leading to a faster decrease in carbon assimilation. Also, a decrease in total sucrose was observed in the leaves and roots as a consequence of photosynthesis inhibition associated with a reduction in sucrose translocation from the leaves to the roots.

In a previous work, Inomoto et al. (1998) has observed higher concentration of soluble sugars in the leaves of arabica coffee parasitized by *P. brachyurus* and *P. coffeae*. Furthermore, while evaluating the effects of different *P. coffeae*-population densities on the photosynthesis of arabica coffee, Kubo et al. (2003) determined that populations above 900 nematodes *per* plant decreased the photosynthesis.

4.8 Concluding Remarks

Given the economic importance of coffee for the countries that cultivate it, and the serious damage caused by the root-lesion nematodes, the scarcity of studies on several aspects of the coffee-*Pratylenchus* interaction is surprising, such as the nematode life cycle, population dynamics, feeding behaviour, pathogenicity, host range, and geographic distribution.

Root-lesion nematodes are efficiently cultured *in vitro*, so a large number of specimens can be easily obtained for life cycle studies. The scarcity of studies on the influence of environmental factors (soil temperature and texture, rainfall, etc.) on the life cycle of *Pratylenchus* sp. explain the conflicting data regarding population dynamics in coffee plantations. Data on nematode population dynamics are essential to determine damage and economic thresholds, which benefit the decisions regarding nematode management with nematicide applications.

For a better understanding of the feeding behaviour and the pathogenicity of *Pratylenchus* sp. on coffee, more studies are necessary on the histopathology and ultrastructure of infected roots. Particularly, the mechanisms involved in coffee resistance should be investigated. Such information might be useful for an early selection of resistant germoplasms in breeding programs.

Currently, research efforts are limited to *P. coffee* and *P. brachyurus*, although five other species are known to parasitize coffee. Therefore, the economic importance of *P. goodeyi, P. gutierrezi, P. loosi, P. panamaensis, P. pratensis*, and *P. vulnus* should be further investigated. Additional information is needed from extensive surveys in coffee-producing countries, in order to fully understand the geographic distribution of root-lesion nematodes in coffee plantations. Even though some information regarding *P. brachyurus*-coffee interaction is available, several aspects require further examination, e.g. the effect of the nematodes on the physiology of coffee, as previous analysed for *P. coffeae*.

Finally, several hypothesis postulated here concerning *Pratylenchus* survival and dispersion mechanisms should be further investigated, allowing a better understanding of key factors involved in *Pratylenchus* dissemination in coffee plantations. For example, a comprehensive survey aiming to identify weed species suitable for *Pratylenchus* spp. is badly needed.

References

Acosta N, Malek RB (1981) Symptomatology and histopathology of soybean roots infested by *Pratylenchus scribneri* and *P. alleni*. J Nematol 13:6–12

Asmus GL (2004) Ocorrência de fitonematóides em algodoeiros no estado do Mato Grosso do Sul. Nematol Bras 28:77–86

Cadet P, Thioulouse J (1998) Identification of soil factors that relate to plant parasitic nematode communities on tomato and yam in the French West Indies. Appl Soil Ecol 8:35–49

Castillo P, Vovlas N, Jiménez-Diaz RM (1998) Pathogenicity and histopathology of *Pratylenchus thornei* populations on selected chickpea genotypes. Plant Pathol 47:370–376

Charchar JM, Huang CS (1991) Sobrevivência de *Pratylenchus brachyurus* em fragmentos de raizes de capim gordura *Melinis minutiflora*. Fitopatol Bras 16:22–25

Colbran RC (1954) Problems in tree replacement. II. The effect of certain methods of management on the nematode fauna of orchard soils. Aust J Agr Sci 20:234–237

Corbett DCM (1976) *Pratylenchus brachyurus*. C.I.H. Descriptions of plant parasitic nematodes, set 6, number 89, C.H.I., St. Albans

Duncan LW, Inserra RN, Thomas WK et al (1999) Genetic and morphological relationships among isolates of *Pratylenchus coffeae* and closely related species. Nematropica 29:61–80

Endo BY (1959) Responses of root-lesion nematodes, *Pratylenchus brachyurus* and *P. zeae*, to various plants and soil types. Phytopathol 49:417–421

Feldmesser J, Feder WA, Rebois RV et al (1960) Longevity of *Radopholus similis* and *Pratylenchus brachyurus* in fallow soil in the greenhouse. The Anat Rec 137:355

Feldmesser J, Rebois RV (1965) Temperature and moisture effects on *Pratylenchus brachyurus*. Nematologica 11:37–38

Gadd CH, Loos CA (1941) Observations on the life history of *Anguillulina pratensis*. Ann Appl Biol 29:39–51

Graham TW (1951) Nematode root rot of tobacco and other plants. South Carolina Agr Exp Stn. Technical Bulletin 390

Guiran G (1971) Le problème *Meloidogyne* et autres nématoses sur cultures vivrières tabac, caféier, riz. In: Ritter M (ed) Les Nematodes des Cultures. ACTA-FNGPC, Paris

Herve G, Bertrand B, Villain L et al (2005) Distribution analyses of *Meloidogyne* spp. and *Pratylenchus coffeae* sensu lato in coffee plots in Costa Rica and Guatemala. Plant Pathol 54:471–475

Inomoto MM, Oliveira CMG, Mazzafera P et al (1998) Effects of *Pratylenchus brachyurus* and *P. coffeae* on seedlings of Coffeae arabica. J Nematol 30:362–387

Inomoto MM, Beluti D, Siqueira KMS et al (2004) Efeito de *Pratylenchus coffeae* e *Meloidogyne incognita* no crescimento de cafeeiro 'Catuai Vermelho'. Nematol Bras 28:143–148

Koen H (1967) Notes on the host range, ecology and population dynamics of *Pratylenchus brachyurus*. Nematologica 13:118–124

Kubo RK, Inomoto MM, Oliveira CMG (2000) Report of damage caused by *Pratylenchus brachyurus* in coffee plantation of São Paulo State. Fitopatol Bras 25:338

Kubo RK, Oliveira CMG, Antedomênico SR et al (2004) Ocorrência de nematóides do gênero *Pratylenchus* em cafezais do estado de São Paulo. Nematol Bras 28:159–166

Kubo RK, Silva RA, Tomazini MD et al (2003) Patogenicidade de *Pratylenchus coffeae* em plântulas de cafeeiro cv. Mundo Novo. Fitopatol Bras 28:41–48

Kumar AC (1982) Studies on nematodes in coffee soils of South India. 7. Histopathology and host parasitic relationship of *Pratylenchus coffeae* and two species of coffee. J Coffee Res 12:23–30

Kumar AC (1984a) Investigations on 'Cannoncadoo dieback' in coffee. J Coffee Res 14:85–103

Kumar AC (1984b) The symptoms and diagnosis of the disorder 'spreading decline' (Cannoncado dieback) with a note on spread and control of the causal agent, *Pratylenchus coffeae*. J Coffee Res 14:156–159

Kumar AC, Samuel SD (1990) Nematodes attacking coffee and their management- a review. J Coffee Res 20:1–27

Küpper A (1981) Fatores climáticos e edáficos na cultura cafeeira. In: Malavolta E, Yamada T and Guidolin JA (eds) Nutrição e Adubação do Cafeeiro. Instituto da Potassa, Piracicaba

Lordello LGE (1972) Nematode pests of coffee. In: Webster JM (ed) Economic Nematology. Academic Press, London

Lordello LGE (1986) Plant-parasitic nematodes that attack coffee. In: Nematodes of Bananas, Citrus, Coffee, Grapes, and Tobacco. Union Carbide Agricultural Products Company, North Carolina

Lordello LGE, Mello Filho AT (1969a) O capim pangola difunde nematóides. Rev Agr 44:122

Lordello LGE, Mello Filho AT (1969b) Capins gordura e jaraguá, hospedeiros novos de um nematóide migrador. O Solo 61:27–28

McClure MA (1977) *Meloidogyne incognita*: a metabolic sink. J Nematol 9:89–90

Mazzafera P, Kubo RK, Inomoto MM (2004) Carbon fixation and partitioning in coffee seedlings infested with *Pratylenchus coffeae*. Eur J Plant Pathol 110:861–865

Mani A, Al-Hinai MS, Handoo ZA (1997) Occurrence, population density, and distribution of root-lesion nematodes, *Pratylenchus* spp., in the sultanate of Oman. Nematropica 27:209–219

Monteiro AR (1981) Não se deve plantar nematóides. Soc Bras Nematol 5:13–20

Monteiro AR, Lordello LGE (1974) Encontro do nematóide *Pratylenchus coffeae* atacando cafeeiro em S. Paulo. Rev Agr 44:164

Moura RM, Pedrosa EMR, Prado MDC (2002) Incidência de *Pratylenchus coffeae* causando severa nematose em cafeeiro no Nordeste. Fitopatol Bras 27:649

Oliveira CMG, Monteiro AR, Antedomenico SR et al (1999) Host reaction of *Coffea* spp. to *Pratylenchus brachyurus*. Nematropica 29:241–244

Olowe T, Corbett DCM (1976) Aspects of the biology of *Pratylenchus brachyurus* and *P. zeae*. Nematologica 22:202–211

Radewald JD, O'Bannon JH, Tomerlin AT (1971) Temperature effects on reproduction and pathogenicity of *Pratylenchus coffeae* and *P. brachyurus* and survival of *P. coffeae* in roots of *Citrus jambhiri*. J Nematol 3:390–394

Reis AJ (1965) Nematóide prejudica muda de café. Divulgação Agronômica 16:16–17

Salas LA, Echandi E (1961) Nematodos parasitos en plantaciones de cafe de Costa Rica. Turrialba 3:21–24

Schenck S, Schmitt DP (1992) Survey of nematodes on coffee in Hawaii. J. Nematol 24:771–775

Schieber E (1968) Nematode problems on coffee. In: Smart, GC and Perry VG (eds) Tropical Nematology. University of Florida Press, Gainesville.

Siddiqi MR (1972) *Pratylenchus coffeae*. C.I.H. Descriptions of plant parasitic nematodes, set 1, number 6. C.I.H., St. Albans

Silva RA, Inomoto MM (2002) Host-range characterization of two *Pratylenchus coffeae* isolates from Brazil. J Nematol 34:135–139

Sivapalan P (1972) Nematode pests of tea. In: Webster JM (ed) Economic Nematology. Academic Press, London

Souza JT, Maximiniano C, Campos VP (1999) Nematóides parasitos encontrados em cafeeiros em campo e em viveiros de mudas do Estado de Minas Gerais. Summa Phytopathol 25:180–183

Stradioto MF, Ferraz, LCCB, Pitelli RA (1983) Dinâmica populacional de *Pratylenchus brachyurus* em cultura do milho (*Zea mays* L.) infestada por plantas daninhas. Soc Bras Nematol 7:99–115

Tomazini MD, Silva RA, Oliveira CMG et al (2005) Resistência de genótipos de cafeeiros a *Pratylenchus coffeae* e *Meloidogyne incognita*. Nematol Bras 29:193–198

Townshend JL, Stobbs L, Carter R (1989) Ultrastructural pathology of cells affected by *Pratylenchus penetrans* in alfafa roots. J Nematol 21:530–539

Trudgill DL (1991) Resistance and tolerance of plant parasitic nematodes in plants. Annu Rev Phytopathol 29:167–192

Vaast P, Caswell-Chen EP, Zasoski RJ (1998a) Effects of a root-lesion nematode, *Pratylenchus coffeae*, and two arbuscular mycorrhizal fungi, *Acaulospora mellea* and *Glomus clarum* on coffee (*Coffea arabica* L.) Biol Fertil Soils 26:130–135

Vaast P, Caswell-Chen EP, Zasoski RJ (1998b) Effects of two endoparasitic nematodes (*Pratylenchus coffeae* and *Meloidogyne konaensis*) on ammonium and nitrate uptake by Arabica coffee (*Coffea arabica* L.) Appl Soil Ecol 10:171–178

Villain L, Anzueto F, Hernández A et al (1999) Los nematodos parasitos del cafeto. In: Bertrand B and Rapidel B (eds) Desafios de la Caficultura en Centroamerico. IICA Promecafe, San Jose

Whitehead, AG (1969) Nematodes attacking coffee, tea and coocoa, and their control. In: Peachey JE (ed) Nematodes of Tropical Crops. CAB, St. Albans

Wilcken SRS, Inomoto MM, Ferraz LCCB et al (2002) RAPD of *Pratylenchus populations* from coffee, banana, ornamental plant and citrus in Brazil. Nematology 4:179–180

Zimmermann A (1898) De nematoden der Koffiewortels. Med.'s Lands Plantentin 27:1–64

Zunke U (1990) Observations on the invasion endoparasitic behavior of the root lesion nematode *Pratylenchus penetrans*. J Nematol 22:309–320

Chapter 5
Economic Importance, Epidemiology and Management of *Pratylenchus* sp. in Coffee Plantations

Luc Villain

Abstract As coffee-parasites, root-lesion nematodes (RLNs), *Pratylenchus* spp., have been underestimated in terms of their importance to coffee production. Indeed, their migratory behavior and the symptoms they induce – non-specific root necrosis – have not caught the attention of nematologists, extensionists and growers until recently. Nowadays, RLNs are being recognized as damaging to arabica and robusta coffees in Guatemala, El Salvador, Indonesia and Vietnam, among others. This awareness has arisen from studies conducted on several aspects, such as population fluctuation, epidemiology, assessment of damage threshold and management through chemical, biological, cultural and genetic approaches. This chapter focuses on discussing in detail all these aspects.

Keywords Root-lesion nematodes · epidemiology · chemical control · cultural control · biological control · genetic control

5.1 Introduction

In some coffee-producing countries or regions, root-lesion nematodes (RLNs), *Pratylenchus* spp., are considered major parasites of arabica and robusta coffees (*C. arabica* L. and *C. canephora* Pierre ex Froehner, respectively). This review complements Chapters 3 and 4, for it deals with management of RLNs. Initially, the chapter emphasizes that these nematodes are likely to be more important to coffee production worldwide than previously estimated. The available literature on RLN population fluctuation is discussed, with emphasis on aspects related to production systems. A thorough discussion is made on the difficulties in establishing and using damage thresholds for RLN-management. Different approaches for controlling these nematodes – chemical, biological, genetic and cultural – have their advantages and disadvantages examined. At the end, research needs are outlined in order to

L. Villain
Centre de Coopération Internationale en Recherche Agronomique pour le Développement TA A-98/IRD, Montpellier, France
e-mail: luc.villain@ird.fr

R.M. Souza (ed.), *Plant-Parasitic Nematodes of Coffee*,

address two main goals: assessing the role played by RLNs on coffee production worldwide and developing strategies for their efficient, durable and environment-friendly management.

5.2 Economic Importance

The economic importance of RLNs to coffee production worldwide has probably been underestimated. Indeed, unlike root-knot nematodes (*Meloidogyne* sp.) which induce root galls or swellings, RLNs induce non-characteristic necroses in the cortex of coffee roots, which correlate with secondary detrimental alterations in the plant's physiology and above-ground symptoms (see Chapter 4). The root symptoms induced by RLNs can easily be taken as death of coffee roots caused by normal physiological changes during the plant's phenological cycle or by unfavorable abiotic, telluric conditions (water saturation, physical and/or chemical factors, etc). Therefore, parasitism by RLNs and the related yield loss (see below) often pass unnoticed unless field samplings and laboratory analyses are performed. Such analyses are particularly necessary when coffee plants are parasitized by *Meloidogyne* sp., whose symptoms easily mask the presence of RLNs.

Under these circumstances, it is quite difficult to estimate the economic importance of specific coffee-parasitic *Pratylenchus* species, all the more considering the uncertainties on the taxonomic status of several amphimitic RLN populations (see Chapter 3).

Because of its pantropic distribution, *P. coffeae* (Zimmerman) Filipjev and Schuurmans Stekhoven is the most reported species on coffee (Villain et al., 2002; Campos and Villain, 2005) and on other tropical or sub-tropical crops such as banana (Gowen et al., 2005) and yam (*Dioscorea* sp.) (Bridge et al., 2005). Recent morphological, biological and molecular studies have raised doubts on the taxonomic status of several amphimitic coffee-parasitic RLN isolates from Central America and Brazil (Hervé, 1997; Villain et al., 1998; Duncan et al., 1999; Villain et al., 2000; Siciliano-Wilcken et al., 2002a,b; Silva and Inomoto, 2002). Particularly, some populations from Guatemala, El Salvador and Costa Rica seem to belong to species morphologically close to but different from *P. coffeae* because of their reproductive isolation and genetic distance. Furthermore, these populations show considerable variability in their root penetration dynamics and reproductive fitness on arabica coffee (Villain et al., 2000; Villain et al., 2001a,c).

In conclusion, an indeterminate proportion of the reports dealing with coffee-parasitic *P. coffeae* could probably be related to other *Pratylenchus* species, or even to undescribed taxa. A similar situation has recently occurred with the description of a new species closely related to citrus-parasitic *P. coffeae* (Inserra et al., 2001).

Bridge et al. (1997) suggested that *P. coffeae*, originally described from coffee roots, could be native to the Pacific islands and the Pacific Rim countries, and that it could have been spread worldwide through banana (*Musa* spp.) planting materials. In Central America, *P. coffeae* has been reported as economically important for coffee cultivation in Guatemala (Chitwood and Berger, 1960; Schieber and

Sosa, 1960; Schieber, 1966; 1971) and El Salvador (Abrego and Holdeman, 1961; Whitehead, 1969; Gutierrez and Jimenez, 1970).

As detailed in Chapter 15, *P. coffeae* has also been reported in Indonesia causing serious damage to plantations of arabica and robusta coffees (Wiryadiputra, 1990 cited by Toruan-Mathius et al., 1995; Toruan-Mathius et al., 1995). In the latter, the yield losses ranged between 29 and 78%. Also, as detailed in Chapter 15, *P. coffeae* seems to be widely distributed in some of the robusta-producing provinces of Vietnam, where it is considered one of the most important nematodes on this crop. In India, *P. coffeae* is considered the most destructive nematode for arabica coffee (Palanichamy, 1973; see Chapter 16).

In Brazil, *P. coffeae* has been reported causing serious damage to some coffee plantations in the States of São Paulo (Monteiro and Lordello, 1974; Kubo et al., 1999; 2001; 2002a) and Pernambuco (Moura et al., 2003). The pathogenicity of *P. coffeae* isolates from São Paulo on arabica coffee has been demonstrated through controlled inoculation assays (Inomoto et al., 1998; Silva et al., 2001; Silva and Inomoto, 2002; Kubo et al., 2002b).

P. coffeae has also been reported on coffee in many different countries in Latin America, the Caribbean region, Asia, Africa and in the North-American State of Hawaii, but without details on its economic significance (Campos and Villain, 2005).

P. gutierrezi Golden, López and Vilchez and *P. panamaensis* Siddiqi, Dadur and Barjas, two amphimitic species that are morphologically similar to *P. coffeae*, have been described from Costa Rica and Panama, respectively (Siddiqi et al., 1991; Golden et al., 1992). Nonetheless, no information was given on their pathogenicity or economic importance on coffee.

P. brachyurus (Godfrey) Filipjev and Schuurmans Stekhoven seems to be the most widely distributed RLN on coffee in Brazil, at least in its main producing States, Minas Gerais and São Paulo (Lordello, 1972; Gonçalves et al., 1978; D'Antonio et al., 1980; Kubo et al., 2002a). Its pathogenicity to seedlings of arabica and robusta (cultivar 'Apoatã') coffees has been demonstrated under controlled inoculation assays (Inomoto et al., 1998; Oliveira et al., 1999). However, no figures are available on this species' damage to coffee worldwide.

Some other *Pratylenchus* species have been reported on coffee locally, such as *P. pratensis* (de Man) Filipjev in south India and *P. loosi* Loof in Sri Lanka (Whitehead, 1968), *P. goodeyi* Sher and Allen in Tanzania (Bridge, 1984) and *P. zeae* Graham in Brazil (Ferraz, 1980; Monteiro et al., 2001). Apparently, these species do not have any economic importance to coffee cultivation.

5.3 Epidemiology

5.3.1 Estimate of Population Damage Thresholds for RLNs

Very few studies have been performed relating RLN population levels to quantitative or qualitative damages to coffee plantations. A field assay carried out in Guatemala revealed a strong negative correlation between the productivity of ungrafted arabica coffee tree plots and their average RLN population (Villain et al., 2000). These authors have found it difficult to correlate coffee yield with nematode population at any specific point in time. Instead, a high correlation was observed between the cumulative population level taken from samplings performed every six months during the plantation's formation and the yield obtained in its third harvest (Fig. 5.1).

The negative correlation between yield loss and cumulative nematode population can be understood if one considers that coffee flowering and subsequent fruit production occur on second year-wood branches. Therefore, for any given production the RLN population will have affected the coffee plants in the previous year, by compromising the growth of productive plagiotropic branches and the production of flowering nodes; in the following year, this will compromise flowering and coffee bean filling, thus reducing productivity.

Another physiological aspect of coffee plants explains why a given production is probably more affected by previous year- than same year-damage caused by RLNs and *Meloidogyne* sp. as well: coffee plants have a peculiar inability to shed excessive fruits in relation to their nutritional status (Cannell, 1985). This can lead to 'die-back' of branches and long-term, dramatic effects on the plantation's productive life span. This explains why Villain et al. (2000) observed a drastic increase in the rate of plant deaths when these began to produce, two years after planting (Fig. 5.2). Four years after planting, after the third harvest, the death rate increased to 50% on average, and it reached 76% in the most RLN-infested plots. In this assay, the plot infested with the lowest population, average of 15 nematodes/g of root, yielded around six tons of coffee berries/hectare (ha), in comparison to the most infested one, 125 nematodes/g of root, which yielded around 0.5 ton/ha (Fig. 5.1).

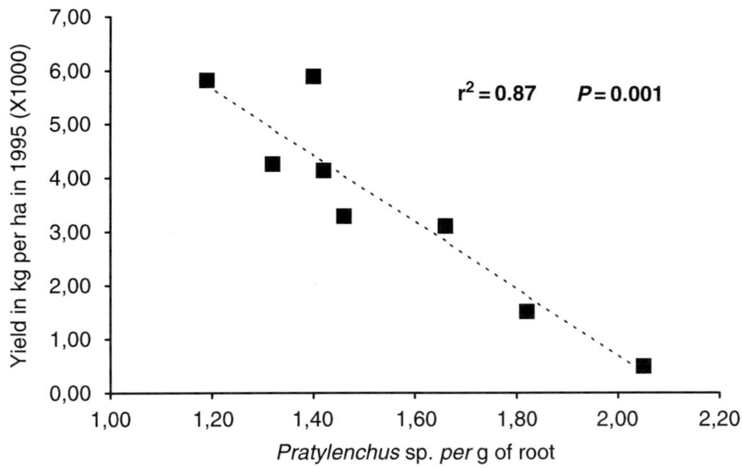

Fig. 5.1 Relationship between coffee berry yield in southwest Guatemala and average of *Pratylenchus* sp. root population during the three years prior to harvest. Nematode numbers are mean of log[x + 1]-transformed original counts in eight plots of 50 ungrafted *Coffea arabica* plants each (from Luc et al., 2000, with permission)

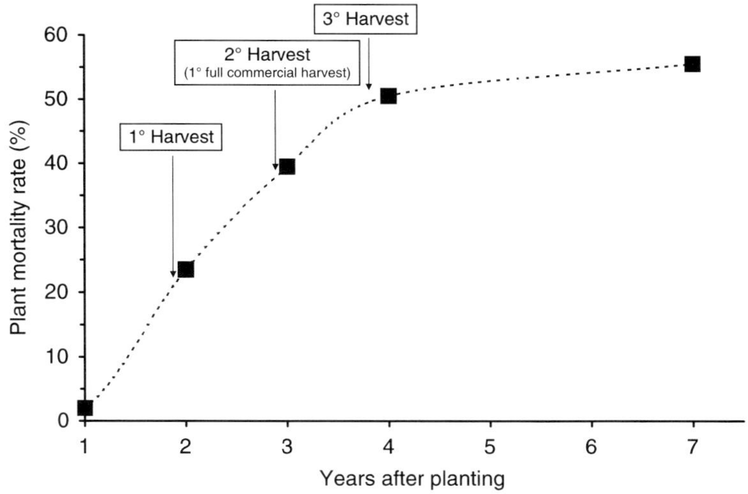

Fig. 5.2 Time-course evolution of average plant mortality rates in plots with 50 ungrafted *Coffea arabica* plants parasitized by *Pratylenchus* sp. in Guatemala. Values are averages of four plots

In addition to this quantitative effect on coffee production, RLN population levels also correlate negatively with the qualitative variable of coffee bean size (Villain et al., 2001b). The share of beans retained in the sieve with an aperture of 17/64 inches or larger was reduced from 95% for the least infested plots to 65% for the most infested ones. These studies show that this RLN population, distinct from *P. coffeae* but still under taxonomic study, and widely present in Guatemala and El Salvador, is highly damaging to arabica coffee. Likewise, other RLN populations or undescribed species could be just as damaging to coffee.

5.3.2 The Effect of Intensive Production Systems on RLN Epidemiology and Damage

The intensification of coffee cultivation began in the 1960s and 1970s with the advent of low-habit cultivars such as ‘Caturra’ and ‘Catuai’, and later of ‘Catimors’ and ‘Sarchimors’ resistant to ‘leaf rust’ caused by *Hemileia vastatrix* Berk and Br., resulted in changes in the agronomic practices employed in this crop. In their turn, these new practices had an impact on coffee-parasitic nematode populations and their damage to plantations. The likely influence of the agronomic practices on RLN-resistance and -tolerance results from their polygenic regulation, which promotes an incomplete protection against the nematodes. As explained later in this chapter, RLN-resistance is likely to be strongly linked to plant metabolism, such as the phenol-related pathways.

In general, the use of modern coffee cultivars has increased the impact of plant-parasitic nematodes on this crop because of their susceptibility to most of the important *Meloidogyne* and *Pratylenchus* species, and because of their lower tolerance to parasitism; such intolerance is linked to the cultivars' high productivity.

Furthermore, the cultivation of more productive coffee cultivars has demanded more intense fertilizations, particularly through nitrogen dressings. Such practices have led to soil acidification and subsequent nutritional imbalance of the plants, such as aluminum toxicity (Bornemiza et al., 1999). This process has occurred in many kinds of soils in many coffee-producing regions, such as the soils of volcanic origin largely present in Central America. Nutritional imbalances faced by the plants increase the impact of nematodes because their parasitic action on the roots negatively affects the plants' potential to uptake nutrients from the soil.

Additionally, in many countries the cultivation of low-habit cultivars has demanded the establishment of plantations with higher densities of plants/ha. This has probably helped spread parasitic nematodes through the plantations because of the more intense mixing of the plants' roots through the soil profile.

Finally, the intensification of coffee cultivation has led to a reduction in coffee shading, a common practice in many producing regions carried out with the help of various tree species, such as *Inga* sp. and *Grevilea* sp. Full-sun coffee plantations are more productive, particularly because of their more abundant blossom under full sunlight, yet they are less tolerant to nematode parasitism. The removal of shade trees eliminates their protection against high diurnal temperatures and water stress, which are particularly serious threats to coffee plantations in regions with well defined, long dry seasons. Hence, shade trees provide a friendlier microclimate for coffee trees.

5.3.3 Population Fluctuation as Related to RLN Epidemiology

Villain (1992) and Villain (2000) studied the seasonal fluctuations undergone by two coffee-parasitic RLN populations located in two regions in Guatemala, at 450 and 1200 masl. Initially, the authors noticed a correlation between the soil and the root populations, with the former presenting continually much lower nematode numbers; hence, only the root populations were monitored during the whole study.

At both altitudes, the same seasonal fluctuation in population was observed: two major population peaks were observed annually, one during the dry season (December through January) and another during a brief period of rain recess within the rainy season (around July). Both population peaks were synchronized to the coffee's periods of root growth, which in turn precede the periods of shoot growth. On the other hand, at both altitudes the lowest population levels were observed during the period of coffee berry maturation, which occurs from August through November at 450 masl and from September through December at 1200 masl.

This nematode population pattern is related to the process of coffee berry maturation, which acts as a priority physiological sink for assimilates and minerals, hence restricting the supply of assimilates to the roots and causing the death of part of the plant's root system (Cannell, 1985). Therefore, it is likely that a decrease in the root's supply of carbohydrates negatively affects nematode feeding and nutrition, hence decreasing their reproduction and population level. It is plausible that a rapid increase in the soil-borne fungi and bacteria populations in the sodden soils accelerates the process of root necrosis following the lesions caused by RLNs. Therefore, although influenced by the rain regime, RLN population fluctuations are more strongly determined by the phenological cycle of the coffee plants, which naturally is influenced by rainfall.

Typically, the fluctuations observed in RLN populations are very ample, with rapid decreases and increases in nematode numbers. This pattern is typical of organisms with an 'r' strategy, which present a high potential for colonization of new ecological niches. Nonetheless, coffee-parasitic RLNs are sexually reproduced. The high reproductive potential of two RLN populations was demonstrated by Villain et al. (1998), who observed as much as 30 thousand nematodes 14 weeks after carrot disks reared monoxenically *in vitro* had been inoculated with just two nematodes at the juvenile stage.

A field assay carried out in Guatemala showed that pruning the coffee plants in December, just after harvest and during the dry season, causes a rapid and severe decrease in RLN population due to the death of a large portion of the root system (L. Villain, unpublished results). One year later, during the next dry season, the plants regenerate their root system and the nematode population increases strongly. It would seem crucial to protect the roots during this regeneration stage so as to guarantee vigorous growth for the recently pruned coffee plants, but employing this strategy is difficult since granulate nematicides do not work properly during the dry season.

5.4 Management of RLNs

5.4.1 The Importance of Nematode Diagnosis and the Difficulties in Establishing and Using Damage Thresholds

Since RLNs do not cause easily recognizable symptoms, laboratory diagnosis is very important for the awareness of coffee growers and the subsequent application of management measures. The detection of RLNs, mainly from root samples, is essential (i) in nurseries, to ensure that seedlings are free of nematodes; in such cases, the acceptable infestation threshold should be zero, and (ii) in the field, to identify the *Pratylenchus* species involved and to obtain a rough estimate of the infestation level.

In microplot or field experiments, one can obtain correlations between RLN population levels and coffee yield loss. However, it seems very risky to manage plantations based on hypothetical damage thresholds. Indeed, population estimates of plant-parasitic nematodes, including RLNs, are influenced by several methodological factors. For example, the following sampling details may strongly influence the outcome of the population estimate: (i) sampling size is important since plant-parasitic nematodes generally present an aggregated spatial distribution in the soil. Such spatial distribution has been observed for RLNs and *Meloidogyne* sp. in coffee plantations in Central America (Cilas et al., 1993; Hervé et al., 2005), (ii) the sampling pattern employed in the field should be rigorous, avoiding the growers' tendency to sample preferentially coffee plants with advanced symptoms, whose nematode population has began to decline, and (iii) the sampling time is of paramount importance, since RLN populations vary rapidly and drastically during the seasons, as discussed above. Finally, the efficiency of nematode extraction from the soil varies with the method employed e.g., centrifugal flotation *vs* mistier technique, and the choice of precision sieves adopted; these variations can influence population estimates as well.

Another difficulty in employing damage thresholds as a platform for RLN management refers to the diversity of ecological and agronomic conditions in coffee-producing regions. As discussed above, the economic loss caused by RLNs is a function of yield loss and plant mortality, which in its turn is a function of the root damage suffered by the plants during their lifetime. These functions are influenced by environmental factors such as soil fertility, amount of exposure to the sun that plants receive in the full-sun or shaded cultivation systems, climate and the plants' genotype, which determines their resistance or susceptibility to the nematodes.

Because of such complex interactions and the large diversity of ecological conditions observed in the coffee-growing regions, it seems difficult to establish a standardized damage threshold and apply it in the process of taking decisions concerning RLN control. Finally, as discussed below, the priority in plant protection is now given to genetics and other control approaches as an alternative to using environmentally hazardous synthetic chemicals, such as nematicides. These approaches focus on alleviating the parasitic action of nematodes on coffee plants. Therefore, the simple presence of the most important RLNs in the coffee field is reason enough to initiate a management program; hence, the nematode population level is not taken into consideration for such a decision.

5.4.2 The Limitations of Chemical Control

In coffee nurseries, conventional granulate nematicides show some efficacy to reduce nematode populations (Abrego, 1974), but they do not guarantee the production of nematode-free seedlings to prevent their dissemination. Therefore, the only efficient approach is to use nematode free-substrates. Since methyl bromide is certain to be globally banned, solarization is the method of choice to kill nematodes infesting the soil (Gaur and Perry, 1991; Ghini and Bettiol, 1991).

In Brazil, there is a tendency among cooperative or private nurseries to adopt soil-fee substrates, which are composed of organic composts and inert substrates, such as vermiculite. Such commercial, nematode-free substrates are somewhat costly, but their use is feasible if the plant seedlings are produced in small, plastic containers, usually called cells. Unfortunately, such commercial soil-free substrates are not available in many coffee-producing countries.

Nematode-free seedlings can only be produced if the irrigation water is also kept free of nematodes. Hence, collecting water from deep wells or treating it before use is mandatory to prevent the production of nematode-infected seedlings, as reported by Ferraz (1980) and Gnanapragasam and Jebamlai (1982).

At the same time, all efforts should be undertaken to have the seedlings at an optimum physiological status at the moment of transplanting to the field. This will allow the coffee plantation to start with resistant plants, either rootstocks or own-rooted cultivars, at their maximum defense level. This aspect is important in the case of polygenic and complex plant resistance to nematodes, as in the case of RLNs, which involves the metabolism of phenolic compounds. For example, it has been demonstrated that arbuscular mycorrhizal fungi enhance plant resistance to nematodes by acting as a physical barrier against nematode penetration and through the nutritional benefits that those symbiotic organisms offer to coffee seedlings (Vaast and Zasoski, 1991; 1992; Vaast, 1996; Vaast et al., 1998). In their turn, the nutritional benefits contribute to enhancing the expression of nematode resistance genes. Likewise, all appropriate agronomic practices (fertilization, shading, watering, etc.) will help to optimize this resistance.

In established coffee plantations, it seems that the efficacy of conventional granulate nematicides against RLNs is indeed limited. To be efficient for a given harvest, these products need to be used for the whole two-year period that determines it. Also, the product dosage and application frequency required make this control method incompatible with the economic constraints on coffee production and with the environmental concerns that involve pesticides with a high level of toxicity.

As discussed before, even amphimitic *Pratylenchus* species present a high reproductive potential. Since nematicides do not eradicate nematodes, RLNs parasitizing susceptible coffee cultivars can quickly recover to high population levels after the nematicide has been washed out of the soil and/or degraded, or after a massive root death caused by pruning.

For example, it has been observed in Guatemala that applications of terbuphos twice a year at a rate of 1 g a.i./plant during the coffee plantation's first three years, and at 2 g during the following year, suppressed RLN populations until only the second year after planting, therefore becoming ineffective even before the coffee plants began to yield (Villain et al., 2000). The early nematicide effect on RLNs delays the 'die-back' process and the rise in the mortality rate observed in susceptible ungrafted 'Caturra' plants, but no significant effect on coffee yield is observed.

Under the same conditions, 'Caturra' plants grafted onto resistant *C. canephora* rootstocks did not show any significant yield increase with terbuphos applications (Fig. 5.3). In Costa Rica, Figueroa (1978) showed that carbofuran applications decreased *P. coffeae* populations for a period of four months only when applied in a three year-old arabica coffee plantation. This author showed that a dosage of 1.5 g a.i./plant applied twice a year increased the yield by 28%, but two years of nematicide application was necessary to obtain a significant yield increase.

Regarding the use of nematicides, it is also important to consider that most quality arabica coffee plantations are located in highlands, in areas that play a major role in the water cycle, with strong surface water runoffs because of the topography and

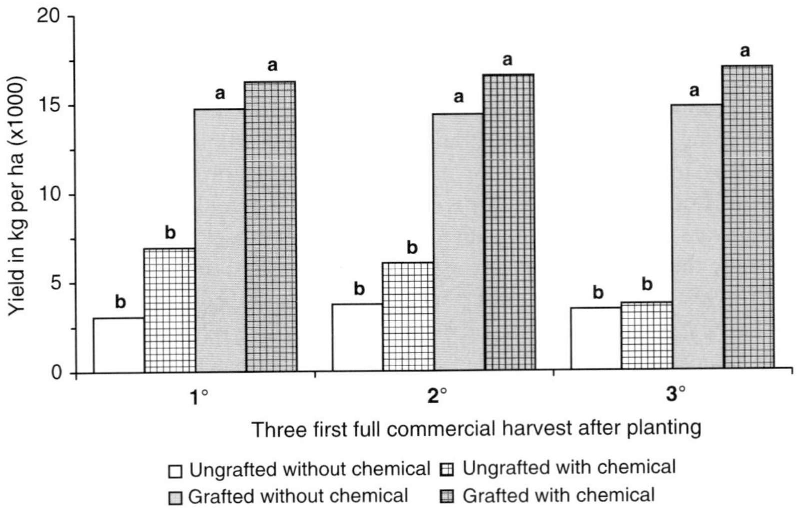

Fig. 5.3 Average coffee berry yield of plots with 50 *Coffea arabica* plants each, according to a factorial statistical design: with or without grafting onto *C. canephora versus* with or without chemical treatment. Values are average of four replicates. For each harvest, yields marked with the same letter are not different according to Newman and Keuls's test at $P = 0.05$ (from Luc et al., 2000, with permission)

intense underground water infiltration, particularly in the volcanic soils present in Meso America, in South American Andean countries, in the Caribbean, in Southeast Asia and the Pacific Islands. Hence it is clear why water contamination with nematicides, most of which are highly water soluble, could have a serious repercussion on the environment. It is important to consider that synthetic nematicides are wide spectrum biocides, so they can have an impact on biological activity in the soil.

As discussed before, high RLN populations occur during the dry season, at least in certain regions such as Central America. Such population peaks are very difficult to control chemically since nematicides need a certain level of soil humidity to act properly. Drastic nematode population decreases occur during the second half of the rainy season in Central America. Thus nematicide applications during this period, as sometimes practiced by coffee growers, are absolutely unjustified under such climatic conditions.

Finally, nematicide applications represent a high cost to coffee growers, at least for brand products, since many generic products can be found on the market today. Such an additional cost may not be acceptable considering the present coffee market, which has been suffering from low or at best medium prices for more than 15 years (see Chapter 2).

As seen above, the control of RLNs with the presently available chemical compounds seems to be of little efficiency as a long term strategy, considering the serious hazards to the environment as well as to humans during the productive lifetime of coffee plantations, at least 15–20 years. Therefore, alternative methods for nematode control must be proposed in order to develop a sustainable coffee cropping system. One of the most promising methods for controlling RLNs and other nematodes is the use of resistant germplasm.

5.4.3 Genetic Control

C. arabica is the most cultivated coffee species in Latin America. As discussed in Chapter 9, its most cultivated cultivars are based on a very narrow genetic pool (Charrier and Eskes, 1997; Anthony et al., 1999), and they are all susceptible to most coffee-parasitic *Pratylenchus* and *Meloidogyne* species (Hernández, 1997; Bertrand et al., 1999; Villain et al., 1999). Consequently, sources of RLN-resistance have been investigated among wild or semi-wild germplasm from the two main centers of *C. arabica* genetic diversity: Yemen, where this species was first cultivated and Ethiopia, where the species originated (Anthony et al., 2001). Eighteen introductions from Yemen and eleven from Ethiopia have been evaluated at the seedling stage for RLN-resistance through controlled inoculations, and they all appeared highly susceptible to a population from Guatemala (Anzueto, 1993; Villain et al., 2004). Hence, it seems unlikely that a source of RLN-resistance will be encountered in *C. arabica*.

On the other hand, RLN-resistance sources have been found in *C. canephora* germplasm. A hipocotyledonary method to graft arabica coffee onto *C. canephora* has been employed in Guatemala for 40 years to control RLNs (Reyna, 1968), and it is now widely used in Guatemalan areas infested with this nematode, ensuring an effective control of it even when non-selected rootstocks are used (Villain et al., 2000; 2001b). Grafting onto *C. canephora* has been also recommended to control *P. coffeae* in Indonesia (Palanichamy, 1973), where highly resistant robusta clones have been selected (see Chapter 15). In this country, breeding of resistant clones has two goals: to control *P. coffeae* in arabica coffee plantations through grafting onto resistant rootstocks, and to control these nematodes in plantations of own-rooted robusta cultivars.

Initially, the grafting of arabica coffee onto *C. canephora* was based on the idea that the latter would be at least tolerant to RLNs (Schieber, 1966; Reyna, 1968). Recent studies have revealed actual resistance factors to these nematodes. The *C. canephora* rootstock cultivar ‘Nemaya’, which is genetically close to ‘Apoatã’ selected in Brazil for its resistance to races one, two and three of *Meloidogyne incognita* (Kofoid and White) Chitwood (Fazuoli, 1986), has been selected for resistance to different *Meloidogyne* species present in Central America (Anzueto et al., 1996; Bertrand et al., 1999; 2000).

A study on the root penetration dynamics of two Guatemalan RLN populations showed that very few nematodes penetrated the roots of ‘Nemaya’, in comparison to a massive penetration in the roots of arabica coffee ‘Catuai’ within 24 h of inoculation (Villain et al., 2001b; 2004; 2006). A histological analysis showed no structure likely to prevent or hinder nematode penetration; therefore, the lower nematode penetration could be linked to an unattractive or repulsive property of the 'Nemaya' roots.

At the post-infectious stage, resistance factors to a Guatemalan RLN population have been observed in an open-pollinated progeny of one of the parents of 'Nemaya' (Villain et al., 2001b; 2004; 2006). The resistance seemed to be linked to the abundance of polyphenols in the roots of 'Nemaya' seedlings, which was not observed in 'Catuai'. The presence of numerous storage cells for phenolic compounds in the roots of 'Nemaya', even in the absence of nematodes, suggests that the plant's defense mechanisms are probably constitutive, i.e. their on set is independent of parasitism.

Studies performed in Indonesia revealed a correlation between resistance level to *P. coffeae* and root polyphenol concentration in different *C. canephora* clones (Toruan-Mathius et al., 1995). If phenolic metabolism is a major component of RLN-resistance, one can expect this resistance not to be very specific (Dalmasso et al., 1992). Such resistance would therefore provide coffee plants with an acceptable level of resistance to different *Pratylenchus* species.

A field assay carried out in Guatemala showed that grafting of arabica coffee 'Catuai Vermelho' onto free-pollinated progenies of *C. canephora* provided an efficient control of RLNs, with a maximum of 26 nematodes/g of roots in the grafted plants in comparison to 135 nematodes/g of roots in the non-grafted ones (Villain et al., 2000). This level of resistance resulted in significantly lower plant mortality rates, with an average of 6% in the plots with grafted plants in comparison to 25–56% in the ungrafted ones. In the latter, that percentage range was due to variations in the nematode distribution in the soil and in the amount of shade. This study also showed that on average the grafted plants yielded more than three times the ungrafted ones.

Moreover, this study showed that grafting did not affect significantly either the coffee beans' chemical composition of sugars, caffeine, trigonelline, chlorogenic acids and lipids, or their roasting parameters of weight loss and volume increase. Also, the parameters related to beverage quality, harshness, body, acidity, bitterness and astringency were not changed (Villain et al., 2001b). These results agree with work by Melo et al. (1976) who stated that grafting does not affect the coffee beans' caffeine concentration, regardless of the genotypes of both scion and rootstock.

Despite these good results in Guatemala, coffee breeders should remain focused on selecting RLN-resistant *C. canephora* rootstocks and breeding programs should be supported. Such a continuous research effort is warranted by the substantial genetic and RLN-resistance found in *C. canephora* germplasm (Leroy, 1993; Toruan Mathius et al., 1995). To support such a research effort it is worth remembering the serious damages caused by *P. coffeae* to robusta coffee plantations in Indonesia (Toruan Mathius et al., 1995). In Brazil, controlled assays have shown that *C. canephora* 'Apoatã' was susceptible to *P. brachyurus* and that *C. canephora* var *Kouillou* (=var *Conilon*) was susceptible to an isolate of *P. coffeae* from São Paulo State (Oliveira et al., 1999; Tomazini et al., 2003).

Finally, in order to develop durable strategies for nematode management, it is important to consider the whole plant-parasitic nematode community, which means all coffee-parasitic *Pratylenchus* and *Meloidogyne* species. In a plantation, these species compete for feeding on the coffee roots; such competition has been observed between *M. paranaensis* Carneiro, Carneiro, Abrantes, Santos and Almeida and *Pratylenchus* sp. in Guatemala, and between *M. exigua* Göldi and *Pratylenchus* sp. in Costa Rica (Cilas et al., 1993; Bertrand et al., 1998; Hervé et al., 2005). Thus, the use of coffee genotypes with a specific resistance may disturb the equilibrium of a nematode community towards a non-targeted species. This was observed in Costa Rica where the planting of *M. exigua*-resistant cultivars favored RLNs, which reached much higher population levels than they did while competing with *M. exigua* (Alpizar et al., 2005).

5.4.4 Biological Control: An Appealing but Unfeasible Strategy

As seen through the research results cited before, RLN-resistant coffee genotypes are not immune. Thus, when such genotypes are planted it is important to avoid high RLN populations in the field, so as to favor a durable management of these genotypes. Biological control could play an important role in this strategy, by reducing nematode populations when infested coffee fields are replanted with resistant genotypes. To date, biological control of nematodes has not been widely used in coffee cultivation, and nematode-antagonistic organisms have been sought and studied more for the control of *Meloidogyne* sp.

Preplant cover crops with nematicidal properties and ability to suppress plant-parasitic nematode populations have already been used on other tropical crops (Sarah, 1996; Wang et al., 2002). Good control or even suppression of RLN populations in vegetable fields has been achieved by previously cropping marigolds (*Tagetes* spp.) (Oostenbrink et al., 1957; Caubel et al., 1978; Kimpinski et al., 2000). In a tomato field, Hackney and Dickerson (1975) observed a drastic reduction of *M. incognita* and *P. alleni* Ferris populations by previously cropping *T. patula* L. or castor bean (*Ricinus comunis* L.). Another successful example was the control of *M. incognita* and *P. brachyurus* in a tomato field by the combination of six week-fallow and cultivation of *Crotalaria mucronata* Desv. (Brodie and Murphy, 1975). In Indonesia, preplant cultivation of *T. patula* and Guatemala grass (*Trypsacum laxum* Nash) suppressed *P. coffeae* in infested coffee plantations (Wyriadiputra, 1987).

Despite the good results, this nematode control strategy presents some difficulties for implementation in the coffee cultivation system: (i) the seeds of some plant species, such as those of vigorous marigold cultivars, are expensive; furthermore, many of these cover crops do not produce any goods to be sold by the growers, (ii) the seeds of some cover crops are not readily available for purchase, (iii) the need for a cover crop that will suppress all major coffee-parasitic *Pratylenchus* and *Meloidogyne* species and (iv) the climate, soil, topography and/or shaded conditions in some coffee growing regions are not necessarily suitable for some of the most efficient nematode-antagonistic cover crops.

Another difficulty that may interfere with this strategy is that some cover crops are antagonistic to some nematode genera or species, but they favour others. This has been observed on pineapple intercropped with *Crotalaria* sp., which efficiently controlled *Meloidogyne* sp. but increased the population of *P. brachyurus* to levels at least as harmful as those of the former (Luc et al., 2005). On a sugarcane plantation, *C. juncea* L. reduced *M. incognita* and *M. javanica* (Treub) Chitwood populations, but increased *P. zeae* (Ceres da Rosa et al., 2003).

Other antagonistic organisms should be tested for RLN-control on coffee, as has been done against other plant-parasitic nematodes using fungi and the bacteria *Pasteuria penetrans* (Thorne) Sayre and Starr (Naves and Campos, 1991; Campos and Campos, 1997; Campos et al., 1998). However, most of these organisms have a degree of antagonistic, parasitic or predatory specificity to plant-parasitic nematodes (Stirling, 1991). Therefore, it will probably be necessary to use a mixture of different biological agents with complementary types of antagonism, depending on the plant-parasitic nematode species and/or pathotypes that are present in a given field.

5.4.5 Cultural Control

As discussed before, the mechanisms involved in RLN-resistance are likely to be complex, involving phenolic metabolism and perhaps other factors. The polygenic nature of such partial resistance (Nelson, 1978; Parlevliet, 1979) increases the probability of its overall expression being determined by the environment (Rapilly, 1991). For instance, nutritional deficiency of *Camellia sinensis* (L.) O.Kuntze and *Prunus avium* L. rootstocks reduced their levels of partial resistance to *P. loosi* and *P. penetrans* (Cobb) Chitwood and Oteifa, respectively (Gnanapragasam, 1982; Melakeberhan et al., 1997).

Therefore, it is crucial for a grower to implement appropriate agronomic management of the coffee plantations in order to maintain the plants at a near optimum physiological stage, thus optimizing the expression of resistance factors and possibly increasing their overall level of tolerance to RLNs. The basic management routine involves fertilization programs based on soil analysis, control of soil pH, application of organic amendments and, in some regions, the rational use of shade trees.

5.5 Concluding Remarks

Many issues on RLNs remain to be tackled by nematologists and breeders in the forthcoming years.

For example, systematic surveys and proper characterization of coffee-parasitic RLN populations are badly needed. Many coffee-producing regions have not yet been surveyed, which hampers our knowledge of their biodiversity. The taxonomic status of many coffee-parasitic, amphimitic populations must be clarified, and any new taxon must be biologically characterized, particularly for its pathogenicity to the cultivated species *C. arabica* and *C. canephora*. All *Pratylenchus* species that are parasitic on coffee, particularly *P. coffeae*, must be better characterized with an integrated approach using modern tools, including molecular analysis.

Whenever a new taxon is described, its distribution and damage potential need to be assessed and, if necessary, a proper regional control strategy needs to be launched. Such taxonomic and biological characterizations are essential for the selection of resistant germplasm in breeding programs, and for developing biological control methods.

Coffee resistance to RLNs must be better characterized. Since the genetic strategy is one of the most promising for controlling plant-parasitic nematodes, the resistance mechanisms, both pre- and post-infection, and their genetic determinism should be further studied. It is possible that the determinism of RLN-resistance may be more complex than *Meloidogyne*-resistance. Conceivably, the genetic characterization of RLN-resistance would allow the development of molecular markers for assisted selection of resistant genotypes, making it easier to screen coffee germplasm and the 'pyramidation' of resistance genes against different plant-parasitic nematodes and other pathogens.

One issue that deserves special attention from nematologists and breeders alike regards the ensemble of methods to be used and the criteria to be adopted in assessments of RLN-resistance and/or –pathogenicity. Some of these issues are now discussed: (i) as emphasized before, any RLN population to be used in such assessments should first be clearly characterized so as to reveal its taxonomic status, (ii) to produce the inoculum, rearing the nematodes *in vitro* on carrot disks or alfalfa callus seems to be a good method, with no evidence of pathogenicity erosion to coffee plants (Anzueto, 1993; Villain et al., 2000), (iii) since RLN-parasitism leads to the destruction of the plant's root system over time, the assessment of the final nematode population for calculation of reproductive rates should be made before the damage caused to the root system reduces nematode reproduction; also, the fresh root weight should be assessed and a sound equation must be established correlating the number of nematodes in the inoculum, the phenological stage of the coffee plants and the test duration, (iv) since RLN-resistance seems to be linked to the metabolism of some compounds, special care should be taken to maintain the tested coffee seedlings under optimal and homogeneous physiological conditions, particularly regarding the supply of light, water and nutrients, (v) although the centrifugation-flotation method is probably the most precise for extracting RLNs, its laborious procedures suggest that the misting chamber method should also be considered a good alternative for RLNs and other migrating endoparasites, (vi) any coffee genotype identified as RLN-resistant in controlled inoculation tests conducted in greenhouse should be further assessed under field conditions. Since in many coffee-producing regions RLNs and *Meloidogyne* sp. occur together, field assays should challenge the genotypes with soil communities composed of both nematodes, and evaluate their resistance and tolerance.

Finally, it is important to develop alternative, rather than chemical, control methods, which could support the sustainable use of RLN-resistant genotypes. In countries where soil-free substrates are not available, ecologically and economically acceptable alternatives to methyl bromide should be developed for the desinfestation of the soil used in coffee nurseries. These alternatives could be either new molecules with nematicidal properties or techniques, such as soil solarization. For coffee plantations, it is necessary to develop nematode control methods that respect the environment and are economically accessible to coffee growers. These methods should aim to suppress or decrease RLN populations in order to sustain the resistance available in selected coffee cultivars.

References

Abrego L (1974) Ensayos de selectividade de nematicidas en el combate del *Pratylenchus coffeae* en almaciqueras de café. Nematropica 4:17

Abrego L, Holdeman Q (1961) Informe de progresos en el estudio del problema de los nematodos del café en El Salvador. Instituto Salvadoreño de Investigaciones del Café, Bol Inf, suppl #8, El Salvador

Alpizar E, Hervé E, Lashermes P et al (2005) Complete and partial resistance to *Meloidogyne exigua* in *Coffea arabica* modified pre-existing field nematode populations. Proceedings XX Int Conf Coffee Sci:1260–1262

Anthony F, Astorga C, Berthaud, J (1999) Los recursos genéticos: las bases de una solución genética. In: B Bertrand, B Rapidel (eds) Desafios de la caficultura en Centroamerica. IICA-PROMECAFE, San Jose

Anthony F, Bertrand B, Quiros A et al (2001) Genetic diversity of wild coffee (*Coffea arabica* L.) using molecular markers. Euphytica 118:53–65

Anzueto F (1993) Etude de la résistance du caféier (*Coffea* sp.) à *Meloidogyne* sp. et *Pratylenchus* sp. Ecole Nationale Supérieure Agronomique de Rennes, DS thesis, Rennes

Anzueto F, Bertrand B, Peña M et al (1996) Desarrollo de una variedad porta-injerto resistente a los principales nematodos de America Central. Proceedings XVII Symp Caficultura Latinoamericana:7

Bertrand B, Aguilar G, Santacreo R et al (1999) El mejoramiento genético en América Central. In: B Bertrand, B Rapidel (eds) Desafios de la caficultura en Centroamerica. IICA-PROMECAFE, San Jose

Bertrand B, Cilas C, Herve G et al (1998) Relations entre les populations de deux espèces de nématodes, *Meloidogyne exigua* et *Pratylenchus* sp., dans les racines de *Coffea arabica* au Costa Rica. Plante Rech Dév 5:279–286

Bertrand B, Duran MXP, Anzueto F et al (2000) Genetic study of *Coffea canephora* coffee trees resistance to *Meloidogyne incognita* in Guatemala and *Meloidogyne* sp. in El Salvador for selection of rootstock varieties in Central America. Euphytica 113:79–86

Bornemiza E, Collinet J, Segura A (1999) Los suelos cafetaleros de América Central y su fertilización. In: B Bertrand, B Rapidel (eds) Desafios de la caficultura en Centroamerica. IICA-PROMECAFE, San Jose

Bridge J (1984) Coffee nematode survey of Tanzania. Report on a visit to examine plant-parasitic nematodes of coffee in Tanzania. Commonwealth Institute of Parasitology, UK

Bridge J, Coyne DL, Kwoseh CK (2005) Nematodes parasites of root and tuber crops. In: Luc M, Sikora RA, Bridge, J (eds) Plant parasitic nematodes in subtropical and tropical agriculture, 2nd edition. CABI, Wallingford

Bridge J, Fogain R, Speijer P (1997) Parasites et ravageurs des *Musa*: fiche technique n° 2. Les nématodes parasites des bananiers. *P. coffeae* (Zimmermann, 1898) Filip. and Schu. Stek., 1941; *P. goodeyi* Sher and Allen, 1953. INIBAP, Montpellier

Brodie BB, Murphy WS (1975) Populations dynamics of nematodes as affected by combinations of fallow and cropping sequences. J Nematol 7:91–92

Campos HD, Campos VP (1997) Efeito da época e forma de aplicação dos fungos *Arthrobotrys conoides, A. musiformis, Paecilomyces lilacinus* e *Verticillium chlamydosporium* no controle de *Meloidogyne exigua* do cafeeiro. Fitopatol Bras 22:361–365

Campos VP, Souza JT, Souza RM (1998) Controle de fitonematóides por meio de bactérias. Rev An Patol Plantas 6:285–327

Campos VP, Villain L (2005) Nematode parasites of coffee and cocoa. In: Luc M, Sikora RA, Bridge J (eds) Plant parasitic nematodes in subtropical and tropical agriculture, 2nd edition. CABI, Wallingford

Cannell MGR (1985) Physiology of the coffee crop. In: Clifford MN, Wilson KC (eds) Coffee: botany, biochemistry and production of beans and beverage. Croom Helm, New York

Caubel G, Hemery J, Le Bohec F (1978) Lutte contre les nématodes: observations sur l'effet de *Tagetes patula* L. et *T. erecta* L. sur *Pratylenchus penetrans* Cobb. La Défense Végétaux 194:1–10

Ceres da Rosa R, Moura RM, Pedrosa EMR (2003) Efeitos do uso de *Crotalaria juncea* e carbofuran observados na colheita de cana planta. Nematol Bras 27:167–171

Charrier A, Eskes A (1997) Les caféiers. In: Charrier A, Hamon S, Nicolas D (eds) L'amélioration des plantes tropicales. CIRAD/ORSTOM, Montpellier

Chitwood BG, Berger CA (1960) Preliminary report on nemic parasites of coffee in Guatemala with suggested control measures. Plant Dis Reporter 44:841–847

Cilas C, Villain L, Licardie D (1993) Etude de la répartition de *Pratylenchus* sp. dans une plantation de caféiers au Guatemala. In: Proceedings XV Colloq Sci Int Café:843–847

D'Antonio AM, Libeck PR, Coelho AJE et al (1980) Levantamento de nematóides parasitas do cafeeiro que ocorrem no sul de Minas Gerais. In: Proceedings VIII Congr Bras Pesqu Cafeeiras:440

Dalmasso, A, Castagnone-Sereno P, Abad P (1992) Tolerance and resistance of plants to nematodes – knowledge, needs and prospects. Nematologica 38:466–472

Duncan LW, Inserra RN, Thomas WK et al (1999) Molecular and morphological analysis of isolates of *Pratylenchus coffeae* and closely related species. Nematropica 29:61–80

Fazuoli LC (1986) Genética e melhoramento do cafeeiro. In: Rena AB, Malavolta E, Rocha M et al (eds) Cultura do cafeeiro: fatores que afetam a produtividade. Associação Brasileira para Pesquisa da Potassa e do Fosfato. Piracicaba

Ferraz LCCB, Oliveira JC (1980) Água de irrigação como agente disseminador de nematóides em viveiros de mudas. Rev Agric 55:14–19

Ferraz S (1980) Reconhecimento das espécies de fitonematóides presentes nos solos do estado de Minas Gerais. Experientiae 26:255–328

Figueroa A (1978) Effectos de carbofuran 5G en la productividad del café Caturra. Nematropica 8:26–33

Gaur HS, Perry RN (1991) The use of soil solarization for control of plant parasitic nematodes. Nematol Abstr 60:153–167

Ghini R, Bettiol W (1991) Coletor solar para desinfestação de substratos. Summa Phytopathol 17:281–286

Gnanapragasam NC (1982) Effect of potassium fertilization and of soil temperature on the incidence and pathogenecity of the incidence and pathogenecity of the root lesion nematode, *Pratylenchus loosi* Loof, on tea (*Camelia sinensis* L.). Tea Quarterly 51:169–174

Gnanapragasam NC, Jebamlai AS (1982) Nematode infestation of nursery plants through irrigation water. Tea Quarterly 51:31–33

Golden AM, Lopez R, Vilchez H (1992) Description of *Pratylenchus gutierrezi* n. sp. (Nematoda:Pratylenchyda) from coffee in Costa Rica. J Nematol 24:298–304

Gonçalves W, Thomaziello RA, Moraes MV et al (1978) Estimativas de danos ocasionados pelos nematóides do cafeeiro. Proceedings VI Congr Bras Pesqu Cafeeiras:182–186

Gowen S, Quénéhervé P, Fogain R (2005) Nematodes parasites of bananas and plantain. In: Luc, M, Sikora RA, Bridge J (eds) Plant parasitic nematodes in subtropical and tropical agriculture, 2nd edition. CABI, Wallingford

Gutierrez G, Jimenez QMF (1970) Algumas observaciones sobre la infestacion del cafe practicada en Guatemala y El Salvador como medio para el control de nematodes. Rev Cafetera 98:35–47

Hackney RW, Dickerson OJ (1975) Marigold, castor bean and chrysanthemum as controls of *Meloidogyne incognita* and *Pratylenchus alleni*. J Nematol 7:84–90

Hernández F (1997) Etude de la variabilité intra et interspécifique des nématodes du genre *Meloidogyne* parasites des caféiers em Amérique Centrale. Université des Sciences et Techniques du Languedoc. DS Thesis. Montpellier

Hervé G (1997) Caractérisation biologique et moléculaire de populations de nématodes phytoparasites dans les plantations de caféiers d'Amérique Centrale. Ecole Nationale Supérieure Agronomique de Rennes, D.E.A. dissertation, Rennes

Hervé G, Bertrand B, Villain L et al (2005) Distribution analyses of *Meloidogyne* spp. and *Pratylenchus coffeae* sensu lato in coffee plots in Costa Rica and Guatemala. Plant Pathol 54:471–475

Inomoto MM, Oliveira CMG, Mazzafera P et al (1998) Effects of *Pratylenchus brachyurus* and *P. coffeae* on seedlings of *Coffea arabica*. J Nematol 30:362–367

Inserra RN, Duncan LW, Troccoli A et al (2001) *Pratylenchus jaehni* sp. n. from citrus in Brazil and its relationship with *P. coffeae* and *P. loosi*. Nematology 3:653–665

Kimpinski J, Arsenault WJ, Gallant CE et al (2000) The effect of marigolds (*Tagetes* spp.) and other cover crops on *Pratylenchus penetrans* and folowing potato crops. J Nematol 32:531–536

Kubo R, Silva R, Tomazini M et al (2002b). Patogenicidade de *Pratylenchus coffeae* em plântulas de cafeeiro cv. Mundo Novo. Fitopatol Bras 28:41–48

Kubo RK, Inomoto MM, Oliveira CMG et al (1999) Ocorrência de fitonematóides em cafezais das regiões de Bauru e Marilia. Fitopatol Bras 24:346

Kubo RK, Inomoto MM, Oliveira CMG et al (2001) Nematóides associados a cafeeiros do Estado de São Paulo. In: Proceedings XXIII Congr Bras Nematol:91

Kubo RK, Oliveira CMG, Antedomenico SR et al (2002a) Distribuição de *Pratylenchus* spp. em cafezais no Estado de São Paulo. Proceedings XXXV Congr Bras Fitopatol:230

Leroy T (1993) Diversité, paramètres génétiques et amélioration par sélection récurrente réciproque du caféier. Ecole Nationale Supérieure Agronomique de Rennes, PhD thesis, Montpellier

Lordello LGE (1972) Nematode pests of coffee. In: Webster JM (ed) Economic Nematology. Academic Press, London and New York

Luc M, Bridge J, Sikora RA (2005) Reflections on nematology in subtropical and tropical agriculture. In: Luc M, Sikora RA, Bridge J (eds) Plant parasitic nematodes in subtropical and tropical agriculture, 2nd edition. CABI, Wallingford

Melakeberhan H, Bird G, Rebecca G (1997) Impact of plant nutrition on *Pratylenchus penetrans* infection of *Prunus avium* rootstock. J Nematol 29:381–388

Melo M, Carvalho A, Monaco LC (1976) Contribuição do porta-enxerto no teor de cafeina em grãos de café. Bragantia 35:55–61

Monteiro AR, Antedomênico SR, Ferraz LCCB et al (2001) Primeira ocorrência de *Pratylenchus vulnus* Allen and Jensen, 1951, em cafeeiro. Proceedings XXIII Congr Bras Nematol:88

Monteiro AR, Lordello LGE (1974) Encontro do nematóide *Pratylenchus coffeae* atacando cafeeiro em São Paulo. Rev Agric 49:164

Moura RM, Pedrosa EM, Prado MDC (2003) Incidência de *Pratylenchus coffeae* causando severa nematose em cafeeiro no nordeste. Fitopatol Bras 27:649

Naves RL, Campos VP (1991) Ocorrência de fungos predadores de nematóides no Sul de Minas Gerais e estudo da capacidade predatória e crescimento in vitro de alguns de seus isolados. Nematol Bras 15:152–162

Nelson RR (1978) Genetics of horizontal resistance to plant disease. Ann Rev Phytopathol 16:359–378

Oliveira CMG, Inomoto MM, Vieira AMC et al (1999) Efeito de densidades populacionais de *Pratylenchus brachyurus* no crescimento de plantulas de *Coffea arabica* cv. Mundo novo e C. canephora cv. Apoatã. Nematropica 29:215

Oostenbrink K, Kuiper K, S'Jacob JJ (1957) Tagetes als Feindpflanzen von *Pratylenchus* Arten. Nematologica 2:424–433
Palanichamy L (1973) Nematode problems of coffee in India. Indian Coffee 37:99–100
Parlevliet JE (1979) Componenets of resistence that reduce the rate of epidemic development. Ann Rev Phytopathol 17:203–22
Rapilly F (1991) L'épidémiologie en pathologie végétale. INRA, Paris
Reyna EH (1968) La técnica de injerto hipocotiledonar del cafeto para le control de nematodos. Café 17:5–11
Sarah JL (1996) Les nématodes phytoparasites, une composante de la fertilité du milieu. In: Pichot J, Sibelet N, Lacoeuilhe JJ (eds) Environment fertility and peasant strategies in humid tropics. CIRAD-SAR, Montpellier
Schieber E (1966) Nematodos que atacan al café en Guatemala, su distribución, sintomatologia y control. Turrialba 16:130–135
Schieber E (1971) The nematode problems of coffee in Guatemala. Nematologica 1:17
Schieber E, Sosa ON (1960) Nematodes on coffee in Guatemala. Plant Dis Rep 44:722–723
Siciliano-Wilcken SR, Inomoto MM, Ferraz LC et al (2002a) Morphometry of *Pratylenchus* populations from coffee, banana, ornamental plant and citrus in Brazil. Proceedings IV Int Congr Nematol:356
Siciliano-Wilcken SR, Inomoto MM, Ferraz LC et al (2002b) RAPD of *Pratylenchus* populations from coffee, banana, ornamental plant and citrus in Brazil. Proceedings IV Int Congr Nematol:163
Siddiqi MR, Dabur KR, Bajaj HK (1991) Descriptions of three new species of *Pratylenchus* Filipjev, 1936 (Nematoda: Pratylenchidae). Nematol Mediterranea 19:1–7
Silva RA, Inomoto MM (2002) Host-range characterization of two *Pratylenchus coffeae* isolates from Brasil. J Nematol 34:135–139
Silva RA, Inomoto MM, Kubo RK et al (2001) Efeito de dois isolados de *Pratylenchus coffeae* na fotossintese e no crescimento de cafeeiros cv. Mundo Novo. Proceedings XXIII Congr Bras Nematol:87
Stirling GR (1991) Biological control of plant parasitic nematodes: progress, problems and prospects. CABI, Wallingford
Tomazini MD, Ferraz LCCB, Silva RA et al (2003) Resistência de genótipos de cafeeiros ao isolado K5 de *Pratylenchus coffee*. Proceedings XXIV Congr Bras Nematol:110
Toruan-Mathius N, Pancoro A, Sudarmadji D et al (1995) Root characteristics and molecular polymorphisms associated with resistance to *Pratylenchus coffeae* in Robusta coffee. Menara Perkebunan 63:43–51
Vaast P (1996) Intérêt de la mycorhization des cultures pérennes en milieu tropical humide: cas du caféier arabica. In: Pichot J, Sibelet N, Lacoeuilhe JJ (eds) Environment fertility and peasant strategies in humid tropics. CIRAD-SAR, Montpellier
Vaast P, Caswell-Chen EP, Zasoski RJ (1998) Effects of two endoparasitic nematodes (*Pratylenchus coffeae* and *Meloiodogyne konaensis*) on ammonium and nitrate uptake by Arabica coffee (*Coffea arabica* L.). Appl Soil Ecol 10:171–178
Vaast P, Zasoski RJ (1991) Influences de sources azoyées et de la mycorhization avec différentes espèces sur la croissance et la nutrition de jeunes caféiers Arabica. Café Cacao Thé 35:121–132
Vaast P, Zasoski RJ (1992) Effect of VA-Mycorhizae and nitrogen sources on rhizsphere soil characteristics, growth and nutrient asquisition of coffee seedlings (*Coffea arabica* L.). Plant and Soil 147:31–39
Villain L (1992) Dynamique des populations de *Pratylenchus coffeae* sur caféier dans le Sud-Ouest du Guatemala. Proceedings XIV Int Sci Colloq Coffee:527–533
Villain L (2000) Caractérisation et bioécologie du complexe parasitaire du genre *Pratylenchus* (Nemata:Pratylenchidae) présent sur caféiers (*Coffea* spp.) au Guatemala. Ecole Nationale Supérieure Agronomique de Rennes, PhD thesis, Rennes
Villain L, Anzueto F, Hernández A et al (1999) Los nematodos parásitos del cafeto. In: Bertrand B, Rapidel B (eds) Desafios de la caficultura en Centroamérica. IICA-POMECAFE, San Jose

Villain L, Anzueto F, Michaud-Ferrière N et al (2006) Resistance of *Coffea canephora* cv. Nemaya and cv. Apoatã to *Pratylenchus* spp. from Guatemala. Proceedings XXVI Congr Bras Nematol:105

Villain L, Anzueto F, Sarah JL (2004) Resistance to root-lesion nematodes on *Coffea canephora*. In: Cook R, Hunt DJ (eds) Proceedings IV Int Congr Nematol, Nematology Monographs and Perspectives. Brill, Lieden

Villain L, Baujard P, Anzueto F et al (2002) Integrated protection of coffee plantings in Central America against nematodes. Plante Rech Dévop, special issue: research and coffee growing:118–133

Villain L, Baujard P, Molina A et al (1998) Morphological and biological characterization of three *Pratylenchus* spp. populations parasiting coffee trees in Guatemala. Proceedings XXIV Int Symp Eur Soc Nematol:600–601

Villain L, Figueroa P, Molina A et al (2001a) Reproductive fitness and pathogenicity of three *Pratylenchus* spp. isolates on *Coffea arabica*. Nematropica 31:163

Villain L, Molina A, Sierra S et al (2001b) Evaluation of grafting and nematicide treatments for the management of a root lesion nematode, *Pratylenchus* sp., in *Coffea arabica* L. plantations in Guatemala. Proceedings XIX Int Sci Colloq Coffee (on CD)

Villain L, Pignolet L, Michau-Ferrière N et al (2001c) Evidence of resistance factors to *Pratylenchus* spp. on *Coffea canephora*. Nematropica 31:163

Villain L, Molina A, Sierra S et al (2000) Effect of grafting and nematicide treatments on damage by root lesion nematodes (*Pratylenchus* spp.) to *Coffea arabica* L. in Guatemala. Nematropica 30:87–100

Wang KH, Sipes BS, Schmitt DP (2002) Crotalaria as a cover crop for nematode management: a review. Nematropica 32:35–57

Whitehead AG (1968) Nematodea. In: Le Pelley RH (ed) Pests of coffee. Longmans, Green and Co. Ltd., London and Harlow

Whitehead AG (1969) Nematodes attacking coffee, tea and cocoa and their control. In: Peachey JE (ed) Nematodes of tropical crops. Technical communication #40. Commonwealth Bureaux of Helminthology. St Albans, Herts

Wiryadiputra S (1987) The use of antagonistic plants to control parasitic nematodes in the coffee plantation. Proceedings IX Cong and Nat Sem Indones Phytopathol Soc:484–490

Zimmermann A (1898) De nematoden der Koffiewortels. Med.'s Lands Plantentin 27:1–64

Part III
The Root-Knot Nematode, *Meloidogyne* spp.

Chapter 6
Taxonomy of Coffee-Parasitic Root-Knot Nematodes, *Meloidogyne* spp.

Regina M.D.G. Carneiro and Elis T. Cofcewicz

Abstract *Meloidogyne* species are characterized primarily on morphological features of females, males and second-stage juveniles. Based on these characters, identifying the 17 coffee-parasitic *Meloidogyne* species is a difficult task even for well-qualified taxonomists. This chapter outlines the most diagnostic morphological and morphometric features for *Meloidogyne* taxonomy, and presents the useful characters for identification of those 17 species. In recent years, esterase phenotyping has become a practical and reliable taxonomic tool for this genus. Unfortunately, only 12 out of the 17 coffee-parasitic species have had their phenotypes characterized; *M. africana, M. decalineata, M. kikuiensis, M. megadora* and *M. oteifae* can only be identified by morphological features. Recently, a new identification tool has been developed: the multiplex PCR (SCAR primers) allows unambiguous differentiation of *M. exigua, M. incognita* and *M. paranaensis* from Brazil, with prospects for extending this method to other species. This chapter concludes by outlining studies and initiatives that should be undertaken to facilitate and improve the reliability of coffee-related *Meloidogyne* taxonomy.

Keywords Morphology · esterase phenotyping · SCAR markers · races · intraspecific variability · distribution

6.1 Introduction

Root-knot nematodes (RKNs) are classified in the genus *Meloidogyne*, which was established by Göldi (1887) and includes 17 coffee-parasitic valid species. *Meloidogyne* species are characterized primarily on morphological features of females, particularly the perineal pattern. Features of males and second-stage juveniles (J2) are complementary. Nonetheless, reliable identification of *Meloidogyne* species based on morphology is a formidable task, even for well qualified taxonomists with expertise in the genus.

R.M.D.G. Carneiro
Embrapa-Recursos Genéticos e Biotecnologia, Brasilia, Brazil
e-mail: recar@cenargen.embrapa.br

R.M. Souza (ed.), *Plant-Parasitic Nematodes of Coffee*,

In most RKN-surveys conducted in coffee (*Coffea* sp.) plantations and nurseries worldwide (summarized by Campos and Villain, 2005), the perineal pattern was the main taxonomic feature used for species identification. Nonetheless, species identification based exclusively on this feature is difficult and uncertain for some coffee-parasitic populations, since it requires observation and subjective judgment of morphological aspects and comparison with figures presented in the species' original descriptions. Furthermore, different species may have similar perineal patterns; such is the case for *M. paranaensis* Carneiro, Carneiro, Abrantes, Santos and Almeida, *M. konaensis* Eisenback, Bernard and Schmitt, *M. izalcoensis* Carneiro, Almeida, Gomes and Hernandez, *M. inornata* Lordello and *M. mayaguensis* Rammah and Hirschmann, whose perineal patterns are similar to *M. incognita* (Kofoid and White) Chitwood.

Therefore, cases of misidentification are probably numerous. For example, reports of coffee-parasitic *M. incognita* populations in Guatemala and El Salvador, which had been based on perineal patterning, should be regarded with caution because recent surveys conducted in those countries, with the aid of enzyme phenotyping, have not detected *M. incognita*; instead, *M. paranaensis* and *M. izalcoensis* have been found (Carneiro et al., 2004; 2005b).

Conversely, perineal patterning can be a complementary tool for taxonomy based on enzyme phenotyping and other biochemical or molecular methods. Indeed, species-specific esterase phenotypes have been characterized for 12 of the 17 coffee-parasitic *Meloidogyne* species. Furthermore, Randig et al. (2002) have developed a polymerase chain reaction (PCR)-based assay to identify RKNs associated with coffee in Brazil. Three RAPD markers have been transformed into sequence-characterized amplified region (SCAR) markers, which are specific for *M. exigua* Göldi, *M. incognita* and *M. paranaensis*. Currently, only five coffee-parasitic *Meloidogyne* species from Africa have not had their enzymatic phenotypes characterized; for these species, identification remains based on morphological features only.

This chapter aims to assist nematologists, plant pathologists and other scientists in identifying the 17 coffee-parasitic *Meloidogyne* species. Initially, the basic RKN-morphology is presented, and the taxonomic reliability of several morphological and morphometric features is discussed. The diagnostic features for each of the 17 species are presented, as well as drawings from their original descriptions (some of them have been published without scale bars). Advances in biochemical and molecular taxonomy are outlined as well.

6.2 Morphology and Morphometry in *Meloidogyne* Taxonomy

Because of the morphological and morphometric similarities between *Meloidogyne* species, the most appropriate approach is to ponder a combination of differential characters of nematode females, males and J2.

Females ($L = 380 - 1348\,\mu m$) are pear-shaped to spheroid, with a short (see *M. kikuyensis* de Grisse) to elongated (see *M. coffeicola* Lordello and Zamith) neck.

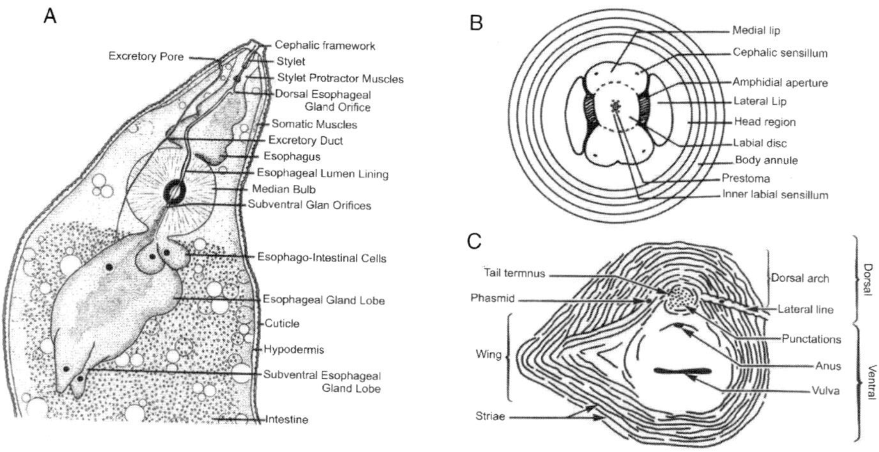

Fig. 6.1 Female morphology of root-knot nematodes (*Meloidogyne* sp.). (**A**) Anterior region. (**B**) Head morphology as revealed by SEM, in face view. (**C**) Perineal pattern (from Eisenback and Triantaphyllou, 1991, with permission)

Their body is white and not transformed into a cyst-like structure upon death. The cephalic region ('head') presents a cuticular framework (Fig. 6.1A), a labial region with six lips, median lips fused into two pairs, and one asymmetrical or symmetrical postlabial annule. The amphid apertures are slit-like (Fig. 6.1B). The stylet is robust (10–25 μm long), with three basal knobs. The positioning of the dorsal oesophageal gland orifice in relation to the base of the stylet knobs (DEGO position) is about 2–10 μm, but this character is variable within populations and species. The excretory pore is located anterior to the median bulb, usually 15–25 annules posterior to the lip region; nonetheless, this positioning varies a lot within and among *Meloidogyne* species, which makes it a poor diagnostic character. The oesophageal glands are usually five-lobed, and they overlap the intestine. The body cuticle presents simple cross annulations, which form a variable, somewhat circular pattern around the vulva and anus, which is called the perineal pattern (Fig. 6.1C). The phasmids are situated on either side of and dorsal to the anus. The eggs are not retained in the body; instead, they are deposited in a gelatinous matrix which is extruded through the anus. The females are usually endoparasitic, inducing the formation of galls ('knots') on the roots of most hosts. A more detailed morphological description of RKNs can be read in Jepson (1987) and Eisenback and Triantaphyllou (1991).

Males are vermiform, with their length (700–2,000 μm) varying according to the environmental conditions during their development. Therefore, the character body length and morphometric ratios relating it to oesophagus and tail lengths or body width are nearly useless for taxonomy. The head (Fig. 6.2A,C) presents a labial cap with six lips, and the median lips are more or less fused into two pairs, assuming a dumb-bell shape; these features provide several good diagnostic features. The amphid apertures are slit-like, conspicuous, leading to broad pouches in the lateral lips. Usually there is only a single postlabial annule, although additional, incomplete

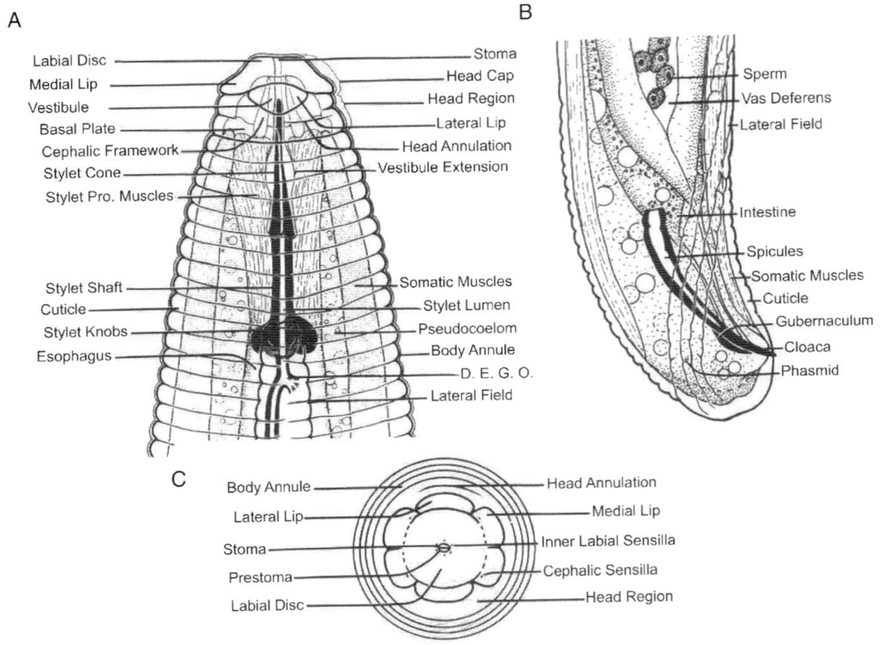

Fig. 6.2 Male morphology of root-knot nematodes (*Meloidogyne* sp.). (**A,C**) Anterior region in lateral and face views, respectively. (**B**) Posterior region (from Eisenback and Triantaphyllou, 1991, with permission)

annules can be present, which can be used to distinguish species and populations. The stylet has well-developed basal knobs; the stylet can be 13–30 μm long, although most *Meloidogyne* species have it in the range of 18–24 μm; this character presents a coefficient of variability of only 4%, which makes it a good character to differentiate species. The size and shape of the stylet cone, shaft and knobs are also excellent supporting characters for species identification. Males have a strong cephalic framework. The DEGO position is 2–13 μm; in general, this character exhibits much variation, although some species can be distinguished from it. The position of the excretory pore varies widely within species, being of limited value as a differential character. The hemizonid is usually located anterior to the excretory pore; thus, its positioning may help in identifying species that have it posterior to the excretory pore. Normal males present one gonad, whereas sex-reversed males have two. Males have gubernaculum (Fig. 6.2B). Spicule length ranges from 19 to 40 μm across the genus, with much overlap in its length among species. Slight differences in spicule structure have been described for some species, but in general spicule morphology is not of diagnostic value. The male tail is bluntly rounded and short, with little variation among the species.

Second-stage juveniles vary in body length from 290 to 912 μm across the genus. In many species this character ranges from 300 to 500 μm, which makes it inadequate for species identification. Due to J2's small size, discerning details precisely

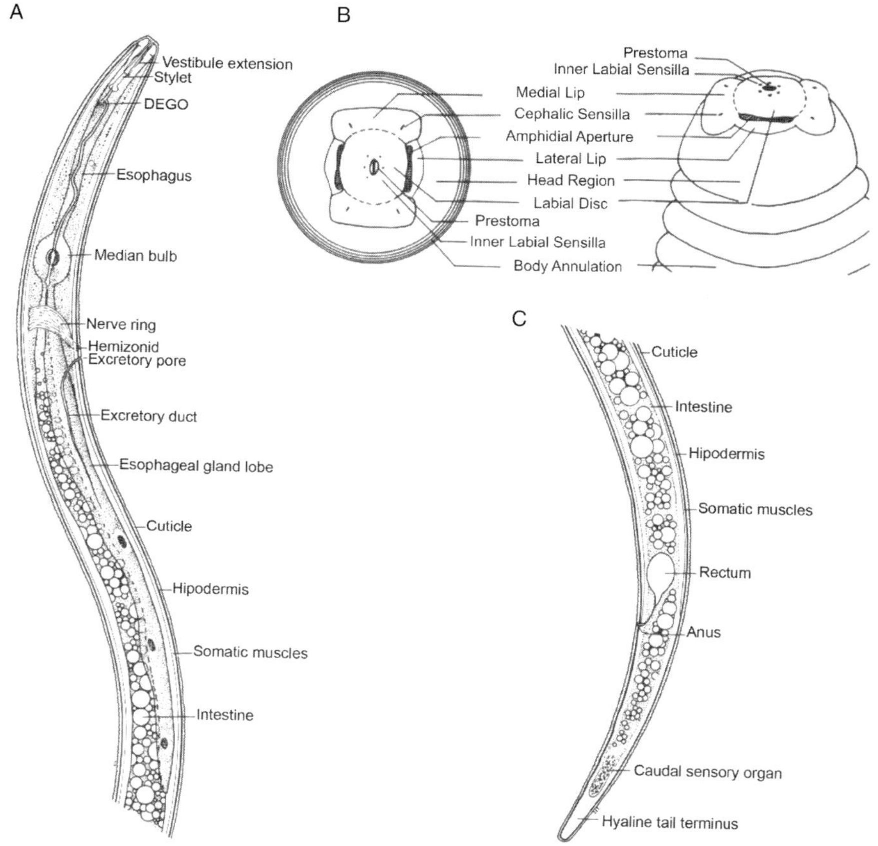

Fig. 6.3 J2 morphology of root-knot nematodes (*Meloidogyne* sp.). (**A,B**) Anterior region in lateral and face views, respectively. (**C**) Posterior region (from Eisenback and Triantaphyllou, 1991, with permission)

in the nematode's head (Fig. 6.3A,B) can only be done with the aid of a scanning electron microscope. Furthermore, head morphology is quite similar among most species, although some differ in the shape of the labial disk, details in the lateral and medial lips, format, size and positioning of labial and cephalic sensilla, and presence of head annulations. Second-stage juveniles have a delicate stylet that ranges in length from 8 to 18 μm across the genus. This character shows low variability among species, although it may be helpful in identifying certain species. The DEGO position is 2–8 μm, and it seems a good differentiating feature, with groups of species being distinguished based on it. The position of the excretory pore is variable. Hemizonid positioning can be a fairly useful diagnostic feature in those species in which it is located posterior to the excretory pore. Tail length varies considerably among species, from 15 to 100 μm. Due to its small intraspecific variation, it is a very useful measurement. In J2, the tail ends in a hyaline terminus (Fig. 6.3C), which can be considered to identify those species in which it is always short or long. Whitehead (1968) and Jepson (1987) have grouped *Meloidogyne* species according to J2 tail lengths and shapes. The latter author has also stated that differences in tail measurements from populations of a single species can be larger than between species. Nevertheless, differences in mean tail length and/or mean length of the tail's hyaline terminus are large enough to distinguish species in groups.

6.3 The Status of Coffee-Parasitic *Meloidogyne* Taxonomy

Meloidogyne sp. comprises more than 90 species. Nineteen have been associated with coffee in many countries worldwide, including very damaging ones that cause great losses to coffee growers and to the economy of developing countries.

In this review 17 species are recognized as valid (see below). *M. thamesi* (Chitwood in Chitwood et al.) Goodey has been synonimized to *M. arenaria* (Neal) Chitwood by Jepson (1987), and confirmed by Eisenback and Triantaphyllou (1991). These authors have also synonymized *M. inornata* to *M. incognita*, but the former has been revalidated by Carneiro et al. (2008).

M. göldii has been described by Santos in his DS thesis (1997); nonetheless, this species' description and diagnosis have never been published. According to the International Code for Zoological Nomenclature, any publication that mentions *M. göldii* Santos, 1997 should refer to it as a *nomen nudum*.

6.3.1 Nominal List of Coffee-Parasitic Meloidogyne *Species*

6.3.1.1 Valid Species

M. exigua Göldi, 1887, type species
M. africana Whitehead, 1960
M. arabicida López, 1989
M. arenaria (Neal, 1889) Chitwood, 1949
  Syn. *M. thamesi* (Chitwood in Chitwood et al., 1952) Goodey, 1963
M. coffeicola Lordello and Zamith, 1960
M. decalineata Whitehead, 1968
M. hapla Chitwood, 1949
M. incognita (Kofoid and White, 1919) Chitwood, 1949
M. inornata Lordello, 1956
M. izalcoensis Carneiro, Almeida, Gomes and Hernandez, 2005
M. javanica (Treub, 1885) Chitwood, 1949
M. kikuyensis de Grisse, 1960
M. konaensis Eisenback, Bernard and Schmitt, 1994
M. mayaguensis Rammah and Hirschmann, 1988
M. megadora Whitehead, 1968
M. oteifae Elmiligy, 1968
M. paranaensis Carneiro, Carneiro, Abrantes, Santos and Almeida, 1996

6.3.1.2 Nomen Nudum

M. göldii Santos, 1997

6.4 Diagnostic Features and Distribution of Coffee-Parasitic *Meloidogyne* Species

6.4.1 *M. exigua*

The females are small (L = 387.5 − 496 μm), being characterized by the perineal pattern round to hexagonal, with the dorsal arch varying from low and rounded to somewhat high and squarish, with striae coarse and widely spaced (Fig. 6.4K, L, M). In the perineal pattern, the lateral fields are usually inconspicuous and only indistinctly forked; however, the inner lateral line regions may have coarse, raised, looped, and folded striae which also cover the anus (Chitwood, 1949; Lordello and Zamith, 1958; Cain, 1974; Jepson, 1987). The female stylet is 12–14 μm long, its shaft is cylindrical, but occasionally it narrows at the junction with the knobs. The DEGO position is usually 4–8 μm (Fig. 6.4F). In males, the head contour accompanies the contour of the body's first cuticle annules, thus being called not offset (Fig. 6.4A). The medial lips are often divided medially by a shallow groove. Stylets are 18–20 μm long; the shaft is straight and cylindrical, and it narrows at the junction with the knobs. The DEGO position is variable (3–5 μm). In J2, the moderately long tail (44–46 μm) ends in a bluntly rounded tip (Fig. 6.4I). A few narrow constricting annulations close to the tail terminus are typical of this species (Eisenback and Triantaphyllou, 1991). Although *M. exigua* populations are very similar morphologically (Lima and Ferraz, 1985), recent molecular studies have showed a high genetic variability among coffee-parasitic populations (Muniz et al., 2008).

M. exigua can be distinguished by its esterase phenotypes (Est E1 and E2, Fig. 6.21) (Carneiro et al., 2000; 2005b) and PCR-SCAR markers (Randig et al., 2002; 2004). It reproduces by meiotic parthenogenesis, with haploid chromosomal number (n) equal to 18 (Tryantaphyllou, 1985).

Coffee-parasitic populations of *M. exigua* have been reported from Brazil, Guatemala, Dominican Republic, Nicaragua, Costa Rica, Puerto Rico, Colombia, Peru, El Salvador, Venezuela, Bolivia, Honduras and Panama (Campos and Villain, 2005).

6.4.2 *M. africana*

Females are 660–910 μm long, and present a typical perineal pattern which is roughly circular, without punctations (Fig. 6.5B). The dorsal arch is low and the phasmids are located close to the wide tail terminus, which is often marked by short disordered striae. The wide lateral fields are unmarked by incisures, but they present tiny, disordered striae. The female stylet is 15 μm long and the DEGO position is

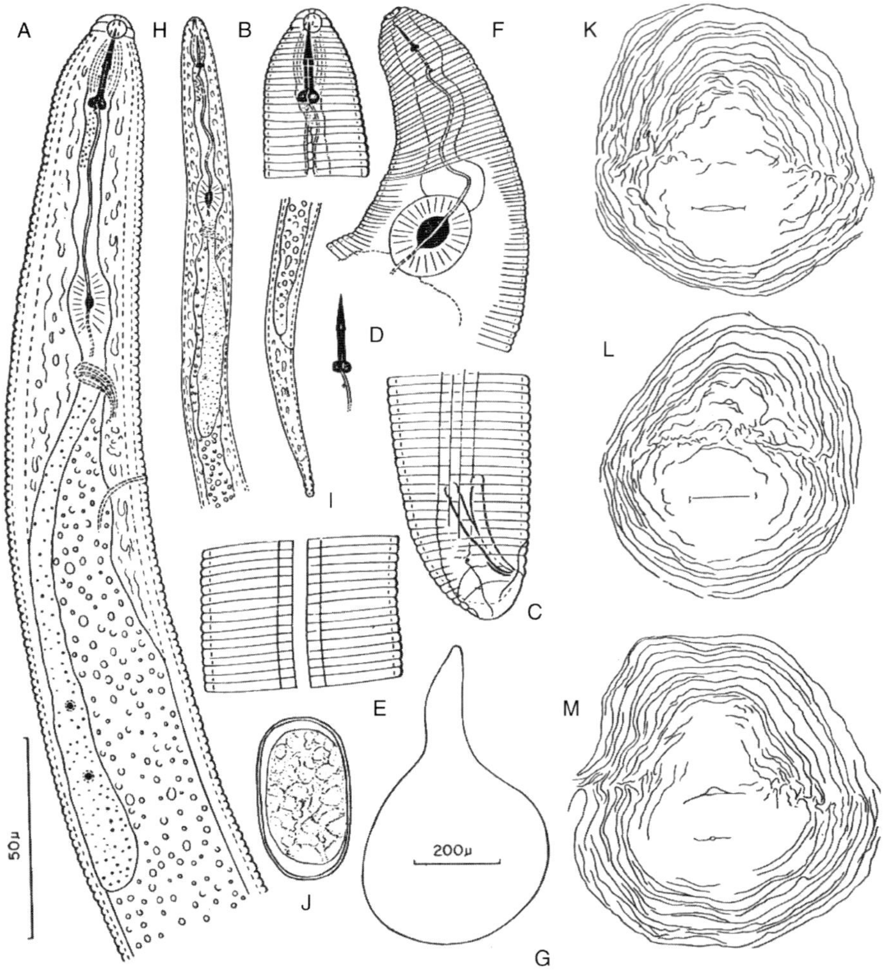

Fig. 6.4 *Meloidogyne exigua*. (**A–E**) Male anterior and posterior regions, stylet and lateral field, respectively. (**F,G**) Female anterior region and body shape. (**H,I**) J_2 anterior and posterior regions, respectively. (**J**) Egg. (**K–M**): Perineal patterns (from Lordello and Zamith, 1958, with permission)

4–9 μm. Males are 1,200–1,850 μm long, presenting one head annule behind the head cap; their stylet knobs are spherical and prominent (Fig. 6.5C,D). In males, the stylet is 19–22 μm, and the DEGO position is 4–6 μm. The spicules have a medial flange; the gubernaculum is crescent in lateral view (Fig. 6.5E). The J2 are 380–470 μm long and their stylet measures 12–18 μm; they present a fairly broad tail (Fig. 6.5A), which gradually tapers to a blunt, rounded terminus, generally without any cuticular constrictions in the hyaline region; instead, their tail presents fine striae extending close to the tail terminus (Whitehead, 1960; Jepson, 1987).

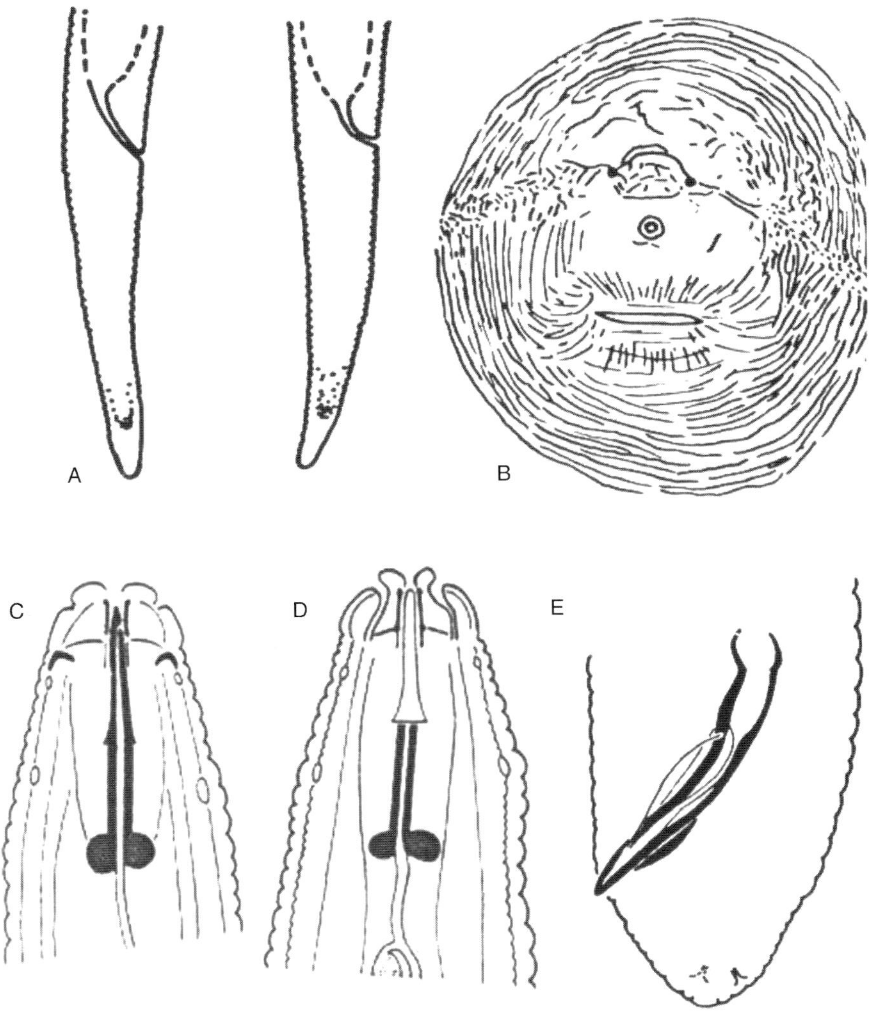

Fig. 6.5 *Meloidogyne africana*. (**A**) J_2 tails. (**B**) Perineal pattern. (**C,D**) Male anterior region. (**E**) Male posterior region (from Whitehead, 1968, with permission)

No esterase phenotype has been characterized for this species; its mode of reproduction and chromosome number are not known. On coffee, *M. africana* is known to occur in Kenya and Zaire (Campos and Villain, 2005).

6.4.3 M. arabicida

This species presents females 543–1,206 μm long, whose perineal pattern is very peculiar: it shows relatively angular contours with thick striae in the center and

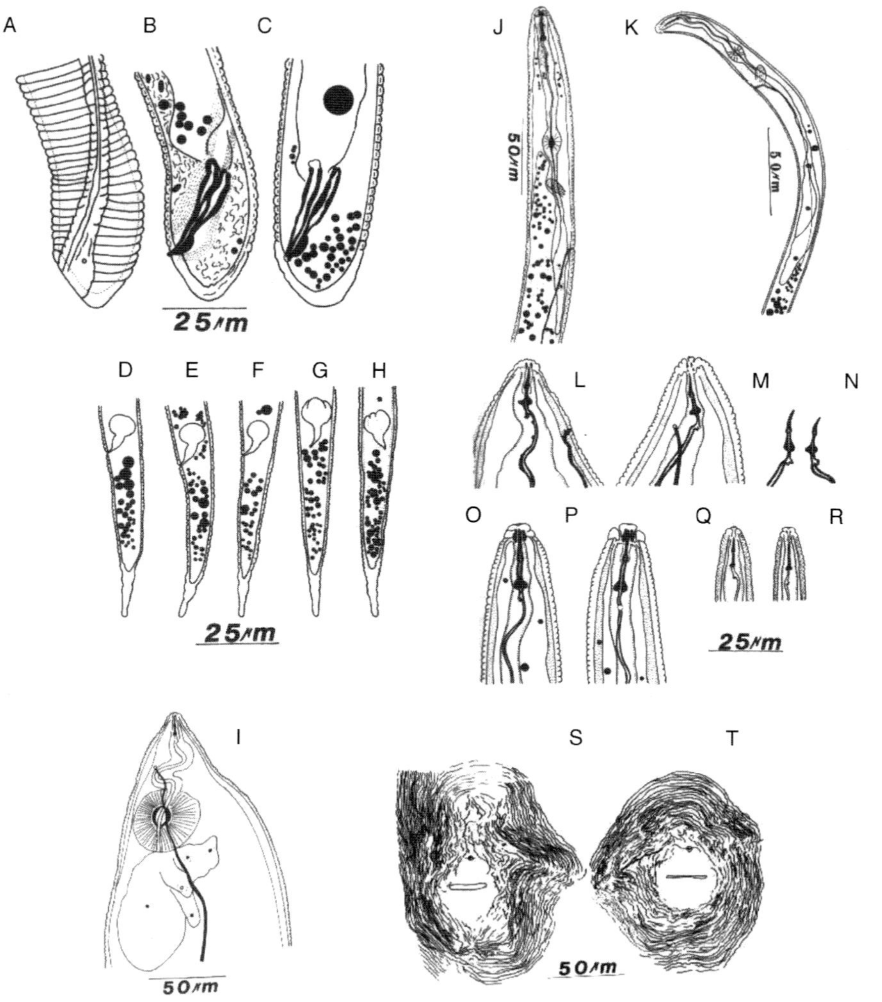

Fig. 6.6 *Meloidogyne arabicida*. (**A–C**) Male posterior end. (**D–H**) J_2 tails. (**J,K,Q,R**) J_2 anterior region. (**I,L,M**) Female anterior region. (**N**) Female stylet. (**O,P**) Male anterior region. (**S,T**) Perineal patterns. (from Lopez and Salazar, 1989, with permission)

thinner ones on the periphery; the dorsal arch is relatively high and rectangular (Fig. 6.6S,T). Most patterns have striae lateral projections ('wings'), which can be present on both sides or on just one. The vulva is elongated and smooth, without prominent striae originating from it. The female medial labial lips are separated by a small indentation in the center. Males are 905–1,881 µm long, with a smooth head region presenting just one annule ring (Fig. 6.6O,P) and areolated lateral fields (Fig. 6.6A). The 372–480 µm long J2 have a smooth head region with narrow lateral lips, slightly arcuate; one relatively short, incomplete striae is found in the lateral area of the head region; the J2 present a dilated rectum (Fig. 6.6D–H) (López and Salazar, 1989).

This species can be diagnosed by its esterase phenotype (Est AR2, Fig. 6.21) (Carneiro et al., 2004; Hernandez et al., 2004). Its mode of reproduction and chromosome number are unknown. On coffee, *M. arabicida* has been reported from Costa Rica (Campos and Villain, 2005).

6.4.4 *M. arenaria*

This species is characterized by its female (510–1,000 μm long) perineal pattern, which is flattened to rounded (Fig. 6.7F). The striae in the arch are slightly indented at the lateral lines; often the dorsal and ventral striae meet at an angle at the lateral lines, and generally form a 'shoulder' on the arch. Some striae fork and are short and irregular near the lateral lines. The striae are smooth to wavy and some may bend towards the vulva. The pattern may also have striae that extend laterally to form one or two 'wings'. Some populations of *M. arenaria* present variant females which present perineal pattern similar to *M. incognita*'s. *M. arenaria* females have unique stylets: in general, their stylet is very robust, 13–17 μm long; the DEGO position is 3–7 μm (Fig. 6.7D,E). Stylet cone and shaft are broad. The shaft increases in width posteriorly and gradually merges with the stylet knobs; these are wide and rounded posteriorly. The males' head region is low and slopes posteriorly. It forms a smooth and continuous structure that is almost as wide as the head region (Fig. 6.7A,B). Two or three incomplete annulations are present on the head region. The stylet is 20–25 μm long, with the posterior portion of its cone much wider than the anterior portion of its shaft. The shaft is generally cylindrical, and it gradually merges with the very large stylet knobs. Typically, the J2 (398–605 μm long) present no annulations in the head region, although some specimens may have two or three annulations. The tail (44–69 μm long) is narrow, tapering to a subacute terminus (Fig. 6.7H).

M. arenaria can be distinguished by its esterase phenotypes (Est A2 and A3, Fig. 6.21) (Carneiro et al., 2000; 2004) and PCR-SCAR markers (Zijlstra et al., 2000a). It reproduces by mitotic parthenogenesis, with 36, 45 or 51–56 chromosomes. Coffee-parasitic *M. arenaria* populations have been found in Jamaica, Cuba and El Salvador (Campos and Villain, 2005).

6.4.5 *M. coffeicola*

This species is diagnosed by its brownish, very elongated females (992–1,348 μm), which have long necks (Fig. 6.8F). The stylet is 15.3–17.6 μm long and the DEGO position is 3.8–4.6 μm. The characteristic perineal pattern shows a low arch, which has very faint striae closely spaced, smooth to slightly wavy in the dorsal sector (Fig. 6.8G). Close to its tip, the tail is rather wide, being marked by faint striae

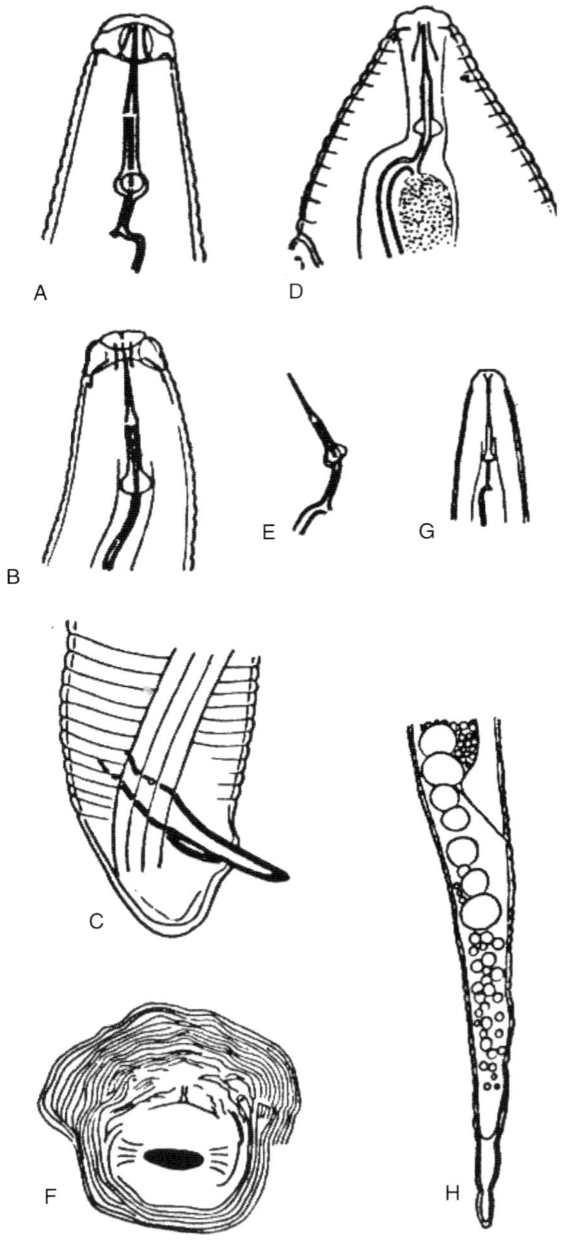

Fig. 6.7 *Meloidogyne arenaria*. (**A–C**) Male anterior and posterior regions. (**D**) Female anterior region. (**E**) Female stylet. (**F**) Perineal pattern. (**G**) J_2 anterior region. (**H**) J_2 posterior end (from Chitwood, 1949, with permission)

and surrounded by concentric circles; the phasmids are located close to the tail tip. The perineal pattern's lateral fields are very poorly defined; in some specimens, it is marked only by slight irregularities in the striae. Males ($L = 1,279 - 1,595\ \mu m$) present four aerolated lateral field incisures (Fig. 6.8D); the head is cupolate, and its contour extends beyond the body's contour (offset) (Fig. 6.8A), having one annule

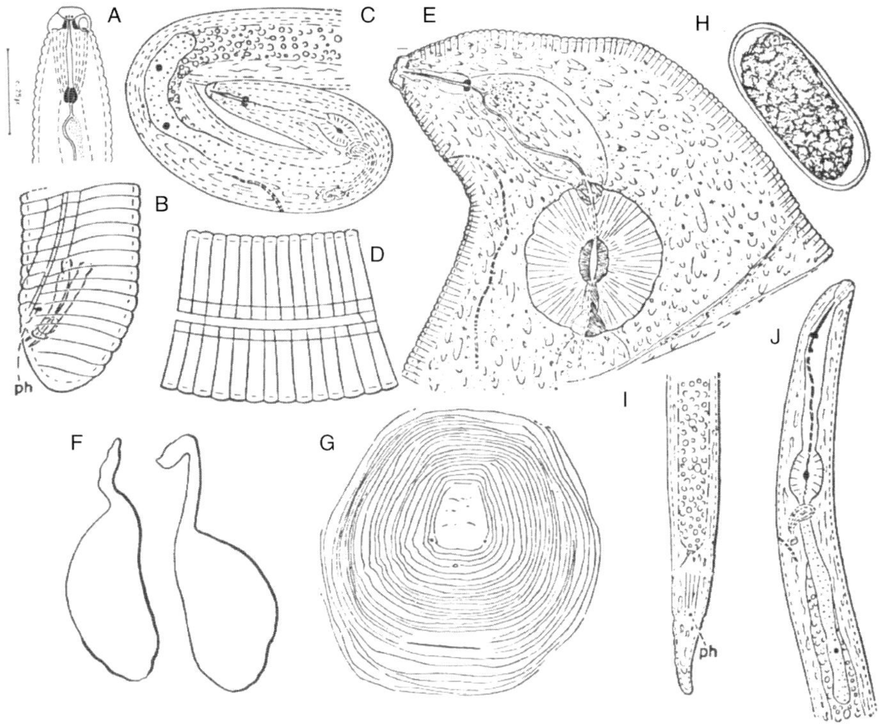

Fig. 6.8 *Meloidogyne coffeicola*. **(A–D)** Male. **(E)** Female anterior region. **(F)** Female body shapes. **(G)** Perineal pattern. **(H)** Egg. **(I,J)** J2 posterior and anterior regions, respectively. ph = phasmid (from Lordello and Zamith, 1960, with permission)

behind the head cap. The stylet knobs are longitudinally ovoid, not prominent. Male stylet length is 23–26 μm and the DEGO position is 3.8–4.6 μm. Phasmids are located before the cloaca (Fig. 6.8B). The J2 ($L = 336.6 - 423.8$ μm) present 9.2–10.7 μm long stylets, with weak, ovoid knobs; their tail is fairly short (29.1–33.6 μm) and bluntly rounded (Fig. 6.8I).

Care should be taken to differentiate *M. coffeicola* from *M. decalineata*, because these species may present similar perineal patterns. *M. decalineata* has smaller females ($L = 649 - 1,041$ μm); males and J2 of these species are quite distinct (Whitehead, 1968).

M. coffeicola may be characterized by its esterase phenotype (Est C2, Fig. 6.21) (Carneiro et al., 2000). Its mode of reproduction and chromosome number are unknown. This species has only been reported in Brazil (Campos and Villain, 2005).

6.4.6 *M. decalineata*

This species is characterized by the length of female body (649–1,041 μm) and stylet (12–17 μm); the DEGO position is 3–4 μm. It also has a peculiar perineal

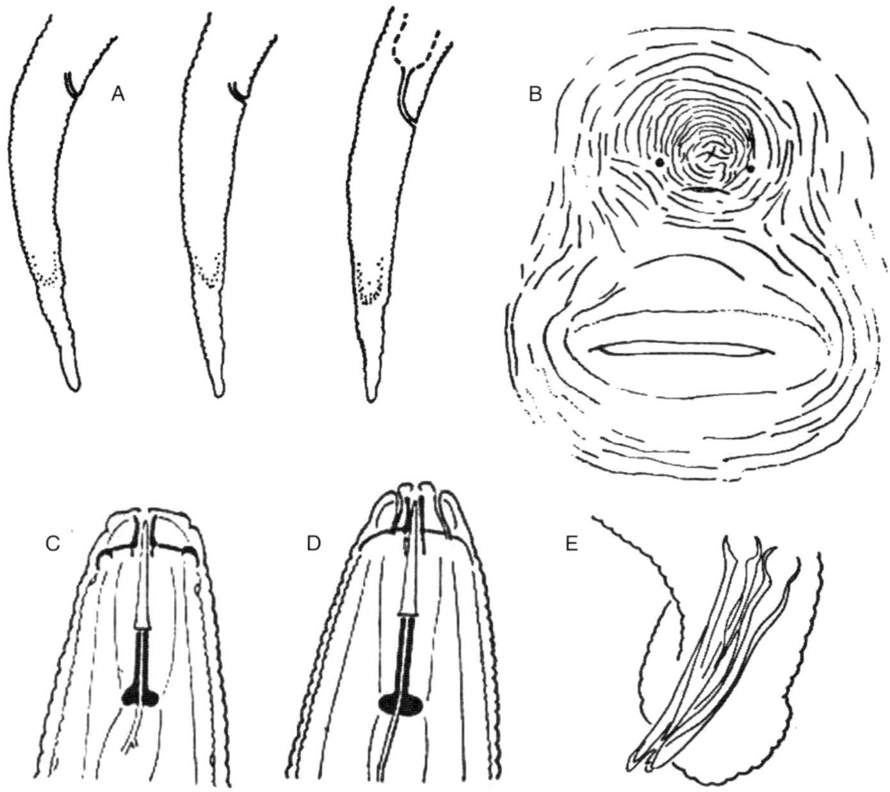

Fig. 6.9 *Meloidogyne decalineata*. (**A**) J_2 tails. (**B**) Perineal pattern. (**C,D**) Male anterior region. (**E**) Male posterior end (from Whitehead, 1968, with permission)

pattern, which shows striae fairly close and evenly spaced, which are often broken, especially at the lateral sides of the pattern (Fig. 6.9B). A distinct tail whorl is present, fairly distant from the vulva; the tail terminus is marked by short, disordered striae; numerous striae can be seen between the tail whorl and the vulva. Rudimentary lateral fields can be seen in some patterns, occasionally with minute disordered striae within the fields. Phasmids are located close to tail terminus. The body cuticle is often folded in the pattern's ventral region. Males are 649–1,041 μm long; their stylet is 12–17 μm long and the DEGO position is 3–4 μm. Males present head not offset, which in lateral view seems fairly low and shaped as a truncate cone, with a small head cap followed by a very short first head annule (Fig. 6.9C,D). Males present ten lateral field incisures. The J2 are 471–573 μm long, their stylet measure 10.7–13.7 μm long and they present their head slightly inflated, with three or four annules behind the head cap. The J2 present a narrow tapering tail, which is 44–52 μm long and ends in a broadly rounded terminus (Fig. 6.9A). The tail hyaline terminus is 15.5 μm long (Whitehead, 1968).

No esterase phenotype has been characterized for *M. decalineata*. The mode of reproduction and chromosome number are unknown. On coffee, this species has been found in Tanzania and São Tome and Principe (Campos and Villain, 2005).

6.4.7 *M. hapla*

The females are 550–790 μm long, and the DEGO position is 5–6 μm. The perineal pattern (Fig. 6.10J–N) is round hexagon to flattened oval, often with punctations in the tail terminal area. Lateral lines are indistinct. Some striae may extend laterally and form one or two 'wings'. Striae are smooth to wavy. The female stylet is short (10–14 μm), and its knobs are round and distinctly offset from the shaft. The stylet cone is slightly curved dorsally, and the shaft is broadest posteriorly (Fig. 6.10F,G). Males are 791–1,432 μm long and their stylet is 17.3–22.7 μm long. Their head is neither annulated nor offset from the body. The stylet is narrow and short, with round knobs, which are offset from the shaft (Fig. 6.10C,V–Z). The DEGO position is 4–6 μm. The J2 measure 312–355 μm long and their stylet is 10–12 μm long. The J2 head present a truncate cone shape, and a head cap that is small and circular. The tail is 33–48 μm long, tapering uniformly to a tip which is variable in shape, usually subacute but sometimes bifid (Fig. 6.10T,U).

This species can be distinguished by its esterase phenotype (Est H1, Fig. 6.21) and PCR-SCAR markers (Carneiro et al., 2000; Zijlstra et al., 2000). It reproduces by meiotic parthenogenesis (race A) or by mitotic parthenogenesis (race B). Race A has $n = 13 - 17$, while race B has $2n = 30 - 31$, although most populations present polyploidy and have $3n = 43 - 48$ (Tryantaphyllou, 1985). On coffee, *M. hapla* has been reported from Brazil, Tanzania, Zaire, India, Kenya, Congo, Guatemala and El Salvador (Campos and Villain, 2005).

6.4.8 *M. incognita*

The lengths of female body and stylet are 510–690 and 15–16 μm long, respectively. The DEGO position is 2–4 μm. This species is diagnosed by its perineal pattern, which has a high dorsal arch composed of smooth to wavy striae (Fig. 6.11F,G,M,R,S). Some striae fork near the lateral lines, but distinct lateral lines are absent. Striae that bend toward the vulva can often be seen. The female stylet cone is distinctly curved dorsally, and the shaft is slightly wider posteriorly. The stylet knobs are broadly elongated, offset from the shaft, and anteriorly indented. Males are 1,200–2,000 μm long. The male head shape is very characteristic, having a centrally concave labial disc, which is raised above the medial lips (Fig. 6.11A,K,J,N,O,P). The medial lips are as wide as the head region, which is generally marked by two or three incomplete annulations. The DEGO position is 1.4–2.5 μm. The stylet is 23–26 μm long, with a tip that is blunt and wider than the medial portion of the cone. The shaft is generally cylindrical and it often narrows near the stylet knobs. The stylet knobs are offset from the shaft, anteriorly indented,

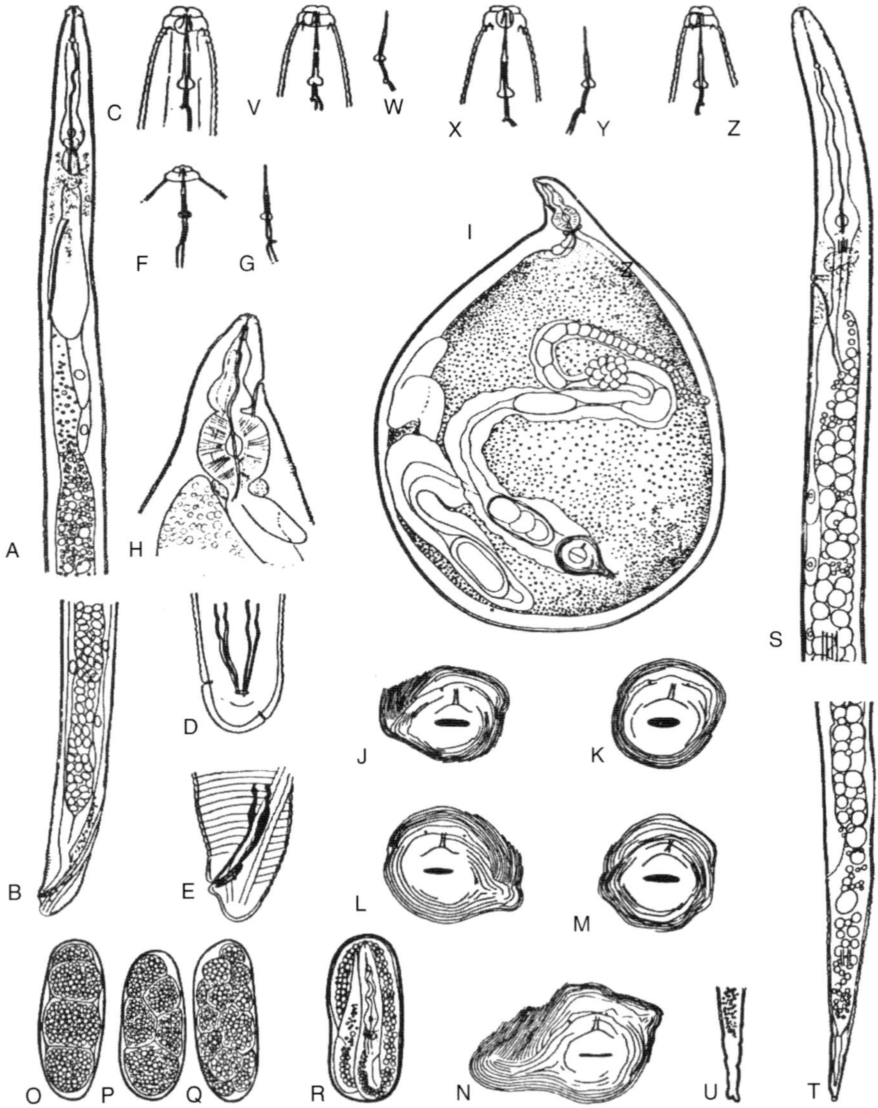

Fig. 6.10 *Meloidogyne hapla*. (**A–E,V,X,Z**) Male. (**F,G**) Female stylets. (**H–N**) Female anterior region, body and perineal patterns. (**O–R**) Eggs. (**S–U**) J_2 anterior region and tails (from Chitwood, 1949, with permission)

and broadly elongated to round (Fig. 6.11A,K). The J2 are 360–393 μm long, their DEGO position is 2.0–2.5 μm; the stylet is 10–12 μm long. The J2 present dumb bell-shaped labial disc and a medial disc. The labial disc is small and round, slightly raised above the medial lips. Lateral lips lie in contour with the head region, which usually bears two to four incomplete annulations. The J2 tail is 38–55 μm long, and it tapers steadily to a subacute terminus, with coarse posterior striae (Fig. 6.11U).

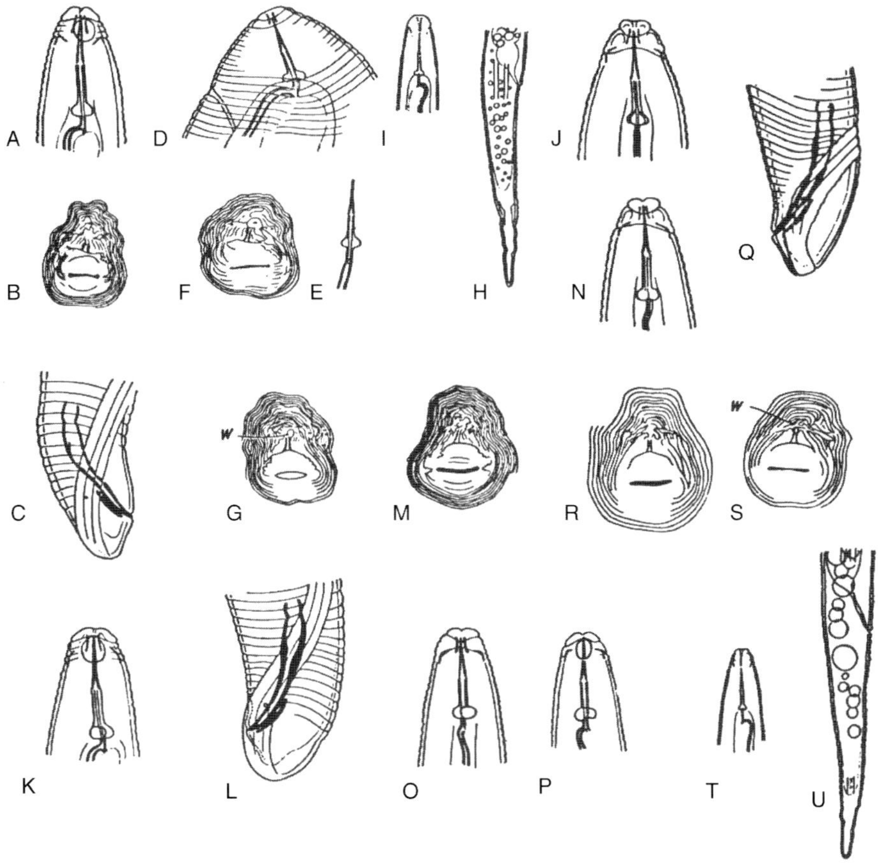

Fig. 6.11 *Meloidogyne incognita*. (**A,J,K,N,O,P**) Male anterior region. (**C,L,Q**) Male posterior end. (**D,E**) Female anterior region and stylet. (**B,F,G,M,R,S**) Perineal patterns. (**H,I,T,U**) J2 anterior and posterior regions. (from Chitwood, 1949, with permission)

M. incognita can be distinguished by its esterase phenotypes (Est I1 and I2) (Carneiro et al., 2000; Fig. 6.21) and PCR-SCAR markers (Zijlstra et al., 2000a; Randig et al., 2002). It reproduces by mitotic parthenogenesis, with 2n = 41 – 48 (Tryantaphyllou, 1985). Coffee-parasitic *M. incognita* populations have been found in Brazil, Tanzania, Jamaica, Venezuela, Guatemala, the Ivory Coast, India, Costa Rica, El Salvador, Nicaragua, Cuba and the U.S.A. (Campos and Villain, 2005).

6.4.9 M. inornata

In its original description and in subsequent taxonomic reviews of *Meloidogyne* sp., *M. inornata* has been considered closely related to *M. incognita* (Whitehead, 1968; Hewlett and Tarjan, 1983). Jepson (1987) and Eisenback and Triantaphyllou (1991) have synonymised *M. inornata* with *M. incognita* based on morphological features. Carneiro et al. (2008) have re-described and revalidated *M. inornata*.

The perineal pattern has a distinct, high dorsal arch composed of smooth to wavy striae, similar to those of *M. incognita* (Fig. 6.12H). The female stylet is 15–17 μm long, with the cone generally slightly curved dorsally and with well developed knobs. The DEGO position is 3.5–4.5 μm. Males have a high, rounded head cap, which is continuous with the body contour; it has a large, round, centrally concave labial disc, raised above the medial lips (Fig. 6.12A,B). The head region is

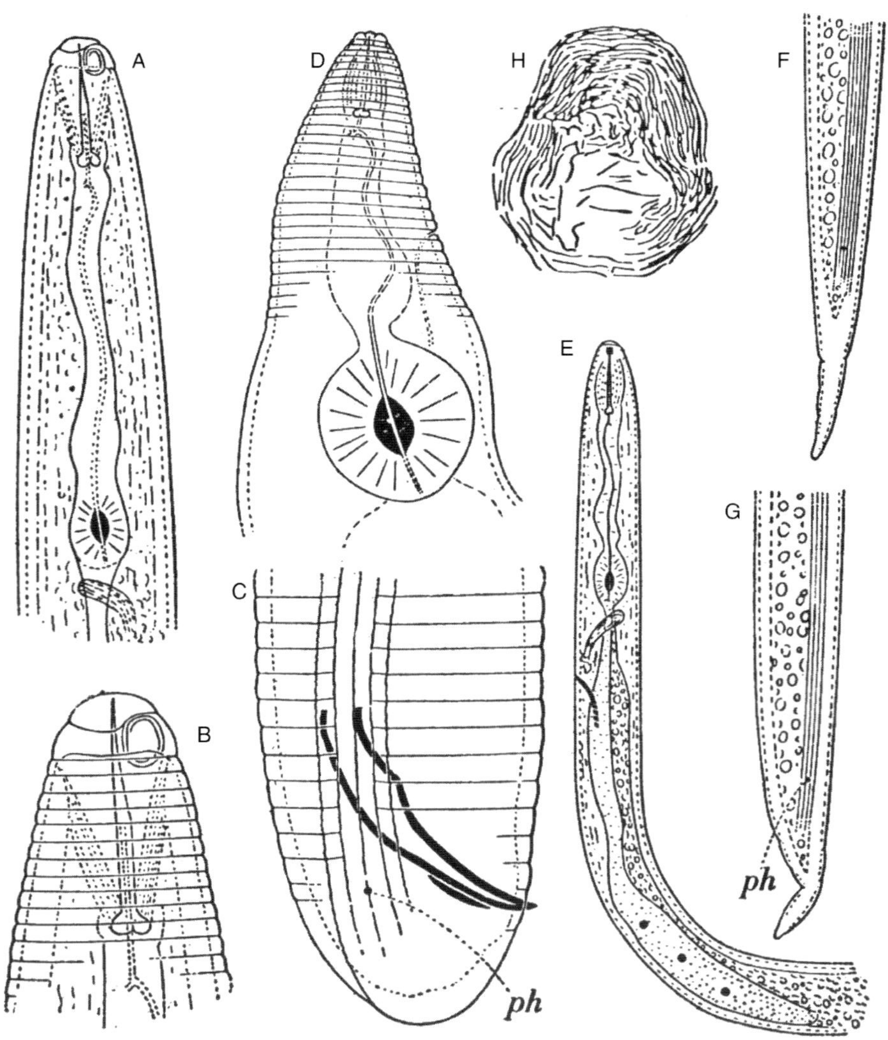

Fig. 6.12 *Meloidogyne inornata*. (**A–C**) Male stylet, anterior and posterior regions. (**D**) Female anterior region. (**E–G**) J_2 anterior and posterior regions. (**H**) Perineal pattern. ph = phasmid (from Lordello, 1956, with permission)

never marked by incomplete annulations. The stylet is robust (20–25 μm long) with a straight cone, cylindrical shaft with several small projections, and pear-shaped, backward-sloping knobs. The male lateral fields are composed of a variable number of crenate incisures in different parts of the body. The J2 stylet is 10–13 μm long and the DEGO position is 2.5–3.5 μm. The lateral fields are composed of four to six straight or undulate incisures (Fig. 6.12F,G), and the tail length is 35–58 μm.

The esterase phenotype I3 (Fig. 6.21) is species-specific, being the most useful character to differentiate *M. inornata* from other species. This species reproduces by mitotic parthenogenesis, with $3n = 54 - 58$ (Carneiro et al., 2008). Coffee-parasitic *M. inornata* has been reported from Guatemala (Campos and Villain, 2005). Nonetheless, a recent survey conducted in Latin America with the aid of esterase phenotyping has not detected this species in Guatemala (Hernandez et al., 2004; Carneiro et al., 2004).

6.4.10 M. izalcoensis

The perineal pattern is similar to *M. incognita* and *M. paranaensis*. It presents a dorsal arch which can be moderately high or high, squarish to round. It also presents striae coarse, smooth to wavy, sometimes zigzaggy, usually without a distinct whorl (Fig. 6.13E). The female head region is offset from the body, sometimes annulated (Fig. 6.13C). The labial disc has two bumps on the ventral side, slightly raised above the medial lips. The female stylet is robust, 15–16 μm long; the DEGO position is 4.5–6 μm. Males have a high, round head cap which is continuous with the body contour (Fig. 6.13B,D). The labial disc is fused with the medial lips to form an elongated lip structure. The head region is never marked by incomplete annulations. The stylet is robust, 23–26 μm long and it has rounded knobs, backwardly sloping (Fig. 6.13B,D); the DEGO position is 4–7 μm. In J2, the stylet length is 12–13 μm and the DEGO position is 3–4 μm. The J2 tail is 45–48 μm long, conoid, with a round terminus (Fig. 6.13G–I).

The esterase phenotype I4 (Fig. 6.21) is unique and is the most useful character to differentiate *M. izalcoensis* from other species (Carneiro et al., 2005a). In molecular analysis, *M. incognita* and *M. izalcoensis* have appeared far apart in majority rule consensus dendrograms, which shows that these species are phylogenetically distant (Carneiro et al., 2004). *M. izalcoensis* reproduces by mitotic parthenogenesis, having $2n = 44 - 48$. This species has been reported from El Salvador (Carneiro et al., 2005a).

6.4.11 M. javanica

The perineal pattern has a round to flattened dorsal arch, with distinct lateral lines which separate the pattern into dorsal and ventral regions (Fig. 6.14AA,BB,C,CC,D, G,N,O,Z). No or few striae cross the lateral incisures, while some striae bend

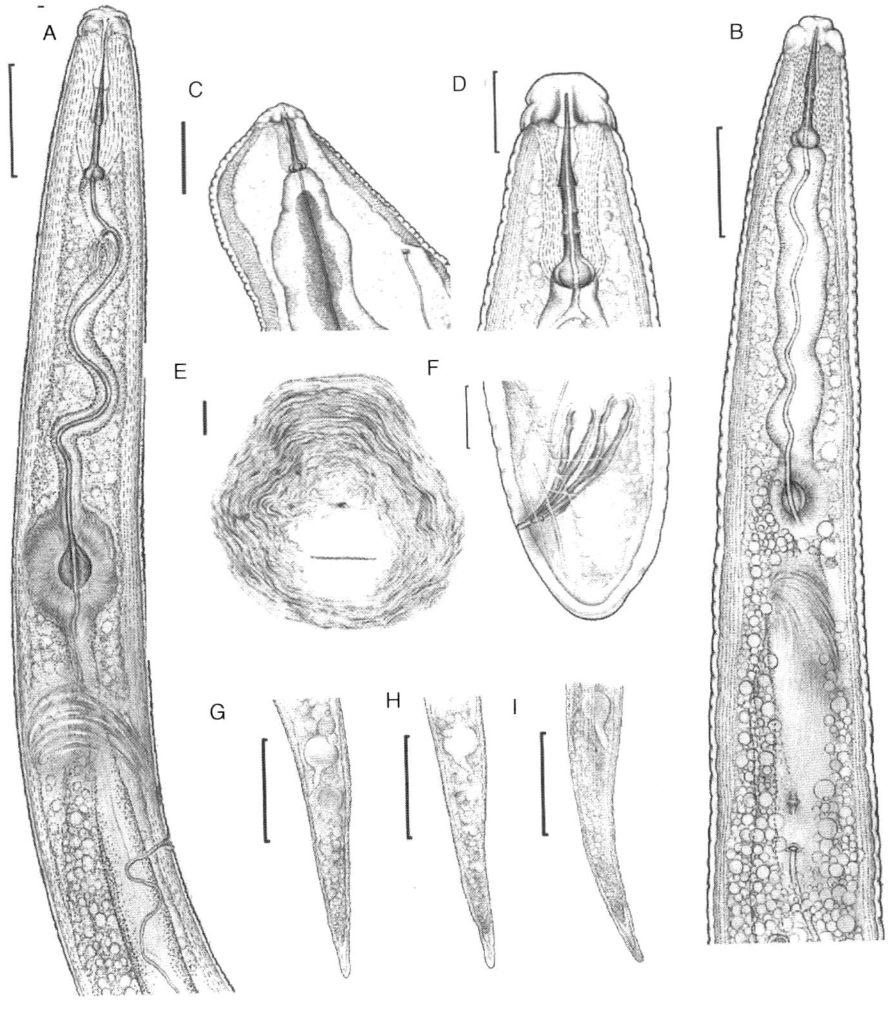

Fig. 6.13 *Meloidogyne izalcoensis*. (**A**) J_2 anterior region. (**B,D**) Male anterior region. (**C,E**) Female anterior region and perineal pattern, respectively. (**F**) male posterior region. (**G–I**) J_2 tails. Scale bars: A, B = 10 μm, C – I = 20 μm (from Carneiro et al., 2005a, with permission)

toward the vulva. Female stylet is 14–18 μm long and similar to *M. incognita*'s, except that its cone is not distinctly curved dorsally, and it gradually increases in width posteriorly (Fig. 6.14A,B,P). The DEGO position is 2–5 μm. Males are 940–1,440 μm long, and the head cap is high and almost as wide as the head region (Fig. 6.14E,H,R,S). The large smooth labial disc and the medial ones are fused. The stylet is 20–21 μm long, with a cone that is narrow anteriorly and very wide posteriorly; its shaft is cylindrical and it often narrows near the junction with the stylet knobs; these are low, wide and offset from the shaft (Fig. 6.14K–M). The DEGO position is 2–3 μm.

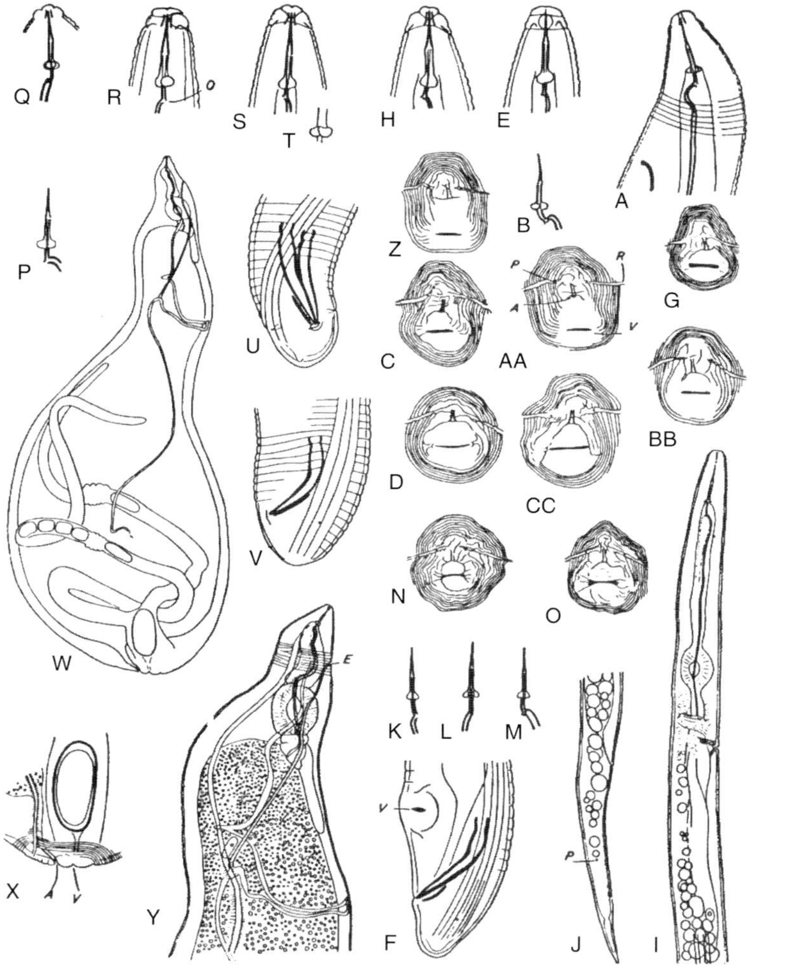

Fig. 6.14 *Meloidogyne javanica*. (**A**) Female anterior region. (**B,K,L,M**) Female stylet. (**AA,BB,C,CC,D,G,N,O,Z**) Perineal patterns. (**E,H,R,S**) Male anterior region. (**F**) Intersex male posterior region with rudimentary vulva. (**I,J**) J_2 anterior and posterior regions, respectively. (**P,Q**) Female stylet. (**U,V**) Male posterior end. (**W,X,Y**) Female body, posterior and anterior regions (from Chitwood, 1949, with permission)

Coffee-parasitic *M. javanica* has been reported from Brazil and other countries (see below). Nonetheless, experimental inoculations on susceptible genotypes have never confirmed that coffee is a suitable host for *M. javanica* (Santos, 1997; Oliveira et al., 1998; Carneiro et al., 2005b).

This species can be distinguished by its esterase phenotype (Est J3, Fig. 6.21) and by PCR-SCAR markers (Carneiro et al., 2000; Zijlstra et al., 2000a). *M. javanica* reproduces by mitotic parthenogenesis, with $2n = 41 - 48$ (Tryantaphyllou, 1985).

On coffee, *M. javanica* has been reported from Brazil, Tanzania, Zaire, El Salvador, India, Cuba and São Tome and Principe (Campos and Villain, 2005).

6.4.12 M. kikuyensis

This species is characterized by females 580–880 μm long, with a peculiar perineal pattern which has a low arch and prominent single lateral lines without incisures. The phasmids are located fairly close to the tail end, and characteristic striae with 'cheek-like' structures are seen on each side of the vulva (Fig. 6.15O,T). The female stylet is 13.5–16 μm and the DEGO position is 3.5–5 μm. Males are 810–1,650 μm long, with hexagonal head cap (Fig. 6.15C,D). The head has three annules behind the head cap. The stylet is 17–20 μm long and the DEGO position is 4.5–6 μm.

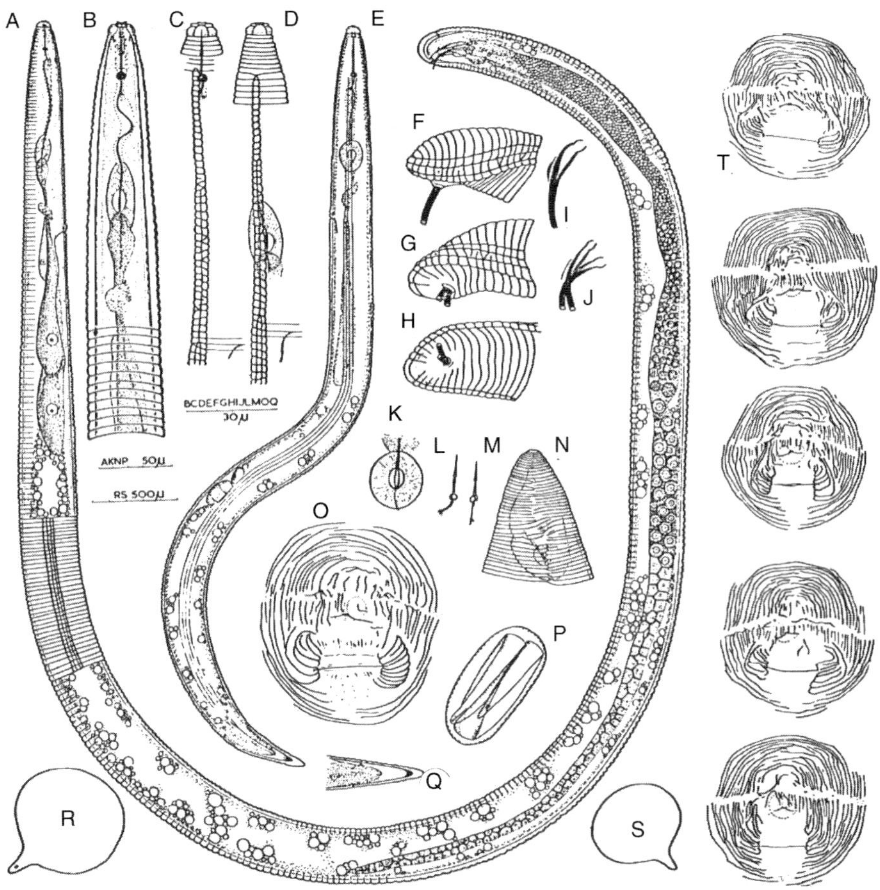

Fig. 6.15 *Meloidogyne kikuyensis*. (**A–D**) Male anterior region. (**E,Q**) J_2. (**F–J**) Male posterior end and spicules. (**K–N**) Female. (**O,T**) Perineal patterns. (**P**) Egg with J_2. (**R,S**) Female body shapes (from De Grisse, 1960, with permission)

The lateral fields present four incisures at mid-body (Fig. 6.15C,D). The J2 are 290–360 μm long, with stylet 12–15 μm long and the DEGO position is 3.5–5 μm. The tail is short (29.1 μm), tapering with a broad, rounded triangular hyaline area (Fig. 6.15E,Q). The short J2 tail differs in this species from all the others, except for *M. africana*. For a detailed morphological description of this species see De Grisse (1960), Whitehead (1968) and Jepson (1987).

No electrophoretic phenotype is available for this species. It reproduces by amphimixis, with n = 7 (Triantaphyllou, 1990). Cytogenetic studies have suggested that despite the small chromosome number, *M. kikuyensis* should be regarded as a true RKN (Triantaphyllou, 1990). The low chromosome number would represent the ancestral *Meloidogyne* condition from which all species have evolved. In comparison to the predominant parthenogenetic mode of reproduction found in *Meloidogyne* sp., the obligatory amphimitic mode of reproduction of *M. kikuyensis* further supports the hypothesis that this species represents the ancestral form of *Meloidogyne* sp. (Triantaphyllou, 1990). On coffee, this species has been reported from Kenya (Campos and Villain, 2005).

6.4.13 *M. konaensis*

In its original description (Eisenback et al., 1994), this species was diagnosed through the morphology of females (L = 531.8 − 1,510 μm) and males (L = 1,149 − 1,872 μm). Its perineal pattern is quite variable and similar to *M. incognita*'s and *M. arenaria*'s (Fig. 6.16M); thus, it is not a good taxonomic character. The morphology of female stylet is similar to *M. arenaria*'s; nonetheless, unlike the latter, the medial lips are divided into distinct lip pairs in *M. konaensis*. The most useful character to identify this species is male stylet morphology, which is 20.2–24.4 μm long, with 6–12 large projections surrounding its shaft (Fig. 6.16D,G); otherwise, the stylet is similar to *M. arenaria*'s. The male head cap is also similar to *M. arenaria*'s; however, the medial lip is often divided into distinct medial lip pairs in *M. konaensis* (Eisenback et al., 1994).

This species presents three different esterase phenotypes (Carneiro et al., 2000; 2004, Sipes et al., 2005), but only populations with the phenotype Est P1 (= Est F1) (Fig. 6.21) reproduce on coffee (Sipes et al., 2005). This species reproduces by mitotic parthenogenesis, with 2n = 44 (Eisenback et al., 1994). *M. konaensis* has only been reported from the USA (Hawaii) (Campos and Villain, 2005).

6.4.14 *M. mayaguensis*

In its original description (Rammah and Hirschmann, 1988), this species was diagnosed by the perineal pattern, which is round to dorso-ventrally ovoid (Fig. 6.17G,H). The dorsal arch is rounded, with striae that are fine, mostly continuous, widely spaced. The pattern's ventral region is rounded, with striae that are fine, closely

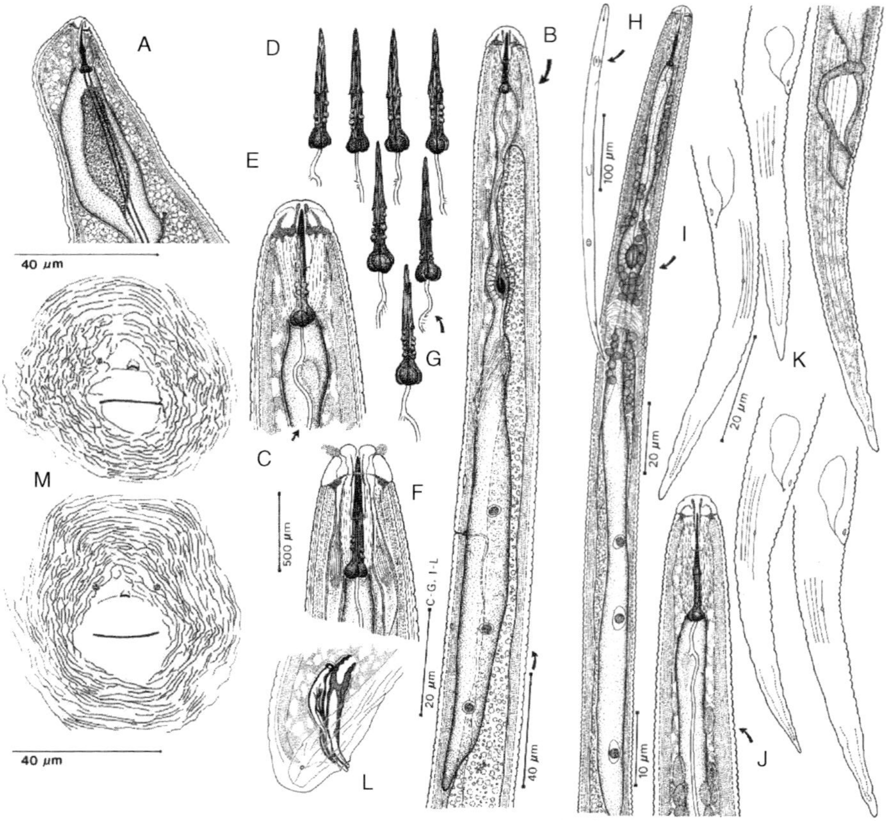

Fig. 6.16 *Meloidogyne konaensis*. (**A**) Female anterior region. (**B,C,F**) Male anterior region. (**D,G**) Male stylet. (**L**) Male posterior region. (**I,J**) J_2 anterior region. (**K**) J_2 tails. (**M**) Perineal patterns (from Eisenback et al., 1994, with permission)

spaced. Lateral lines are only seldom distinguishable; when seen, they break in striae; alternatively, a single lateral line may occur on one side of the pattern, at the junction of the dorsal and ventral arches. The tail tip area is large, circular, and usually free of striae. The female body is 518.4–769.5 μm long. Recently, Brito et al. (2004) have argued that the perineal pattern is not a good character for identification of *M. mayaguensis*, because it presents an accentuated variability and because many specimens show a pattern similar to *M. incognita*'s. The female stylet is 13.8–16.8 μm long, with knobs characteristically reniform in shape. In males, the high head cap is only slightly defined, is not offset from the body, and it lacks annulations. The stylet is 20.7–24.6 μm long, with knobs that are distinctly separated and not longitudinally divided by a groove; the base of the dorsal knob is concave. The stylet shaft is irregular in its diameter, with a wavy lumen, and it narrows near the junction with the stylet knobs. In J2, the tail measures 49.2–62.9 μm, and it tapers

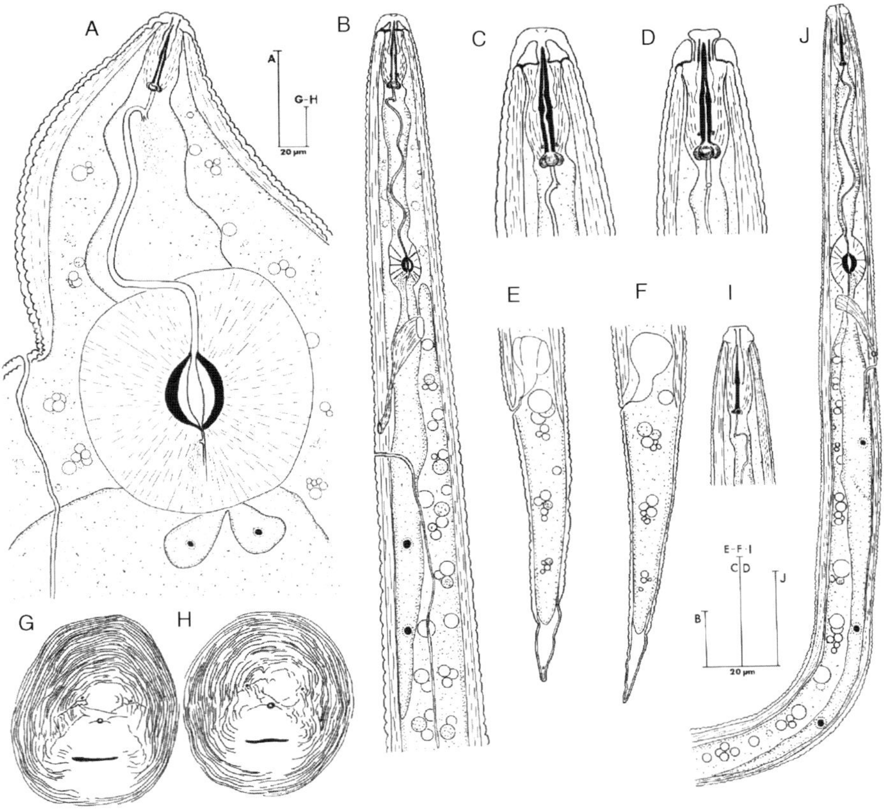

Fig. 6.17 *Meloidogyne mayaguensis*. (**A**) female anterior region. (**B–D**) Male anterior region. (**E,F**) J_2 tails. (**G,H**) Perineal patterns. (**I,J**) J_2 anterior region (from Rammah and Hirschmann, 1988, with permission)

gradually to its tip; the tail terminus is not distinctly narrow (Fig. 6.17E,F; Rammah and Hirschmann, 1988).

Considering the difficulty of characterizing *M. mayaguensis* on morphological grounds, the identification can be based on its esterase phenotype (Est M2, Fig. 6.21) (Carneiro et al., 2000; 2001) and DNA analysis (Block et al., 2002). *M. mayaguensis* reproduces through mitotic parthenogenesis, with $2n = 44 - 45$ (Esbenshade and Triantaphyllou, 1985a). On coffee, its geographical distribution includes Cuba, Costa Rica and Guatemala (Campos and Villain, 2005).

6.4.15 *M. megadora*

This species is diagnosed by its characteristic perineal pattern, which is more or less circular with very low dorsal arch; the pattern is also marked by short, thick

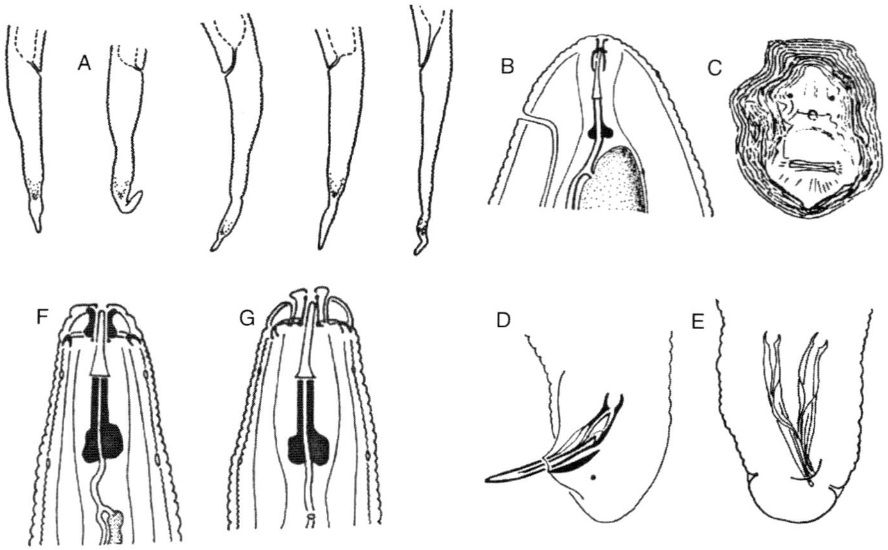

Fig. 6.18 *Meloidogyne megadora*. (**A**) J_2 tails. (**B**) Female anterior region. (**C**) Perineal pattern. (**D,E**) Male posterior region. (**F,G**) Male anterior region (from Whitehead, 1968, with permission)

striae, generally smooth but often broken (Fig. 6.18C). Phasmids are fairly close to tail terminus; the tail end is fairly wide. Lateral lines are not generally visible, but they are marked in the posterior region of the pattern by characteristic short coarse striae. In some patterns the tail whorl is seen distinct from the rest of the pattern. The female stylet is 13–17 μm long and the DEGO position is 4–9 μm. Males present a head that is low, shaped as a truncate cone, with one indented annule behind the head cap (Fig. 6.18F,G). In normal males, which are 905–2,277 μm long, the stylet is strong, 18.3–21.9 μm long, with knobs that are longer than wide, with outer margins longitudinally and transversely grooved (Fig. 6.18F,G). Dwarf males present reduced stylet with more rounded knobs. The DEGO position is 4–8.3 μm. The J2 are 413–548 μm long, with three annules behind head cap. Their tail is 47–58 μm long, subacute; it tapers irregularly in three 'sections', with its tip having various shapes (Fig. 6.18A; Whitehead, 1968).

No electrophoretic phenotype is available for *M. megadora*. Its reproduction mode and chromosome number are unknown. A review on this species has recently been prepared (I. Abrantes, U. Coimbra, personal communication). On coffee, this species' geographical distribution include Angola, Uganda and São Tome and Principe (Almeida and Santos, 2002; Campos and Villain, 2005).

6.4.16 M. oteifae

This species is diagnosed by small females ($L = 520 - 680$ μm) with short neck, and by the perineal pattern with low dorsal arch, very smooth and faint striae which are

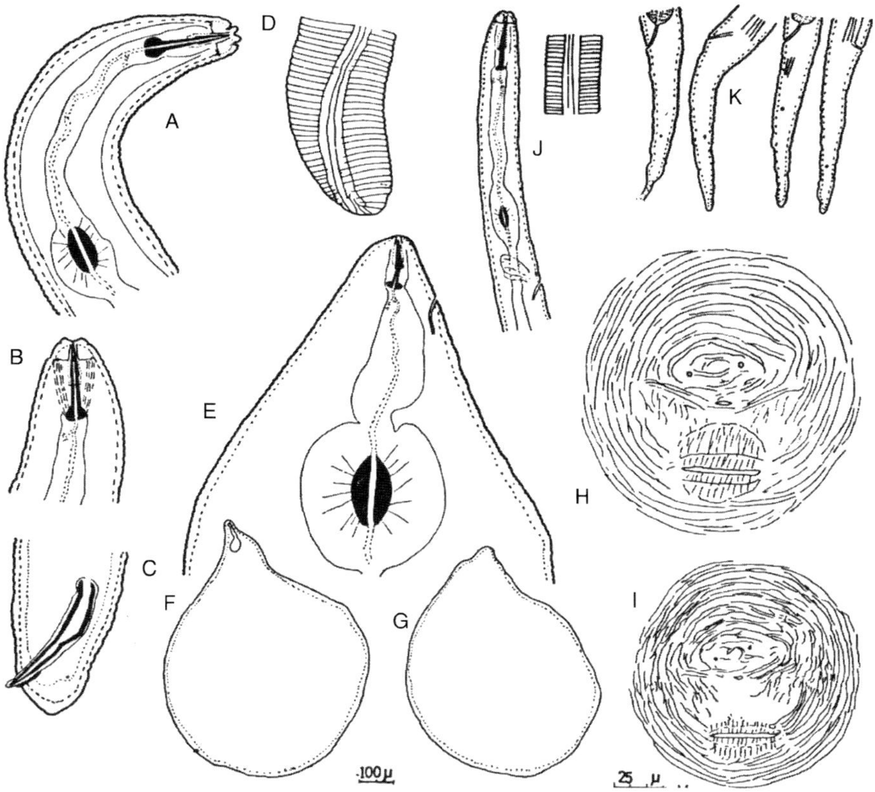

Fig. 6.19 *Meloidogyne oteifae*. (**A–D**) Male anterior and posterior regions, stylet and spicule. (**E–G**) Female anterior region and body shape. (**H,I**) Perineal patterns. (**J,K**) J_2 anterior region, lateral field and tails (from Elmiligy, 1968, with permission)

close together (Fig. 6.19H,I). The tail terminus is wide, covered by short, coarse striae and surrounded by concentric circles of striae, which form a distinct tail pattern that is not raised as a knob. The vulva is wide. *M. oteifa* and *M. africana*'s perineal patterns are similar, but the former has the vulva surrounded by circles of striae, which are themselves crossed by some striations radiating from the vulva; also, *M oteifa* does not have a wide, relatively clear area in the lateral field (Elmiligy, 1968). In *M. oteifa*, large phasmids are present, which are positioned closer than the vulva width. The female excretory pore is located posterior to the stylet knobs (Fig. 6.19E), at 18–23 µm from the anterior end of the body. The stylet is 13–14 µm long, slightly curved, and the knobs are round; the DEGO position is 3–4 µm. Males are 980–1,270 µm long, with one or two postlabial annules. The stylet is strong, 19–23 µm long, with elongated basal knobs (Fig. 6.19A,B). The tail is very short (Fig. 6.19C). The J2 ($L = 320 - 400$ µm) have stylet 11–13 µm long, tail tapering to a round terminus (Fig. 6.19K), and the lateral field is marked by four lines (Fig. 6.19J); the number of lines decrease towards the anterior and posterior ends of the body.

No electrophoretic phenotype is available for *M. oteifa*. Its mode of reproduction and chromosome number are unknown. On coffee, it has been reported only from Zaire (Campos and Villain, 2005).

6.4.17 *M. paranaensis*

This species can be distinguished from others by the combination of the following characters: the females (L = 512 − 780 μm) have labial and medial lips fused, asymmetric and rectangular. Their stylet is 15–17.5 μm long, with broad, distinctly offset knobs, and the DEGO position is 4.2–5.5 μm. The perineal pattern is similar to *M. incognita*'s (Fig. 6.20AA). Males (L = 983 − 2, 284 μm) have high, round head cap continuous with the body contour (Fig. 6.20B–D). The labial disc is fused with the medial ones, forming an elongated lip structure. Sometimes the head region is marked by an incomplete annulation. The stylet is robust (20–27 μm), usually

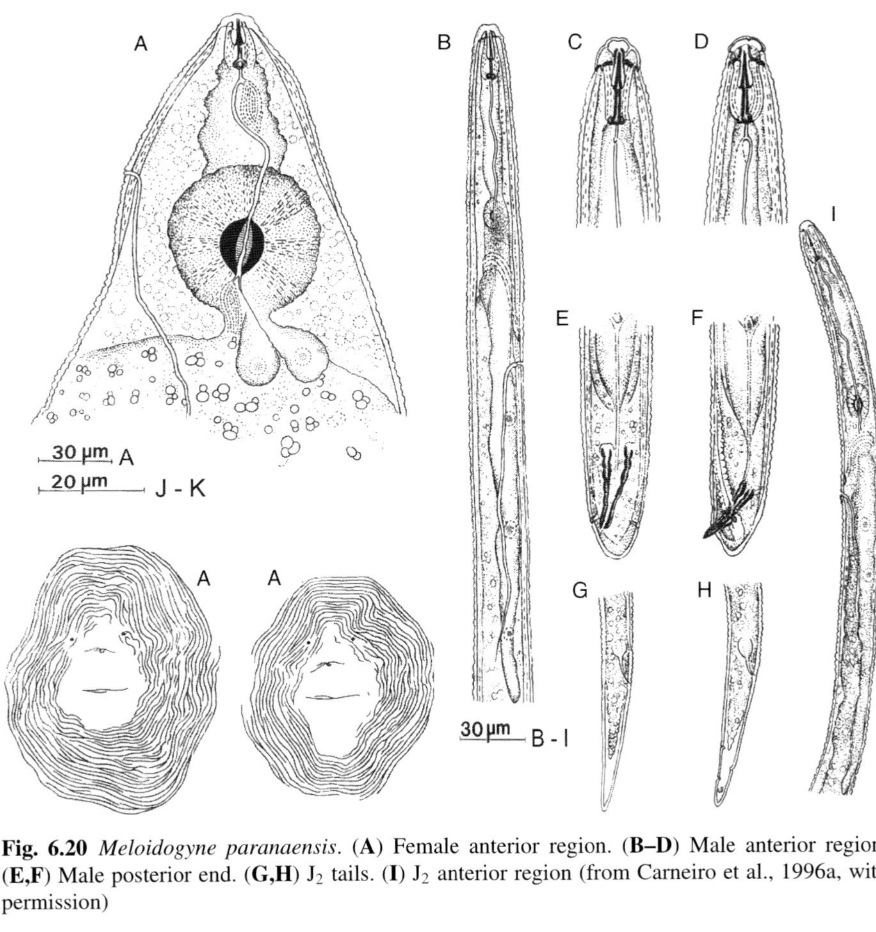

Fig. 6.20 *Meloidogyne paranaensis*. (**A**) Female anterior region. (**B–D**) Male anterior region. (**E,F**) Male posterior end. (**G,H**) J_2 tails. (**I**) J_2 anterior region (from Carneiro et al., 1996a, with permission)

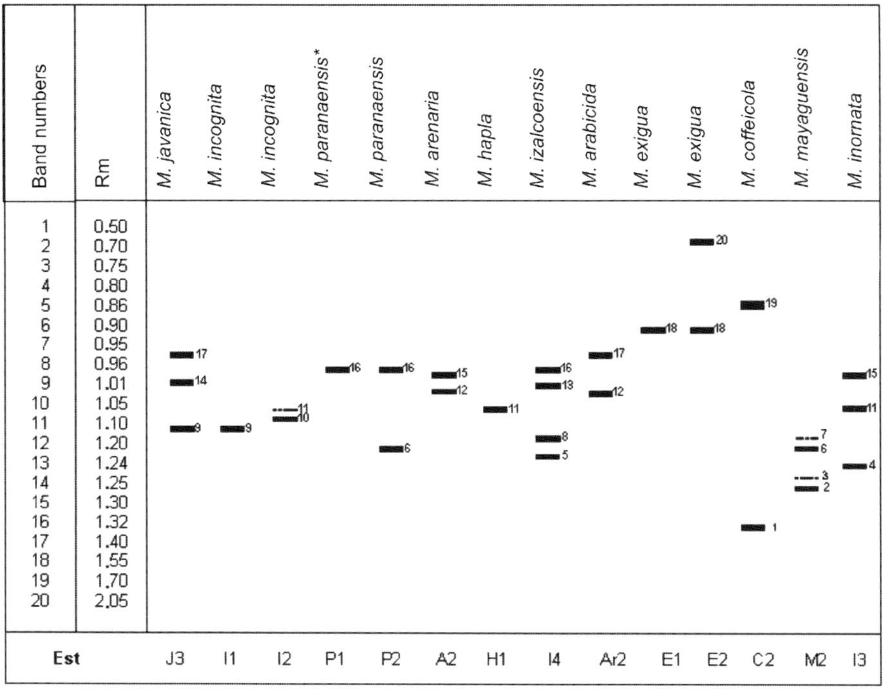

Fig. 6.21 Esterase (Est) phenotypes of coffee-parasitic *Meloidogyne* spp. Rm = ratio of migration in relation to the fastest band of *M. javanica*. Dotted lines indicate weak bands
* phenotype Est P1 (= Est F1) has been detected in *M. konaensis* from coffee

with rounded to transversely elongated knobs (Fig. 6.20C,D), and sometimes with one or two projections protruding from the shaft. The DEGO position is 3.5–5 μm. The J2 stylet is 13–14 μm long, and the DEGO position is 4–4.5 μm. The tail is 48–51 μm long, usually conoid and with a rounded terminus. The hyaline tail terminus is distinct (Fig. 6.20G,H). The rectal dilatation is large and the phasmids are small and located posterior to the anus.

M. paranaensis can also be distinguished by its esterase phenotypes [Est P1 (= Est F1) and P2] (Fig. 6.21; Carneiro et al., 2004) and PCR-SCAR markers (Randig et al., 2002; 2004). It reproduces by mitotic parthenogenesis, with 2n = 50 − 56 (Esbenshade and Triantaphyllou, 1985a; Carneiro et al., 1996a). On coffee, it has been reported from Brazil, Guatemala and the USA (Hawaii) (Carneiro et al., 2004; Campos and Villain, 2005).

6.5 Electrophoresis-Based *Meloidogyne* Species Identification

The difficulties and benefits of identifying *Meloidogyne* species based on electrophoresis have been revealed by studies on about one thousand RKN populations from different crops (Esbenshade and Triantaphyllou, 1985a; 1990; Carneiro et al., 1996b; 2000; 2004; Cofcewicz et al., 2004; 2005). These studies have demonstrated that several *Meloidogyne* species can be identified through enzyme phenotypes (esterase and malatodesidrogenase) revealed through polyacrilamide-gel electrophoresis. Through the methodology outlined by Esbenshade and Triantaphyllou (1985b) and Carneiro and Almeida (2001), the esterase phenotype of as many 20–25 individual females can be compared in the same gel.

Therefore, this biochemical taxonomic approach is a valuable tool in *Meloidogyne* research, specially (i) in extensive surveys, to determine the frequency and relative distribution of *Meloidogyne* species, (ii) to routinely identify RKN populations, and to detect atypical ones, and (iii) to purify RKN populations, prior to studies on DNA analyses, morphological characterization or others that need pure species (Carneiro et al., 1996b; 2000; 2005b; Cofcewicz et al., 2004; 2005; Esbenshade and Triantaphyllou, 1985a; 1990).

Unfortunately, there are no enzymatic phenotypes available for identification of all *Meloidogyne* species. Of the 17 coffee-parasitic *Meloidogyne* species, esterase phenotypes are available for the identification of 11 (Fig. 6.21). For each phenotype, the bands have their ratio of migration (Rm) calculated in relation to the fastest band of *M. javanica*, which is used as a reference.

The phenotypes available are: *M. incognita* (Est I1, Rm = 1.01; Est I2, Rm = 1.05 and 1.10); *M. exigua* (Est E1, Rm = 1.55; Est E2, Rm = 1.55 and 2.05); *M. coffeicola* (Est C2, Rm = 0.50 and 1.70); *M. javanica* (Est J3, Rm = 1.01, 1.25 and 1.40); *M. hapla* (Est H1, Rm = 1.10); *M. arenaria* (Est A2, Rm = 1.20 and 1.30); *M. paranaensis* (Est P1 (= F1), Rm = 1.32; Est P2, Rm = 0.90 and 1.32); *M. arabicida* (Est Ar2, Rm = 1.20 and 1.40); *M. mayaguensis* (Est M2, Rm = 0.70, 0.75, 0.90 and 0.95); *M. izalcoensis* (Est S4 (= I4), Rm = 0.86, 0.96, 1.24 and 1.32); and *M. inornata* (Est I3, Rm = 0.80, 1.10 and 1.30) (Carneiro et al., 2000; 2004; 2005b, 2008).

M. konaensis has been reported as presenting three different esterase phenotypes (Est F1, Est I1 and Est F1-I1), depending on the plant it is parasitizing (Sipes et al., 2005). According to these authors, only the Est F1 isolate parasitizes arabica coffee (*C. arabica* L.). In that publication, the morphological comparisons between Est F1, Est I1 and Est F1-I1 isolates are rather poor, and those authors have not convincingly shown that they all belong to *M. konaensis*. It is quite unusual that the same *Meloidogyne* species should present three esterase phenotypes when parasitizing different plants.

A coffee-parasitic RKN isolate from Hawaii (USA), reportedly belonging to *M. konaensis*, has been examined through morphological, isozyme and molecular approaches (Carneiro et al., 2004). This isolate presented the Est F1 (= P1) esterase phenotype and 90% genetic similarity with *M. paranaensis*. Thus, it is obvious that *M. konaensis* is not a clearly characterized species, as suggested by its variable esterase phenotype. The isolate studied by Carneiro et al. (2004) has indubitably been identified as *M. paranaensis* through a SCAR marker with a specific size fragment of 208 pb (Randig et al. 2004).

6.6 DNA-Based *Meloidogyne* Species Identification

The advent of PCR has allowed recent progress in nematode diagnostics. Through this technique, a single nematode or egg mass can be precisely identified at the species level.

Recently developed SCAR-primer sets have enabled sensitive and rapid identification of *M. incognita, M. javanica, M. arenaria, M. hapla, M. chitwoodi* and *M. fallax* (Zijlstra et al., 2000; Zijlstra et al., 2000). These SCAR primers were deduced from sequences of species-specific RAPD markers.

Randig et al. (2002) have developed a PCR-based assay to identify coffee-parasitic RKNs from Brazil. Three RAPD markers have been further transformed into SCAR markers specific for *M. exigua, M. incognita* and *M. paranaensis*. After the PCR procedure, the SCAR primers allow the initial polymorphism between those species to be retained as presence *vs* absence of DNA amplification. Moreover, multiplex PCR using the three pairs of SCAR primers in a single reaction allowed unambiguous identification of those *Meloidogyne* species, even when they were mixed in relative concentration as low as 1% (Randig et al., 2004).

Recently, 54 RKN populations from coffee fields in São Paulo and Minas Gerais States, Brazil, have been identified through esterase phenotyping and PCR reactions using the six SCAR primers altogether (Carneiro et al., 2005b). The multiplex PCR allowed unambiguous identification of *M. exigua, M. incognita* and *M. paranaensis* when present in the samples alone or in mixture; therefore, the potential of this approach for routine diagnostics has been confirmed. This coffee SCAR kit should be extended to include other important coffee-parasitic *Meloidogyne* species from Latin America, Africa and Asia.

Isolates of *M. mayaguensis* have also been identified through DNA-based methods, such as RFLP (Fargette et al., 1996), RAPD (Blok et al., 1997a), amplification of ribosomal DNA of the intergenic spacer region between the 18S and 5S genes (Blok et al., 1997b) and analysis of mitochondrial DNA with products of 705 bp from *COII* and lRNA region (Blok et al., 2002).

6.7 *Meloidogyne* Intraspecific Variability

The International *Meloidogyne* Project has summarized the response of nearly one thousand populations of the most common *Meloidogyne* species and their races to a list of differential hosts (Table 6.1; Hartman and Sasser, 1985).

As regards *M. incognita*, all four races have been found associated with coffee. In Paraná, one of the most important coffee-producing States in Brazil, race two is prevalent and race four the rarest (R. Carneiro, IAPAR, personal communication). Three *M. exigua* races have been detected in Brazil, two of them parasitizing coffee (Carneiro et al., 2000). No races have been detected on other coffee-parasitic *Meloidogyne* species in Brazil.

Table 6.1 Differential host test for the most common coffee-parasitic *Meloidogyne* species [Adapted from Hartman and Sasser (1985) and Carneiro and Almeida (2000)]

Species and races	Differential host plants[a] and results						Original host
	Cotton	Tomato	Tobacco	Pepper	Watermelon	Peanut	
M. incognita race 1	−[b]	+	−	+	+	−	coffee
M. incognita race 2	−	+	+	+	+	−	coffee
M. incognita race 3	+	+	−	+	+	−	coffee
M. incognita race 4	+	+	+	+	+	−	coffee
M. exigua race 1	−	−	−	+	−	−	coffee
M. exigua race 2		+	−	−	+	−	coffee
M. exigua race 3	−	−	−	−	−	−	rubber tree
M. paranaensis	−	+	+	−	+	−	coffee
M. coffeicola	−	−	−	−	−	−	coffee

[a] Cotton ‘Deltapine’; tomato ‘Rutgers’; tobacco ‘NC95’; pepper ‘Early California Wonder’; watermelon ‘Charleston Gray’; peanut ‘Florunner’.
[b] ‘−’ indicates a resistant host; ‘+’ indicates a susceptible one.

There have been few studies on diversity and phylogenetics of coffee-parasitic *Meloidogyne* species; these studies have focused only on meiotic or mitotic parthenogenetic species (Randig et al., 2002; Carneiro et al., 2004). A high level of intraspecific polymorphism has been detected in *M. arenaria, M. exigua* races two and three and *M. hapla*, in comparison to *M. incognita* and *M. javanica*. Phylogenetic analyses have showed that *M. hapla* and *M. exigua* are more closely related to each other than they are to other species; this suggests an early evolutionary divergence of these meiotically-reproducing species from those that reproduce mitotically, and supports the hypothesis that amphimixis is the ancestral reproductive state of *Meloidogyne* (Triantaphyllou, 1985).

A recent study on 18 RKN populations from coffee fields in Brazil, Central America and the USA (Hawaii) has revealed their diversity with respect to enzyme phenotypes, morphology and genome (Carneiro et al., 2004). An analysis of the dendograms deduced from RAPD data has allowed the definition of different clusters of species with high bootstrap support: (i) *M. paranaensis* and *M. arabicida*; (ii) *M. exigua* and *M. mayaguensis*; (iii) *M. arenaria, M. javanica* and *M. izalcoensis*. Intraspecific groups with a low degree of polymorphism have been observed in *M. paranaensis* (polymorphism of 20.3%) and in *M. incognita* (esterase phenotypes Est I1 and I2) (polymorphism of 11.2%). In *M. exigua*, the two coffee-parasitic races presented a genetic diversity of only 8.6%.

Recent studies by Muniz et al. (2008) using RAPD-PCR have showed a high variability among *M. exigua* populations belonging to different races and enzymatic phenotypes. No relationship was observed between races, enzymatic phenotypes and genetic polymorphism. This high genetic variability had been predicted to occur in *Meloidogyne* species that reproduce by facultative meiotic parthenogenesis, in comparison to mitotic parthenogenesis (Triantaphyllou, 1985). Indeed, previous investigations had showed the monophyly of *M. arenaria* and *M. incognita* races (Cenis, 1993; Baum et al., 1994).

These findings suggest that for a given *Meloidogyne* species, its races do not form monophyletic groups; this indicates that such intraspecific groups may not have a common ancestor. In other words, races do not have a genetic determinism, suggesting that this variability should be considered in breeding programs for RKN-resistance (Muniz et al., 2008).

6.8 Concluding Remarks

There have been considerable advances in recent years in the taxonomy of coffee-parasitic *Meloidogyne* species: misidentifications have been revised, species have been described or revalidated, and new identification methods have been developed or consolidated. Isozyme phenotyping, for example, is now well established for most RKNs associated with coffee, and it has become a fairly simple and inexpensive taxonomic tool. Furthermore, over the last few years nematologists worldwide have become aware of the complexity of *Meloidogyne* taxonomy, and the need for characterizing several morphological and morphometric features of RKN populations to accurately identify them.

Proper procedures should also be followed during surveys conducted in coffee fields and nurseries, so that precious time and resources are not wasted. Indeed, one should collect only non-rotten roots with typical RKN-symptoms (galls, swellings or crackings); in old, rotten roots the RKN females are unlikely to be useful for isoenzyme characterization. Roots should be packed in plastic bags, surrounded by moist soil collected from the same site. If samples cannot be examined and processed immediately, they should be maintained in cold chamber or refrigerator; samples should not be frozen or left in the sun or in hot locations.

Wherever possible, the first step in identifying RKN populations should be characterizing their esterase phenotype(s), according to the methodology outlined by Esbenshade and Triantaphyllou (1985b) and Carneiro and Almeida (2001). For each RKN population, at least 30 females should be individually submitted to esterase phenotyping.

In those nematology laboratories where esterase phenotyping cannot be performed, perineal patterning should be cautiously used for species identification. Morphological characterization will also be needed whenever the RKN population presents an unreported esterase phenotype. In these cases, the RKN population under investigation could be either a new species or a population of those five *Meloidogyne* species for which esterase phenotyping has not yet been performed.

Perineal patterning should be carefully done, making sure only mature, egg-laying females are collected, properly cut and mounted in glass slides for examination under the light microscope. Perineal patterns should be properly cleaned of body residues and carefully mounted to avoid the creation of artifacts that will make observation and judgment of perineal pattern characters difficult; special care should be taken to avoid deformation of the perineal pattern through the pressure (weight) of the coverslip. At least 10 perineal patterns should be examined *per* RKN population.

Perineal patterning should be complemented by observation of male and J2 morphology and/or morphometry, paying special attention to those features and/or measures that are typical of one or just a few *Meloidogyne* species.

Currently, five coffee-parasitic *Meloidogyne* species from Africa are not available in international nematology collections and/or their types are not in good conditions for examination: *M. africana, M. decalineata, M. kikuyensis, M. megadora* and *M. oteifae*. For these and new *Meloidogyne* species to be described, it would be extremely interesting to have live samples shipped to Embrapa/Genetic Resources and Biotechnology (Brasilia, Brazil), where a complete infrastructure is available to maintain RKN populations from across the globe, either alive or cryopreserved (Carneiro et al. 2005c). This collection has allowed morphological, physiological, electrophoretic and molecular studies on many coffee-parasitic RKN populations.

References

Almeida AMSF, Santos MSN (2002) Resistance and host-response of selected plants to *Meloidogyne megadora*. J Nematol 34:140–142

Baum TJ, Gresshoff PM, Lewis SA et al (1994) Characterization and phylogenetic analysis of four root-knot nematode species using DNA amplification fingerprint and automated polyacrylamide gel electrophoresis. Mol Plant-Microbe Interact 7:39–47

Blok VC, Phillips MS, Fargette M (1997a) Genetic variation in tropical *Meloidogyne* spp. as shown by RAPDs. Fundam Appl Nematol 20:127–133

Blok VC, Phillips MS, McNicol JW et al (1997b) Comparison of sequences from the ribosomal DNA intergenic region of *Meloidogyne mayaguensis* and other major root-knot nematodes. J Nematol 29:16–22

Blok VC, Wishart J, Fargette M et al (2002) Mithochondrial DNA differences distinguishing *Meloidogyne mayaguensis* from the major species of tropical root-knot nematodes. Nematology 4:773–781

Brito JA, Powers TO, Mullin PG et al (2004) Morphological and molecular characterization of *Meloidogyne mayaguensis* from Florida. J Nematol 36:232–240

Cain SC (1974) *Meloidogyne exigua*. C.I.H. descriptions of plant–parasitic nematodes. Set 4, # 49, Commonwealth Institute of Helminthology, Herts

Campos VP, Villain L (2005) Nematode parasites of coffee and cocoa. In: Luc M, Sikora RA, Bridge J (eds) Plant parasitic nematodes in subtropical and tropical agriculture, 2nd edition. CABI, Wallingford

Carneiro RMDG, Almeida MRA (2000) Caracterização isoenzimática e variabilidade intraespecifica dos nematóides de galhas do cafeeiro no Brasil. Proceedings I Simp Pesqu Cafes Brasil: 280–282

Carneiro RMDG, Almeida MRA (2001) Tecnica de eletroforese usada no estudo de enzimas dos nematóides de galhas para identificação de especies. Nematol Bras 25:555–560

Carneiro RMDG, Almeida MRA, Carneiro RG (1996b) Enzyme phenotypes of Brazilian populations of *Meloidogyne* spp. Fundam Appl Nematol 19:555–560

Carneiro RMDG, Almeida MRA, Gomes ACMM et al (2005a). *Meloidogyne izalcoensis* n.sp. (Nematoda:Meloidogynidae), a root-knot nematode parasitizing coffee in El Salvador. Nematology 7:819–832

Carneiro RMDG, Almeida MRA, Queneherve P (2000) Enzyme phenotypes of *Meloidogyne* spp. populations. Nematology 2:645–654

Carneiro RMDG, Carneiro RG, Abrantes IMO et al (1996a) *Meloidogyne paranaensis* n. sp. (Nematoda: Meloidogynidae) a root-knot nematode parasitizing coffee in Brazil. J Nematol 28:177–189

Carneiro RMDG, Martins I, Teixeira ACO et al (2005c) Freezing and storing *Meloidogyne* spp. in liquid nitrogen. Nematol Bras 29:221–224

Carneiro RMDG, Mendes ML, Almeida MRA et al (2008) Additional information on *Meloidogyne inornata* Lordello, 1956 (Tylenchida: Meloidogynidae) and its characterisation as a valid species. Nematology 10:123–136

Carneiro RMDG, Randig O, Almeida MRA et al (2005b) Identification and Characterization of *Meloidogyne* spp. on coffee from São Paulo and Minas Gerais States using esterase phenotype and Scar multiplex. Nematol Bras 29:233–241

Carneiro RMDG, Tigano MS, Randig O et al (2004) Identification and genetic diversity of *Meloidogyne* spp. (Tylenchida: Meloidogynidae) on coffee from Brazil, Central America and Hawaii. Nematology 6:287–298

Cenis JL (1993) Identification of four major *Meloidogyne* spp. by random amplified polymorphic DNA (RAPD-PCR). Phytopathol 83:76–78

Chitwood BG (1949) Root-knot nematodes – Part I. A revision of the genus *Meloidogyne* Göldi, 1887. Proc Helminthol Soc Wash 16:90–104

Cofcewicz ET, Carneiro RMDG, Castagnone-Sereno P et al (2004) Enzyme phenotypes and genetic diversity of root-knot nematodes parasitising *Musa* in Brazil. Nematology 6:109–123

Cofcewicz ET, Carneiro RMDG, Randig O et al (2005) Diversity of *Meloidogyne* spp. on *Musa* in Martinique, Guadeloupe and French Guiana. J Nematol 37:313–322

De Grisse A (1960) *Meloidogyne kikuyensis* n. sp., a parasite of kikuyu grass (*Pennisetum clandestinum*) in Kenya. Nematologica 5:303–308

Eisenback JD, Bernard EC, Schmitt DP (1994) Description of the kona coffee root-knot nematode, *Meloidogyne konaensis* n. sp. J Nematol 26:363–374

Eisenback JD, Triantaphyllou HH (1991) Root-knot nematode: *Meloidogyne* spp. and races. In: Nickle WR (ed) Manual of agricultural nematology. Marcel Dekker, New York

Elmiligy IA (1968) Three new species of the genus *Meloidogyne* Göldi, 1887 (Nematoda: Heteroderidae). Nematologica 14:577–590

Esbenshade PR, Triantaphyllou AC (1985a) Use of enzyme phenotypes for identification of *Meloidogyne* species. J Nematol 17:6–20

Esbenshade PR, Triantaphyllou AC (1985b). Electrophoretic methods for the study of root-knot nematode enzymes. In: Barker KR, Carter CC, Sasser JN (eds) An advanced treatise on *Meloidogyne*, vol. 2. Methodology. North Carolina State University Department of Plant Pathology and USAID, Raleigh

Esbenshade PR, Triantaphyllou AC (1990) Isoenzyme phenotypes for the identification of *Meloidogyne* species. J Nematol 22:10–15

Fargette M, Phillips MS, Block VC et al (1996) An RFLP study of relationships between species, populations and resistance breaking lines of tropical species of *Meloidogyne*. Fundam Appl Nematol 19:193–200

Göldi EA (1887) Der kaffeenematode Brasiliens (*Meloidogyne exigua* G.). Zoologische Jahrbücher, abt. Systematik R 4:261–267

Hartman RM, Sasser JN (1985). Identification of *Meloidogyne* species on the basis of differential host test and perineal pattern morphology. In: Barker KR, Carter CC, Sasser JN (eds) An advanced treatise on *Meloidogyne*, vol. 2. Methodology. North Carolina State University Department of Plant Pathology and USAID, Raleigh

Hernandez A, Fargette M, Sarah JL (2004) Characterisation of *Meloidogyne* spp. (Tylenchida: Meloidogynidae) isolated from coffee plantations of Central America and Brazil. Nematology 6:193–204

Hewlett TE, Tarjan AC (1983) Synopsis of the genus *Meloidogyne* Göldi, 1987. Nematropica 13:79–102

Jepson SB (1987) Identification of root-knot nematodes (*Meloidogyne* species). CABI, Wallingford

Lima RDA, Ferraz S (1985) Análise comparativa das variações morfometricas entre diferentes populações de *Meloidogyne exigua*. Rev Ceres 32:362–372

Lopez R, Salazar L (1989) *Meloidogyne arabicida* sp. n. (Nemata: Heteroderidae) nativo de Costa Rica: un nuevo y severo patógeno del cafeto. Turrialba 39:313–323

Lordello LGE (1956) *Meloidogyne inornata* sp.n., a serious pest of soybean in the State of São Paulo, Brazil (Nematoda, Heteroderidae). Rev Bras Biol 16:65–70

Lordello LGE, Zamith APL (1958). On the morphology of the coffee root-knot nematode, *Meloidogyne exigua* Göldi, 1887. Proc Helminthol Soc Wash 25:133–137

Lordello LGE, Zamith APL (1960) *Meloidogyne coffeicola* sp.n., a pest of coffee trees in the state of Paraná, Brazil, (Nematoda, Heteroderidae). Rev Bras Biol 20:375–379

Muniz MFS, Campos VP, Castagnone-Sereno P et al (2008) Diversity of *Meloidogyne exigua* (Tylenchida: Meloidogynidae) populations from coffee and rubber tree. Nematology (*in press*)

Oliveira CMG, Kubo RK, Antedomenico SR et al (1998) Reação de cafeeiros a *Meloidogyne javanica*. Rev Agric 73:307–313

Rammah A., Hirschmann H (1988) *Meloidogyne mayaguensis* n. sp. (Meloidogynidae), a root-knot nematode from Puerto Rico. J Nematol 20:58–69

Randig O, Bongiovanni M, Carneiro RMDG et al (2002). Genetic diversity of root-knot nematodes from Brazil and development of SCAR markers specific for the coffee-damaging species. Genome 45:862–870

Randig O, Carneiro RMDG, Castagnone-Sereno P (2004) Identificação das principais especies de *Meloidogyne* parasitas do cafeeiro no Brasil com marcadores Scar-Cafe em Multiplex-PCR. Nematol Bras 28:1–10

Santos JM (1997) Estudos das principais especies de *Meloidogyne* Göldi que infectam o cafeeiro no Brasil com descrição de *Meloidogyne Göldii* sp.n. Universidade Estadual Paulista, DS thesis, Botucatu

Sipes BS, Schmitt DP, Xu K et al (2005) Esterase polymorphism in *Meloidogyne konaensis*. J Nematol 37:438–443

Triantaphyllou AC (1985) Cytological methods for the study of oogenesis and reproduction of root-knot nematode. In: Sasser JN, Carter CC (eds) An advanced treatise on *Meloidogyne*, vol. 1. Biology and Control. North Carolina State University Department of Plant Pathology and USAID, Raleigh

Triantaphyllou AC (1990). Cytogenetic status of *Meloidogyne kikuyensis* in relation to other root-knot nematodes. Rev Nematol 13:175–180

Whitehead AG (1960) The root-knot nematodes of East Africa I *Meloidogyne africana* n.sp. a parasite of arabica coffee (*Coffea arabica* L.). Nematologica 4:272–278

Whitehead AG (1968) Taxonomy of *Meloidogyne* (Nematoda: Heteroderidae) with description of four new species. Trans Zool Soc Lond 31:263–401

Zijlstra C (2000). Identification of *Meloidogyne chitwoodi, M. fallax* and *M. hapla* based on SCAR–PCR: a powerful way of enabling reliable identification of populations or individuals that share common traits. Eur J Plant Pathol 106:283–290

Zijlstra C, Donkers-Venne DTHM, Fargette M (2000) Identification of *Meloidogyne incognita, M. javanica* and *M. arenaria* using sequence characterised amplified regions (SCAR) based PCR assays. Nematology 2:847–853

Chapter 7
Coffee-Associated *Meloidogyne* spp. – Ecology and Interaction with Plants

Ricardo M. Souza and Ricardo Bressan-Smith

Abstract This chapter reviews the basic biology of coffee-parasitic root-knot nematodes (RKNs), *Meloidogyne* spp., their interaction with environmental factors, epidemiology-related issues and interaction with coffee plants at the cellular, tissue and physiological levels. For most of these topics, the available information is largely restricted to *M. exigua*; some information exists for *M. incognita* and *M. konaensis*. More specifically, this review examines the literature on RKNs' thermal requirements, the influence of soil, host and climate factors on nematode population fluctuation, sampling strategies, damage threshold and epidemiology of RKNs, complex diseases involving *M. arabicida* and *M. incognita*, and physiological alterations caused on parasitized coffee plants.

Keywords Physiology of parasitism · histopathology · epidemiology · life cycle · population fluctuation

7.1 Introduction

As far as we know, all coffee-parasitic root-knot nematodes (RKNs) undergo the basic *Meloidogyne* sp. life cycle: egg masses in the soil and/or within roots are believed to be the nematode's main survival stage; once ecloded, second-stage juveniles (J2) infect the roots and, in susceptible plants, they start feeding and sequentially molt into J3, J4 and adult stages. Eight coffee-parasitic *Meloidogyne* species reproduce by mitotic parthenogenesis; *M. hapla* Chitwood undergoes mitotic and meiotic parthenogenesis; *M. exigua* Göldi undergoes meiotic parthenogenesis and *M. kikuyensis* de Grisse is amphymitic. No information is available for the other six species.

As will be discussed in this chapter, a great many studies remain to be done to reveal life cycle details of most coffee-parasitic RKNs. Furthermore, understanding

R.M. Souza
Universidade Estadual do Norte Fluminense Darcy Ribeiro/CCTA/LEF, Campos dos Goytacazes, Brazil
e-mail: ricmsouza@censanet.com.br

R.M. Souza (ed.), *Plant-Parasitic Nematodes of Coffee*,

how their life cycle is influenced by host suitability, soil biota and environmental cues (root availability, air and soil temperatures, soil intrinsic characteristics and temporary conditions) would be of extraordinary scientific relevance, and possibly relevant to RKN management as well. The same applies to the understanding of nematode-induced alterations in coffee physiology, which certainly are the key to nematode-related yield losses.

As seen below, only *M. exigua, M. konaensis* Eisenback, Bernard and Schmitt and *M. incognita* (Kofoid and White) Chitwood have received much attention from studies to reveal more about these aspects.

7.2 Ecology and Epidemiology of Coffee-Associated RKNs

7.2.1 In-Vitro and Greenhouse Life Cycle Studies

As regards *M. konaensis*, Zhang and Schmitt (1995a) have conducted detailed work on its embryogenesis and post-infection development. These authors reported that nematode eggs kept at 30°C presented fast embryogenesis, being closely followed by those kept at 28, 26 and 35°C. Embryogenesis took longer at lower temperatures and it was not completed at 10 or 40°C. Taking into account egg death and hatching rates, Zhang and Schmitt considered 24°C to be the nematode's ideal temperature for embryonic development. Upon being inoculated in seedlings of arabica coffee (*C. arabica* L.) 'Guatemalan', which were maintained in greenhouse or growth chambers, *M. konaensis* took 48 and 38 days to complete its life cycle under average air temperatures of 26 and 30°C, respectively. At these temperatures, the nematode required 866 and 836 degree-days, respectively, to complete its life cycle.

As regards *M. exigua*, Lima and Ferraz (1985a) have observed a slower embryonic development in vitro at 15°C, in comparison to 20 and 25°C; at 30°C, 50% of the eggs died. Santos and Ferraz (1977) have observed that J2 eclode readily in vitro at 25°C; fewer eclosions occurred at 15, 20 and 30°C. Upon inoculation of seedlings of arabica coffee 'Catuai Vermelho' with *M. exigua*, Tronconi et al. (1986) have observed a positive correlation between number of nematode-induced root galls and air temperature, which was kept constant at 16, 20, 24 or 28°C. Nematode reproduction was greater at 20 and 24°C than at 16 and 28°C.

Lordello and Lordello (1983) have performed a detailed study following the development of *M. exigua* after inoculation in seedlings of arabica coffee 'Mundo Novo', which were maintained in greenhouse, growth chamber or in the field. In the latter (average temperature of 22, 4°C), the nematode completed its life cycle in 38 days, requiring 6,788 heat-units above the minimum temperature for its development, which was calculated to be 15°C. In an excellent study on the postembryonic development of *M. exigua* inoculated on 'Mundo Novo' coffee seedlings, Lima and Ferraz (1985b) have performed morphometrics and description of some life cycle aspects; at constant air temperature of 28°C, the life cycle lasted 32–42 days. In Colombia, Baeza (1977) [cited by Villalba-Gault et al. (1983)] have observed *M. exigua* complete its life cycle on arabica coffee 'Caturra' in 58–62 days.

As regards *M. incognita*, Villalba-Gault et al. (1983) have conducted detailed observations on the embryonic and postembryonic developments of coffee-parasitic *M. incognita* race five. Upon inoculation in 'Caturra' coffee seedlings, the nematode took 48–52 days to complete its life cycle.

Jaehn (1990) followed the development of *M. incognita* race two on 'Mundo Novo' coffee seedlings, under different constant air temperatures. More J2 infected roots at 20 and 24°C, in comparison to 28 and 32°C. The life cycle was completed in 48, 40, 32 and 32 days at 20, 24, 28 and 32°C, respectively. In another study, Jaehn (1991a) inoculated *M. incognita* races one, two and four separately in 'Mundo Novo' coffee seedlings, keeping them under constant air temperature in a growth chamber or in the field. Jaehn concluded that temperatures ranging between 28 and 32°C were the most suitable for all races assessed. He also inferred that day/night thermal oscillations affect nematode oviposition more than any other phase of the nematode life cycle. By assuming 10°C as the minimum temperature for nematode development, Jaehn calculated that *M. incognita* races one, two and four would need $534+/-63$, $580+/-92$ and $718+/-109$ degree-days to complete their life cycle.

In an interesting study, Jaehn (1991b) calculated the number of generations undergone *per* year by *M. incognita* races one, two and four in the different climate regions in the State of São Paulo, Brazil. He built a State map from which he predicted that between five and 11 generations occur *per* year, depending on the race and region involved. Consequently, life cycles would take between five and eleven weeks.

Collectively, these investigations suggest that the *M. exigua* populations studied are adapted to an upland, tropical temperature regime. This would probably hold true for most populations found across Latin America, which are typically associated with upland coffee cultivation. Lowland populations could be better adapted to higher temperatures. Accordingly, the *M. incognita* populations studied by Jaehn are adapted to higher temperatures, typical of the central and western regions of São Paulo, which present mean maximum temperatures in the 27–30°C interval (Anonymous, 2007). As regards *M. konaensis*, it also seems adapted to high temperatures. The higher degree-days required to complete its life cycle on coffee, in comparison to *M. incognita*, probably result from the fact that coffee is not a particularly good host to *M. konaensis*; indeed, this nematode required nearly twice as many degree-days to complete its life cycle on coffee in comparison to tomato; accordingly, twice the number of days were required for life cycle completion on coffee in comparison to tomato (Zhang and Schmitt, 1995a).

7.2.2 Field Population Fluctuation as Related to Environmental and Cultivation Conditions

While in vitro and greenhouse studies lay the foundations of nematode life cycle, field studies which are often time-consuming and arduous are necessary to reveal how nematodes interact with diverse environmental cues, such as host suitability, root availability, season-related climate changes and soil characteristics. This kind of information is valuable in many aspects; for example, it may help research and extension personnel to plan actions according to nematode distribution. For example, Villain et al. (1999) have reported that in Guatemala coffee-parasitic RKNs are more often found at low altitudes (50% of the infested farms are located below 800 masl) and in regions of more rainfall (80% of the positive samples have been collected in localities submitted to 2,000 mm/year). Those authors reported that soil type is not a limitation to RKNs in Guatemala. In contrast, in Panama RKN populations found in coffee plantations (but not parasitic to coffee) have been found to decline when monthly rainfall exceeds 500 mm/month (Pinochet et al., 1986). Nonetheless, this study was conducted during a single year, with no further confirmation of this trend.

In relation to *M. konaensis*, Zhang and Schmitt (1995c) have followed the fluctuation of J2 soil population in a naturally infested field planted with 10 coffee genotypes, either susceptible or resistant to this nematode. Unfortunately, the authors conducted only four unevenly-spaced samplings during the 16 months of the study; the data presented were restricted to the 0–15 cm-deep soil zone, although the authors stated that the nematode was more abundant in the 16–45 cm-deep zone; and climate variables were provided for only part of the period covered by the study. Their results show considerable variation in the J2 population on the genotypes assessed, and the J2 distribution in the soil profile (mostly at 16–45 cm deep) warrants further studies since coffee roots typically remain concentrated in the top soil zones, especially in irrigated plantations (Rena and Guimarães, 2000).

Serracin and Schmitt (2000) have studied the effect of soil type on coffee-parasitism by *M. konaensis*. In all four soil types assessed the nematode reduced root growth of seedlings of arabica coffee var *Typica*, with a tendency for more damage to be inflicted in the sandiest type of soil. Although the nematode reproduced readily in all soil types, significant differences occurred between them. Soil moisture content (constant vs fluctuating, with periods of water stress) did affect root galling and nematode reproduction, which were lower under the latter irrigation regime. A similar study was conducted in greenhouse by Tronconi and Ferraz (1985), who assessed the influence of four soil types on root galling by and reproduction of coffee-parasitic *M. exigua*. The authors considered humic latosol to be somewhat unsuitable for the nematode, while red-yellow latosol provided it with the best infective and reproductive conditions.

Soil types and their intrinsic properties and typical biota may possibly, play a major role in the distribution of coffee-parasitic *Meloidogyne* species. For example, if on the one hand, *M. exigua* is widespread across coffee-growing regions in the Americas, on the other hand *M. paranaensis* Carneiro, Carneiro, Abrantes, Santos and Almeida remains restricted in the States of Minas Gerais and Espirito Santo (Brazil), but more widely present in the State of São Paulo. Furthermore, there have been reports of entire regions in which *M. incognita* populations simply do not parasitize coffee (e.g., Barbosa et al., 2004), while other regions have suffered most from this species. In Chapter 14, Villain et al. report a similar association between soil type and distribution of *Meloidogyne* spp. in Central America.

As regards *M. exigua*, Huang et al. (1984) monitored four nematode epidemiological variables for 12 months in a non-irrigated coffee plantation in Minas Gerais State, Brazil. These authors found the nematode population to vary widely during the rainy and dry seasons. In nearby areas, Almeida et al. (1987) have obtained results that contradicted those of Huang et al. (1984), and Maximiniano et al. (2001) have found no statistical correlation between the number of J2 in the soil and mean air temperature and rainfall.

These apparently contradictory results led Souza et al. (2008a) to conduct an epidemiological study in an upland, non-irrigated coffee plantation naturally infested by *M. exigua*. Through 32 sampling dates three weeks apart, those authors observed the numbers of J2/100 cc of soil and J2/5 g of roots to fluctuate seasonally. This trend was not clearly observed in the number of nematode-induced root galls/5 g of roots. Their results do not support the widely accepted notion that in southeast Brazil the higher temperature and rainfall that occur in mid-spring trigger an epidemic of *M. exigua*; indeed, the numbers of J2 *per* unit root and *per* unit soil actually decline during late spring and summer; the number of galls *per* unit root does not respond to summer inputs.

With regard to *M. exigua* survival in the absence of host, Moraes et al. (1977) found no J2 in the soil six months after eradicating a heavily infested coffee plantation. This suggested that it would be safe to replant coffee one year after eradication, if the soil is maintained free of weed hosts. This confirmed previous greenhouse studies by Alvarenga (1974), who had concluded that *M. exigua* does not survive beyond six months in the soil without a suitable host.

Almeida and Campos (1991) have confirmed these studies by noting that *M. exigua* survived less than six months when the soil was cultivated with soybean 'Doko', *Crotalaria spectabilis* Roth, sorghum 'BR-12' or *Stilozobium aterrimum* Piper and Tracy. Those authors tested other crops for rotation; under rice cultivation, the nematode lasted up to 17 months in the soil. In yet another study, Almeida and Campos (1993) concluded that uprooting parasitized coffee plants sharply decreases *M. exigua* soil population, although occasional J2 were found in the soil up to 17 months after uprooting.

A steep decline in the *M. exigua* population has also been documented after coffee plants are drastically pruned, since this practice leads to death of most of the root system. Drastic pruning followed by proper agronomic practices has been proposed as a management strategy against *M. coffeicola* Lordello and Zamith (Rebel et al., 1976, cited by Gonçalves et al., 1998). Drastic pruning combined with nematicide applications is under investigation for management of *M. exigua* (Barbosa, 2008).

M. coffeicola has also been reported to survive briefly in the soil (Rebel et al., 1976; Carneiro Filho and Yamaguchi, 1995, cited by Gonçalves et al., 1998), while a single short-term study concluded that crop rotation was not a feasible strategy for *M. incognita*-infested areas because of this species' long survival (Jaehn and Rebel, 1984). Considering that nematode survival is not the same across different soil types and biota, it would be interesting to assess *M. incognita* survival in soils submitted to plowing and discing coupled with irrigation [to stimulate J2 eclosion (Campos, 2007)], followed by fallowing; this strategy could be of use in regions where nematode-resistant rootstocks of robusta coffee (*C. canephora* Pierre ex A. Froehner) cannot be used because of their inadaptability to mild climate.

In conclusion, although a reasonable body of knowledge exists on environmental factors influencing *M. exigua* and *M. konaensis* life cycles on coffee, very little has been experimentally assessed for other important species, such as *M. paranaensis* and *M. incognita*. Virtually nothing seems to be know for recently described or geographically restricted species, such as *M. izalcoensis* Carneiro, Almeida, Gomes and Hernandez or *M. mayaguensis* Rammah and Hirschmann, among others.

7.2.3 Interaction Between Coffee-Parasitic Nematode Species

Although there have been many studies on plant parasitism by concomitant nematode species (see reviews by Eisenback, 1993; Abawi and Chen, 1998), there seems to be just one study on coffee (Herve et al., 2005). These authors have examined the spatial distribution of *M. paranaensis, M. exigua* and *Pratylenchus coffeae sensu lato* in coffee plantations in Costa Rica and Guatemala. Those authors found signs of competition between *P. coffeae* and those RKNs for the coffee roots; this competition was more evident when involving *M. exigua*, which was more abundant and evenly distributed in the plantation than *M. paranaensis*.

7.2.4 Coffee Complex Diseases Involving RKNs

For coffee complex diseases involving RKNs, reports only appear regarding *Fusarium oxysporum* (Schltdl.) W. C. Snyder et H. N. Hansen. According to Cardoso (1986), Garcia (1945) was the first to report coffee wilt induced by *Fusarium* sp. in Puerto Rico; since then, many other reports and studies have been published, driven mainly by the damage caused by this fungus to coffee cultivation in the African continent.

In a follow-up to field observations, which had suggested a complex disease involving *F. oxysporum* f.sp. *coffeae* and *M. incognita* in Puerto Rico, Negron and Acosta (1989) conducted greenhouse experiments during which they observed chlorosis, wilting and root necrosis in seedlings of arabica coffee var *Bourbon* six months after inoculation with 16 thousand eggs and J2 of *M. incognita per* plant plus *F. oxysporum* f.sp. *coffeae*. Seedling height and dry root and shoot weights were significantly lower when the fungus was inoculated two or four weeks after the nematode, in comparison with simultaneous inoculations or inoculation with the fungus alone. These results are interesting, but as the authors did not assess the damage caused by the nematode alone, this study may be considered inconclusive as far as stating that a complex disease does exist in this case. Furthermore, the excessive nematode inoculum used further compromises the results. The need for a careful experimental design to confirm complex diseases involving nematodes has been put forward by Sikora and Carter (1987).

A distinct disease named 'corchosis' was first reported in Costa Rica by Lopez and Salazar (1989) (cited by Bertrand et al., 2000). According to the latter authors, diseased plants show a progressive decline characterized by leaf chlorosis, flower and fruit falling and poor root system which develops corky tissues in the main and secondary roots; death occurs within two to three years. Field observations associated 'corchosis' with parasitism by *M. arabicida* Lopez and Salazar. Greenhouse and field studies conducted by Bertrand et al. (2000) confirmed that 'corchosis' results from concomitant parasitism by *F. oxysporum* and *M. arabicida*, but not *M. exigua*, and that the fungus alone is not capable of invading and damaging the plants. Those authors were nonetheless unable to detect any additional damage to coffee plants inoculated with both pathogens in greenhouse, in comparison to plants inoculated with *M. arabicida* alone. Efforts are underway to control this disease through genetic resistance (see Chapter 9).

7.2.5 The Potential of Damage Thresholds as Guidelines for RKN Management

Throughout the literature, management of coffee-parasitic RKNs is proposed as a set of practices, either prophylactic or to be adopted after confirming the plantation infestation (Campos, 1997; Villain et al., 1999; Campos and Villain, 2005; Chapter 8). Many factors interact to determine the damage and consequent yield losses caused by RKNs, such as the nematode and coffee species involved, the agronomic requirements and nematode susceptibility of the cultivar or variety planted, and the region's edaphic and climate conditions. In some cases, additional plant-pathogens may aggravate damage, such as *F. oxysporum* in the presence of *M. incognita* or *M. arabicida*. To prescribe a management strategy, the nematologist or extension official must juggle with yet more aspects, such as the local traditions of coffee cultivation, the grower's monetary means to invest in the crop and the reselling prospects for the future harvests.

If, on the one hand, the literature on RKN management overstresses the need to consider all the above factors when devising a management strategy, on the other hand very few studies offer guidelines on how this should be done. For example, several reports exist on the aggravated damage caused by *M. incognita* and *M. paranaensis* in areas of sandy soil, in which coffee plants suffer concomitant abiotic stresses. Nonetheless, no guidelines exist for growers located in areas of soil with medium texture, infested with other *Meloidogyne* species or that seek prospects of revenues from their investment in RKN management. Even less information is available for management of *Pratylenchus* sp. and other nematodes for which the economic relevance has not been well established or that occur in restricted regions.

A common misunderstanding on the merit of establishing nematode damage thresholds (DTs) comes from the notion that the knowledge gained in one region on the relationship between nematode population and yield losses would not be readily applicable to other regions, thus reducing the applicability of DTs for nematode management. According to this view, the specificities of each plantation or region would be so great that nematode management strategies would need to be tailored for each locale.

By emphasizing the differences, this criticism denies the benefits of DTs; indeed, DTs could be instrumental to coffee-parasitic RKN management because of this crop's commonalities: (i) the agronomic practices employed in this crop are not so diverse as to hamper their categorization into 'cultivation systems', for which different DTs could be developed; (ii) just a handful of cultivars and varieties are largely cultivated across several countries, markedly in the Americas; from what we know today, these genotypes have an enormous genetic similarity as far as nematode susceptibility or resistance go; (iii) currently, just a handful of *Meloidogyne* species are of economic importance, thus reducing the need for studies with different RKN species and 'races'.

Hence, through a mid-term concerted initiative involving nematologists from different regions and/or countries, RKN-DTs could be developed for the main *Meloidogyne* species, coffee genotypes and 'cultivation systems'. For example, a *M. exigua*-DT developed for upland 'Catuai' plantations in Minas Gerais State, Brazil, would certainly be informative in nearby States with similar climate and agronomic characteristics, such as Rio de Janeiro and Espirito Santo. Combined, these three States have over 50% of Brazil's hectarage of arabica coffee, and at least the first two States are largely infested by *M. exigua*. Hence, establishing such a DT could have enormous scientific and economic relevance.

Naturally, DTs would need to be established considering key variables that interfere with the coffee-RKN interaction, such as major soil types and major climate types. Also, there would be no sense in establishing a DT for situations (nematode species and regions) in which an irreversible plant decline occurs, leading to plantation decimation within months.

Despite the inherent complexity of the subject, one should remember the prospects for the coffee industry worldwide, which indicate that growers will increasingly need to optimize their production system if they are to remain in business while preserving sustainability and profitability (see Chapter 2). The management of pests and diseases, nematodes included, is part of the equation. Under these circumstances, nematologists will have to go beyond vague statements, such as that on average nematodes cause coffee yield losses of around 10–15%, but that depending on the circumstances they may reach 100%.

In arabica coffee, any given harvest is partially linked to the plant's vegetative growth in the previous year. Hence, as a perennial crop, several edaphic, climatic and biotic factors have a dynamic effect on production, which has a typical biannual fluctuation. Therefore, assessing the nematode's role in coffee production requires well controlled field experiments, in which all other biotic and abiotic factors are minimized. Because of coffee's natural biannual cycle, experiments should probably cover at least four harvests, and different statistical approaches should be tested to consistently relate productivity and nematode population. Quantifying soil nematode population is not an easy matter (McSorley, 1987; Barker, 1985; Been and Schomaker, 2006), and only a single study has been conducted to assess different sampling strategies for quantitative sampling of coffee-parasitic nematodes (Souza et al., 2008b).

7.2.5.1 Greenhouse Estimates of DTs

Indubitably, determining nematode DTs in greenhouse can only characterize damage caused to seedlings under these experimental conditions; the relation obtained between nematode numbers in the soil or roots and the reduction in the seedling's vegetative growth would hardly have any predictive value when set against the complexity of commercial coffee production.

A few greenhouse experiments have been conducted in recent years. For example, Negron and Acosta (1987) observed a significant *M. incognita*-induced reduction in the height and dry root and shoot weights of *Bourbon* coffee seedlings; this effect occurred at all inoculum levels assessed, starting at four thousand eggs and J2/plant.

Vovlas and Di Vito (1991) applied different inoculum levels of *M. incognita* race one or *M. javanica* (Treub) Chitwood on seedlings of arabica coffee 'São Tome', observing chlorosis and a marked reduction in shoot growth at the inoculum level of 16 eggs/cm^3 of soil. The tolerance limit calculated through Seinhorst's equation for the variables total and shoot fresh weights was around two eggs or J2/cm^3 of soil for *M. javanica* and *M. incognita*; nematode damage to roots was more pronounced.

Zhang and Schmitt (1995b) observed a correlation between *M. konaensis* inoculum density (150, 750, 3,750 or 18,750 eggs/seedling) and the variables shoot height, dry shoot and root weights and percentage of root necrosis; the severity of damage varied according to the genotype tested (arabica coffees 'Guatemalan', 'SL28', 'Guadalupe', 'Mundo Novo' and 'Red Bourbon').

Rodrigues and Crozzoli (1995) inoculated *M. exigua* on coffee seedlings of 'Caturra Amarillo' and 'Catimor P4', using inoculum levels from 0.125 up to 64 eggs/cm^3 of soil (intermediate levels in geometric steps). Those authors observed a reduction in shoot growth starting at 16 eggs/cm^3 of soil; the Seinhorst tolerance limit for the variables shoot and seedling total weights was 0.25 eggs/cm^3 of soil. Di Vito et al. (2000) conducted a similar experiment, now using coffee 'São Tome'. The reduction in the seedlings' shoot growth started at the inoculum level of eight eggs and J2/cm^3 of soil, and the tolerance limit for the variable shoot weight was 1.2 eggs and J2/cm^3 of soil. When these experiments are compared, it is impossible to infer whether the two-fold difference in the damage threshold for shoot growth and the five-fold difference in the tolerance limit for shoot weight result from differences in the genotypes used or in the experimental conditions; furthermore, no inference can be made as to the damage that may occur under field conditions.

7.2.5.2 Field Determinations of DTs

To determine root-parasitic nematode DTs, a necessary first step is understanding nematode distribution in the soil, vertically and horizontally, and how it changes from season to season. This information is crucial to determine how and when (by season or plant phenological stage) soil sampling should be performed in order to obtain reliable nematode counts, i.e. nematode counts that are accurate and precise. Alternatively, nematode population estimates can be obtained through root sampling, with the same need for prior assessment of the best strategy.

While attesting that a coffee plantation is infested by RKNs or *Pratylenchus* sp. poses no major challenge for a nematology laboratory, there have been almost no studies aiming to understand nematode distribution in coffee plantations, nor to devise a sampling strategy to quantify infestation.

Herve et al. (2005) studied the spatial distribution of *M. exigua, M. paranaensis* and *Pratylenchus coffeae sensu lato* in two coffee plantations in Costa Rica and Guatemala; the nematodes were quantified in the coffee roots. The authors found those species to have an aggregated distribution in the fields (*k* value equal to or less than 1.576). In a similar study, Bertrand et al. (1998) had found the same tendency. In the two plantations, Herve et al. associated *k* values with nematode population levels – lower population, lower *k* –, suggesting that this might indicate that the plantations had been established using infected seedlings, and that the initial nematode foci had not spread throughout both fields. Since samplings were performed on a single occasion in both fields, it would have been interesting to assess nematode distribution in a different season and in the soil as well.

In Brazil, Souza et al. (2008b) conducted a two-year study in a commercial coffee plantation to examine *M. exigua* distribution in the soil and roots, and to determine the best strategy for quantitative sampling; the sampling patterns evaluated combined five different sampling core locations around coffee plants and four different epidemiological variables. Statistical analysis concluded that *M. exigua* was evenly disseminated in the plantation, thus not presenting an aggregated distribution; also, the sampling strategy routinely used for RKNs, i.e. sampling at the coffee canopy's edge to quantify J2/100 cc of soil, was by far the worst. The best strategy was sampling coffee roots under the coffee canopy and at 0–20 cm deep soil zone to assess the number of root galls/5 g of roots.

Upon definition of an appropriate sampling strategy, it is necessary to verify the relation between nematode population levels in the soil and/or roots and productivity. If all other biotic and abiotic factors that influence productivity are minimized, one should be able to establish a nematode DT. Such a study has been conducted for the last five years in a commercial coffee plantation infested by *M. exigua* (Barbosa, 2008).

A few other studies have been conducted to determine DTs under field conditions. In Colombia, Leguizamon-Caycedo (1976) has associated the level of *M. incognita* and *M. javanica* soil infestation with symptoms on the shoot (nutrient deficiency and leaf falling) and roots (abundance of suberous tissues, cracking and overall reduction of the root system). The author found the nematodes to be more abundant at the 0–20 cm deep soil zone and at 0–25 cm from the tree trunk. Also in Colombia, Leguizamon-Caycedo (1997) has associated the percentage of the root system affected by *M. incognita* and *M. javanica* with yield losses of coffee 'Caturra'. Based on production over a four-year period, the author calculated that each 1% of root infection would correspond to a yield reduction of 78 g. Unfortunately, the author did not explain how the root infection rate was calculated, the epidemiological variable assessed (presumably root swelling) or the sampling pattern adopted.

In the field, Zhang and Schmitt (1995b) tried to correlate *M. konaensis* J2 soil population in four samplings three months apart with the variables tree height, canopy width and trunk diameter of coffee 'Guatemalan' and 'Guatemalan' grafted onto 'Deweveri' rootstock. Only plant height was significantly related to J2 soil population for both genotypes. Indirectly, these authors calculated the DT to be around 10 eggs/plant.

7.3 RKN-Induced Cell and Tissue Alterations in Coffee Roots

The first study on RKN-coffee interaction at cell and tissue levels was conducted by Mendes et al. (1976; 1977). These authors performed detailed histological observations on the compatible (= susceptible) interaction between *M. exigua* and seedlings of 'Mundo Novo' until the thirtieth day after nematode inoculation. Although their micrographs were not published in high quality, these authors outlined all the main cellular and tissue alterations induced by the nematode. Some interesting features that were observed include: (i) *M. exigua* J2 penetrate the roots preferentially at the meristematic region, (ii) the invasion of this region by many J2 results in the induction of terminal root galls coupled with cessation of root elongation, (iii) J2 migrate either inter- or intra-cellularly through root tissues and (iv) adventitious roots are often differentiated within root galls, but they do not emerge presumably because of physical barriers, such as giant cells, thickened cell walls and female nematode bodies. The first three features are distinct from the broad concept of RKN-feeding behavior, which was built from studies conducted mostly on *M. incognita* feeding on roots of *Arabidopsis thaliana* (L.) Heynh. (von Mende, 1997; Wyss, 2002; Gheysen and Jones, 2007). The preferential J2 penetration at the meristematic region has been confirmed by Nakasono et al. (1980).

Some other studies followed, adding relatively little to the subject. For example, well-developed giant cells and associated tissue alterations were described by histological studies performed by Negron and Acosta (1987) and Vovlas and Di Vito (1991). These authors worked on *Bourbon* and 'São Tome' coffee seedlings, susceptible to *M. incognita* and *M. exigua*, respectively. Vovlas and Di Vito also reported undersized giant cells induced by an isolate of *M. incognita* race two that was unable to parasitize and reproduce successfully on the coffee seedlings. Anthony et al. (2005) also observed features suggestive of intracellular migration of *M. exigua* J2 in the roots of susceptible coffee 'Caturra'; occasionally J2 were found within the differentiated vascular tissues. An array of cell and tissue alterations were described which are consonant with a compatible interaction.

A comprehensive ultrastructural study has been conducted on the compatible interaction between *M. exigua* and rubber tree seedlings until the forty-fiftieth day after nematode inoculation (Fonseca et al., 2003a,b). Apart from anatomical differences between seedling roots of rubber tree and coffee, the tissue and cellular alterations revealed by this study are certainly informative of the alterations induced on the latter.

In an ultrastructural study conducted up to the sixth day after inoculation of *M. exigua* on seedlings of arabica coffee ‘Catuai Amarelo’, Rodrigues et al. (2000) have observed J2 migrating through the root cortex intra- and inter-cellularly, with the feeding site being induced in parenchymatic cells adjacent to developing xylem elements. Early cell and tissue alterations seemed similar to those observed in other compatible RKN-plant interactions. When the same coffee cultivar was inoculated with *M. megadora* Whitehead, the authors observed cell and tissue alterations suggestive of an incompatible (= resistant) interaction, which included changes in cell membranes and abundance of electron-dense vesicles. When both *Meloidogyne* species were inoculated on seedlings of coffee ‘Catimor’, a typical hypersensitive (= resistant) reaction was observed, which included necrosis of cells around and in the feeding site induced by J2.

Anthony et al. (2005) have paid special attention to the incompatible interaction *M. exigua*-arabica coffee ‘Iapar-59’, which harbors the resistance gene *Mex*-1. In this study, fewer J2 seemed capable of invading the roots, while cells stained dark and seemed disorganized or necrotic around those J2 that had successfully invaded the roots. Giant cells were occasionally noticed, but they seemed altered or collapsed. Their results suggested that a hypersensitive reaction is involved in the *Mex*-1-mediated resistance to *M. exigua*.

7.4 *Meloidogyne*-Coffee Interaction: A Physiological Approach

A comprehensive understanding of the mechanisms involved in coffee yield losses caused by RKNs can only emerge from experiments that, on the one hand, consider the particularities of the plant’s physiology and, on the other hand, are designed to isolate variables and allow data to be properly interpreted.

In this section, a brief review is presented on coffee physiology, as a platform for examining the available literature on the mechanisms of nematode-related yield losses. Although RKNs are important parasites of arabica and robusta coffees, in this section most data and analysis are focused on the former.

7.4.1 Coffee Climate Requirements

7.4.1.1 Temperature

Collectively, studies on the effect of temperature on arabica coffee plants present results that are highly variable; this is a consequence of variations in experimental conditions, including the plants’ genotype and phenology (DaMatta and Ramalho, 2006). Generally speaking, seedlings and young plants are more sensitive to extreme temperatures than adult ones. The optimal thermoperiod for young plants is 26°C at day and 18°C at night, while the optimal one for seedlings is 30°C and 23°C, respectively (Went, 1957; Franco, 1958, cited by Rena et al., 1994; Kumar and Tieszen, 1980, cited by Damata and Ramalho, (2006). The exposure of seedlings to 38°C or 13°C causes the cessation of growth.

For mature arabica coffee plants, the ideal mean temperature range is 18–21°C. Above this, growth is impaired and productivity decreases; beverage quality may be affected. In Brazil, most plantations are located in the States of Minas Gerais, Espirito Santo, São Paulo and Paraná, whose mean temperatures fall into that range. Some studies and field observations have indicated that coffee plants exposed to long periods of high temperature coupled with drought and high irradiance have their growth impaired, followed by abortion of leaves and flowers, and subsequent yield loss. This has prompted the launching of breeding programs aimed at adapting arabica coffee to regions with elevated temperatures (Fahl and Carelli, 2007); new cultivars have enabled this crop to be grown in semi-arid regions in northeast Brazil (States of Bahia, Rio Grande do Norte and Ceará) and Africa.

Extreme temperatures, either high or low, impair plant growth and inhibit reproduction. Under such circumstances, photosynthesis, respiration, nutrient assimilation and other metabolic processes are differently affected by the duration and intensity of the extreme temperature. Coffee plants possess a variety of acclimation mechanisms which are activated under such conditions.

As regards low temperature, arabica coffee plants have their growth compromised below a mean temperature of 17–18°C. This condition is relatively common in coffee-producing upland regions in Minas Gerais and Paraná. Frost may cause significant yield losses or irreversible damage to plantations. Low temperature negatively affects cell metabolism, reducing enzymatic rates and the fluidity of cell membranes, thereby affecting the transport of compounds into and out of the cell (Oliveira et al., 2002; Campos et al., 2003). These effects are observed mostly in organelles such as chloroplasts and mitochondria. In coffee chloroplasts, the photosynthetic rate (A_N) and stomatal conductance (g_s) are reduced almost to zero at 5–10°C or lower temperatures (Larcher, 1981; Oliveira et al., 2002). The destruction of pigment complexes, viz. light harvest complex (LHC) in thylakoids and the reduction in photochemical efficiency of photosynthesis are also attributed to low temperature. Afterwards, the metabolic flux in the Calvin cycle is declined, affecting the overall carbohydrate metabolism in the chloroplast.

Coffee physiology is also altered when plants are exposed to high temperatures, with negative effects being observed above 26°C (Coste, 1992). In this case, a decrease occurs in photosynthesis, because a reduction in g_s and the damage to mesophyll and chloroplasts cause a reduction in carbon carboxylative efficiency (Nunes et al., 1968; 1973).

Most of the understanding of the effect of extreme temperatures on coffee physiology has been acquired from experiments conducted with seedlings grown in small pots. Therefore, the imposed temperature stress is quite precise and does not represent weather conditions observed in the field. In the field, daily irradiance levels and air humidity are very variable, particularly in the tropics. To cope with these variations plants use acclimation mechanisms, such as changes in the metabolic flux of photosynthesis and respiration. For example, DaMatta et al. (2001) and Da Matta (2004a) observed that a long period of acclimation allowed plants maintained at 30°C to keep A_N at the same level of efficiency as that in plants maintained at 24°C; in addition, the maximum photosynthetic rate (or potential photosynthesis) was reached at 35°C. This demonstrates the great capacity of arabica plants to adjust their metabolism in regions where temperature stays around 30°C for several hours a day. Therefore, mechanisms of acclimation are essential to the success achieved in breeding coffee for cultivation in regions once thought inappropriate.

The process of acclimation can also be observed in relation to high irradiance conditions. Although arabica plants originated in a shaded environment – the forests in northeast Africa – their cultivation under full sun is common throughout the world. In cultivars not adapted to full sun, an overcharge of energy in the photosynthesis process, commonly called photoinhibition, may be observed (Gilmore and Govindjee, 1999). This condition affects several structures and metabolic processes in the chloroplast, such as the water splitting complex and the repairing capacity of photosystems I and II (Long et al. 1994). Consequently, the electron transport chain in the thylakoids is damaged, resulting in bleaching of photosynthetic pigments, notably chlorophylls.

Although in most instances photoinhibition causes reversible damages, it may be irreversible in some coffee cultivars (Oliveira et al., 2002). In tropical countries, most breeding programs have unintentionally bred for tolerance to high irradiance and drought, even though these were not the original goals. Since coffee has been bred for higher productivity in different climate regions, its physiological plasticity has enabled its expansion to areas with high light intensity and has made it tolerant of drought (DaMatta, 2004; De los Santos-Briones and Hernández-Sotomayor, 2006; Fahl and Carelli, 2007).

7.4.1.2 Water Availability

Many studies have been conducted on the effect of drought on coffee physiology. Nonetheless, greenhouse experiments in which sudden water stress is imposed do not accurately reproduce the climate conditions faced by plantations. According to DaMatta and Ramalho (2006), such studies have the following limitations: (i) root growth is usually restricted by the reduced size of the pots in which seedlings are cultivated; (ii) in pots, soil presents low water conductivity; (iii) in greenhouse, the atmosphere is different from in the field; and (iv) in tropical countries it is difficult to control temperature and air humidity in a greenhouse environment; therefore, experimental plants may present an artificial increase in their evaporative demand due to high air temperatures, which may compromise data and their interpretation. Under natural conditions, drought arrives slowly and concomitantly with other stressful factors, such as extreme temperature, high irradiance and low air humidity, which creates a multidimensional stress (DaMatta et al., 2003; DaMatta and Ramalho, 2006).

In coffee plants, relative water content (RWC) varies during the course of a day and is highly dependent on soil moisture. Stomata and leaf cuticle play a determinant role in controlling water loss (Akunda and Kumar, 1981). A useful index to assess water status in plants is the water potential (Ψ_w), a pressure measurement whose maximum value reaches zero MPa (Pascal, a unit of pressure). Negative values of Ψ_w indicate that the plant is facing water deficit, i.e., the cell turgor begins to decrease from its maximum capacity. In this aspect, it is important to mention that leaves lose water to the atmosphere because air Ψ_w is lower than leaf Ψ_w.

Water deficit occurs frequently in coffee plants growing in the tropics, independently of soil moisture; it can reach −2.2 MPa before a loss of turgor occurs (Pinheiro et al., 2005). Interestingly, this corresponds to approximately 90% of the RWC, a value considered high and largely related to low cell wall elasticity (DaMatta et al. 2003). Different from most robusta cultivars, arabica ones such as 'Catuai', 'Catimor', 'Mundo Novo' and 'Catucai', present a high volumetric modulus of elasticity (ϵ), irrespective of the stress caused by drought (DaMatta, 2004). This is evidence that arabica coffee plants have an efficient control of stomatal transpiration, since they evolved under drier conditions, in comparison to robusta ones. This could explain why arabica is less responsive to irrigation than robusta.

Some drought-tolerant species have an ability to regulate their solute potential (Ψ_s, a component of Ψ_w) to stand periods of water stress; these plants accumulate solutes in the vacuole, such as proline, glycinebetain, sucrose and ions. This process, once called osmotic adjustment (OA), allows plants to stand drought without a significant loss of turgor. Nonetheless, the role of OA in maintaining cell turgor in coffee plants grown under field conditions is still a matter of debate (Rena et al., 1994). DaMatta (2004b) postulated that OA is not significant in many coffee cultivars under drought conditions, as proposed by Goldberg et al. (1984). It seems likely that OA is not related directly to the maintenance of stomatal sensitivity to drought. This is supported by observations that drought-stressed arabica plants show considerable leaf gas exchange rates at low or zero levels of turgor (Meinzer et al., 1990b). Thus, even with the production of osmotically active solutes such as proline under water-stressed conditions, no relation has been found between OA and drought tolerance in coffee, as normally occurs in other plant species.

Water availability in the soil is decisive for stomatal control in coffee plants because it is intrinsically associated with the evaporative demand of the atmosphere. In recent years, some investigations have shown the considerable role of vapor pressure deficit (VPD) in stomatal control, having g_s as a variable (Wormer, 1965). VPD increases as air relative humidity (RH) decreases, and this determines a strong driving force for transpiration, followed by an increase in xylem tension in the plant. In coffee, VPD is highly effective in controlling stomata, concomitantly with rapid changes in air RH during the course of a day. On the other hand, slow soil dehydration, which occurs during the dry season, is the main factor influencing stomatal control, defining the pattern of maximum stomatal aperture under such conditions (Nunes and Correia, 1983).

Since reduced water availability in the soil promotes a decrease in g_s, it would be interesting to define a threshold where A_N begins to decrease. This would be particularly important because the maintenance of high values of A_N is desirable, because it provides more biomass production in the plant. However, it is difficult to define when A_N reduction starts because genetic as well as environmental variations occur during experimental observations. According to Kumar and Tieszen (1980), cited by Damata and Ramalho (2006), a decrease from 7.6 to 2.5 μmol $CO_2/m^2/s$ occurred when Ψ_w changed from -1 to -3.5 MPa. These authors suggested that mesophyll conductance and carboxylation efficiency seem to control A_N, since stomata begun to close when Ψ_w reached values below -2.0 MPa.

Regardless of establishing a starting-point for A_N reduction, a strong correlation is normally observed between g_s and A_N (Ronquim et al., 2006). Nonetheless, this correlation may vary when distinct environmental conditions are imposed. For example, under the same value of g_s, A_N was lower in leaves submitted to RH below 50%, in comparison to leaves submitted to 80% (Nunes, 1988). Since RH is an important factor controlling g_s, it is advisable not to irrigate a plantation during the hottest hours of the day, when RH is normally low.

It has been postulated that temperature and water availability are the main factors affecting coffee physiology; however, these factors alone do not explain the growth cycle observed in coffee. Moreover, these two factors may occur jointly or independently in some regions of the world. Under natural conditions, coffee growth follows rainfall, with wet and dry seasons determining periods of rapid and slow shoot growth (Cannel, 1972). Arabica coffee plants that are not submitted to a dry season typically bloom on young plagiotropic branches, although irradiance may induce water internal tensions. As a result, a single branch displays flowers, immature and mature fruits (Haarer, 1962), leading the growers to practice a hand-picking harvest of ripe fruits only. Therefore, high water tensions may be a factor to synchronise blooming and fruit maturation in regions with a defined dry period.

7.4.2 Carbon Metabolism and Nutrition

Since coffee-associated nematodes parasitize roots, the monitoring of physiological processes related to the plant's aerial part is likely to offer little indication of the primary effects of those parasites on the plant's physiology. This chapter's authors have conducted a series of experiments focused on understanding root- and soil-related factors that might unveil the mechanisms involved in *M. exigua*-related yield losses.

More specifically, these studies have focused on determining (i) how nematode parasitism affects arabica coffee's overall photosynthesis and growth; (ii) whether water stress amplifies the negative effect of nematodes on photosynthesis and nutrient uptake and content; (iii) how sugar translocation and partitioning in roots are altered in parasitized plants.

Silva (2005) have conducted a greenhouse experiment in which seedlings of 'Catuai Vermelho' cultivated in large pots were inoculated with 14 thousand eggs and J2 of *M. exigua*. An average of 800 root-knots/plant was observed seven months after inoculation; the plants suffered a significant reduction in shoot and root dry matters, leaf area, plant height and branching. Such a reduction in vegetative growth of susceptible coffee plants have been reported by other studies (e.g., Silva et al., 2007). Nonetheless, Silva (2005) did not observe a relation between nematode parasitism and the photochemical efficiency of photosynthesis, which was evaluated through fluorescence of the chlorophyll *a*. Furthermore, the difference in the maximum quantum yield of photosynthesis (ratio between fluorescence and maximum fluorescence, Fv/Fm) was negligible between nematode-free and nematode-parasitized plants. Accordingly, the author observed no chlorosis in the parasitized plants, which indicates that the chlorophyll content of their leaves was not affected.

In greenhouse, Souza (2006) continued the studies by Silva (2005), observing no differences in A_N, transpiration and Fv/Fm between nematode-free and -parasitized plants, 13–21 months after inoculation with *M. exigua*. When Souza (2006) submitted a subset of plants to water stress during a 10-day period, no relation was observed between nematode parasitism and carbon assimilation, stomatal conductance or transpiration, although water tension in non-irrigated plants reached 160 kPa.

During a dry season, a non-irrigated plantation naturally infested by *M. exigua* was compared to a nematode-free one in northwest Rio de Janeiro, Brazil (Reis et al., unpublished results). Again, no relation was observed between the parasitism by *M. exigua* and photosynthetic variables in the leaves. No differences were noted in stomatal function or gas exchange that could indicate a reduction in water translocation through the xylem, as reported by Dutra and Campos (2000).

Despite these negative results, for other plant-nematode interactions there have been consistent results indicating that RKNs as well other nematodes interfere with the plant's water absorption and/or translocation, and that acute water stress may aggravate nematode damage (see reviews by Hussey, 1985; Wilcox-Lee and Loria, 1987; Melakeberhan and Webster, 1993). For example, on *Pinus* sp., *Bursaphelenchus xylophilus* (Steiner and Buhrer) Nickle induces water stress by damaging the plant's xylem through cavitation of the tracheids; this may lead to acute wilting and death (Ikeda and Suzaki, 1984; Iwasaki et al., 1999).

There have been conflicting reports on the interaction between RKN-parasitism and the nutritional status of coffee plants. For example, Macedo et al. (1974) did not observe a significant relationship between nematode parasitism and leaf concentration of nitrogen (N), phosphorus (P), potassium (K), calcium (Ca), magnesium (Mg), iron (Fe) and manganese (Mn); these authors have not observed a delay in the plants' development due to nematode parasitism.

On the other hand, Santos et al. (1981) conducted greenhouse studies to examine the effect of *M. exigua* inoculum levels on the growth of coffee seedlings, and on their absorption and translocation of N, P, K, Ca and Mg. Nitrogen and Ca absorption, as well as plant height and root dry weight, were inversely related to inoculum level. The absorption of P, K and Mg was not altered by nematode parasitism. These authors concluded that nutrient translocation was not affected by nematode parasitism, because no nutrient accumulation occurred in the roots. Accordingly, Boneti et al. (1982) observed a reduction in the absorption, but not in the translocation, of zinc (Zn), copper (Cu), boron (B), Fe and Mn by coffee seedlings parasitized by *M. exigua*.

Similar studies have been conducted with coffee plants parasitized by *M. incognita, M. konaensis* and *P. coffeae* (Zimmerman) Filipjev and Schuurmans Stekhoven. Gonçalves et al. (1995) inoculated seedlings of 'Mundo Novo' with six thousand eggs of *M. incognita*; 390 days after inoculation, parasitized plants showed delayed development, and the leaf content of Ca and Zn was significantly lower in comparison with nematode-free controls; no differences occurred in the leaf content of N, P, K, Mg, Fe, Mn and B.

In the field, Hurchanik et al. (2003) observed an inverse correlation between *M. konaensis* population level in the soil and the root concentration of P + K + Mg, P, K, Mg, Ca + Mg, Cu and B. A somewhat different pattern was found when Hurchanik et al. (2004) studied nutrient partitioning in the roots of coffee *Typica* parasitized by *M. konaensis* in greenhouse. Twenty-five weeks after nematode inoculation, these authors found that nematode parasitism caused a decrease in the root concentration of Ca, Mg, P and B, and an increase in Mn, Cu, Zn and Ca/B ratio.

Despite the importance of nitrogen to plant physiology and productivity, relatively little attention has been paid to this nutrient. Vaast et al. (1998) inoculated 'Catuai Vermelho' potted-plants with *M. konaensis* and *Pratylenchus coffeae*, separately. *M. konaensis*-parasitism decreased the proportion of feeder roots in the root system by about 50%, and reduced the uptake of NO_3^- and NH_4^+ by 63% and 54%, respectively. This reduction was related to root galling because non-parasitized feeder roots maintained their N uptake. In contrast, migratory and feeding activities of *P. coffeae* seemed to affect nitrogen uptake by the whole root system.

According to Gonçalves et al. (2004), well-nourished coffee plants stand parasitism by RKNs better than plants concomitantly submitted to nutrient deficiency. Also, Gonçalves and Silvarolla (2007) stated that *M. paranaensis*- and *M. incognita*-related damages are more pronounced in areas of sandy, biologically and chemically poor soils, in comparison with areas of more plant-conducive soils. Hence, optimizing fertilization might seem a valid strategy to stimulate plant tolerance and/or resistance, thus decreasing nematode population.

Nonetheless, increasing fertilization of coffee plants increased *M. konaensis* population as well (Schmitt and Riggs, 1989). Accordingly, Jaehn et al. (1983; 1984) supplied plants with extra amounts of nitrogen in the form of ammonium nitrate; they observed an increase in the root density of mature females, average number of eggs/egg mass and overall nematode reproduction. It is possible that through extra nitrogen fertilizations one might alleviate nematode-related symptoms by providing enough nitrogen for the synthesis of rubisco, a major photosynthetic protein which acts in the Calvin cycle; rubisco is the most abundant protein in plant tissues and nitrogen's main sink in the plant (Netto et al., 2005). Despite reports that *M. exigua*-related yield losses can be avoided by extra fertilizations (see Chapter 8), long-term field studies should probably be conducted for major soil types and coffee cultivars for better assessment of this management strategy.

As regards soluble sugar and starch leaf content, Gonçalves et al. (1995) have postulated that a decrease in A_N would result in a decrease in the leaf carbohydrate content. However, no significant difference was found between *M. incognita*-parasitized and nematode-free 'Mundo Novo' plants. Mazzafera et al. (2004) have noted that root sugar content decreased in 'Catuai Vermelho' seedlings parasitized by *P. coffeae*; carbon fixation in the leaves and its partitioning to the roots were also affected. Those authors suggested that the physiological damage caused by *P. coffeae* is readily expressed in the leaves through a reduction in photosynthesis and phloem transport, which are themselves a consequence of the nematode's destructive action in the roots.

Because RKNs present a more subtle feeding habit, one could expect that RKN-parasitized coffee plants present a different pattern of carbon assimilation and partition, in comparison to *Pratylenchus*-parasitized ones. Indeed, investigations conducted by Del Valle et al. at UENF (unpublished results) have shown that in potted-coffee plants parasitized by *M. exigua, M. paranaensis* or *M. incognita*, a decrease in glucose, fructose and sucrose contents occurs in the post-gall region of the rootlets, in comparison to the pre-gall region and the nematode-free control rootlets. Root galls presented the highest concentration of those sugars, which suggested that nematode females draw those nutrients in their benefit at the expense of the root's distal region. Nonetheless, these results were not confirmed in rootlets obtained from plantations naturally infested with *M. exigua*.

7.5 Concluding Remarks

This review demonstrates our relatively poor knowledge of many basic and applied aspects of coffee-parasitic RKNs. Indeed, some of these *Meloidogyne* species are only known from their original descriptions, and no live cultures of them exist. Other species have been described recently and/or their known geographical distribution is restricted. From the 17 coffee-parasitic species recognized by Carneiro and Cofcewicz (see Chapter 6), only *M. exigua, M. incognita* and *M. konaensis* have been examined in a variety of aspects, and this is no coincidence. Indeed, the widespread incidence of *M. exigua* throughout Latin America broke down one of the main constraints to plant nematology worldwide, viz. the low number of nematologists *per* tropical country. Because of widespread decimation of coffee plantations in Brazil in the 1970s, *M. incognita* caught the attention of nematologists and funding agencies alike, who often elect their priorities on the basis of economic importance. In its turn, *M. konaensis* is relatively better known thanks to a research effort that has spanned more than a decade at the University of Hawaii (USA).

Presumably, all coffee-parasitic RKNs present the basic life cycle features of *Meloidogyne* sp. Nonetheless, embryonic and postembryonic details and climate requirements have only been studied for the three best-known species. A reasonable amount of information exists on the environmental factors that influence *M. exigua* population fluctuation; some data exist for *M. incognita* and *M. konaensis*.

Except for *M. exigua*, no systematic study has been conducted to assess sampling strategies for monitoring populations in coffee plantations. Therefore, nematologists lack the most basic tool for studies involving RKN populations! For example, developing an accurate and precise sampling plan is basic for evaluating the effectiveness of any management approach. Also, a relation should be established between mean RKN population level and plantation productivity because this is essential for anyone assessing, on a scientific basis, the effectiveness and economic soundness of any chemical, cultural or biological control approach. Even if genetic resistance is envisaged as the ultimate control approach, estimating damage thresholds can be useful. Indeed, it seems advisable to breed cultivars that present multi-species, horizontal resistance towards RKNs. In this case, it may be recommended to growers to invest in crop management to enhance plant resistance. Therefore, a cost-benefit analysis would be needed to assess the best management approach for nematode-resistant plantations.

As regards cell and tissue alterations related to induction and maintenance of feeding sites, most information is available for *M. exigua*. Throughout the plant-parasitic nematode groups, the features associated with feeding site induction and maintenance are largely constant for each genus or family; therefore, it is not likely that major differences would be found from histopathological and/or ultrastructure studies conducted on all coffee-parasitic *Meloidogyne* species; nonetheless, putting any widely accepted, 'natural' assumption to the test is, in itself, of scientific relevance.

Finally, our understanding of the physiological alterations induced by coffee-parasitic nematodes is still incomplete. Again, assumptions can be made that RKNs negatively affect water uptake and translocation, with all subsequent physiological damage; some results exist from studies on nutritional imbalances caused by *M. exigua, M. incognita* and *M. konaensis*. A broad and sound picture of the physiology of RKN-coffee parasitism can only arise from mid-term studies conducted under field conditions; during such studies, a plethora of physiological variables should be monitored.

As in all aspects of coffee-parasitic nematodes, national and/or international research collaborations would certainly be the best approach for nematologists in tropical countries to overcome the daily difficulties faced by anyone practicing scientific research, and to make substantial advances for nematology.

References

Abawi GS, Chen J (1998) Concomitant pathogen and pest interactions. In: Barker R, Pederson GA (eds) Plant and nematode interactions. Agronomy monograph #36. ASA, CSSA, SSSA, Madison

Akunda EM, Kumar D (1981) A simple technique for timing irrigation in coffee using cobalt chloride disks. Irrig Sci 3:57

Almeida VF, Campos VP (1991) Alternância de culturas e sobrevivência de *Meloidoygne exigua* em áreas de cafezal infestado e erradicado. Nematol Bras 15:30–42

Almeida VF, Campos VP (1993) Sobrevivência de *Meloidogyne exigua* no solo e em raizes de cafeeiro no campo. Fitopatol Bras 18:147–150

Almeida VF, Campos VP, Lima RD (1987) Flutuação populacional de *Meloidogyne exigua* na rizosfera do cafeeiro. Nematol Bras 11:159–175

Alvarenga G (1974) Determinação preliminar da longevidade no solo do nematóide *Meloidogyne exigua*. Proceedings II Congr Bras Pesq Cafeeiras, no pagination

Anonymous (2007) Instituto Nacional de Meteorologia. http://www.inmet.gov.br/, visited on December 30th 2007

Anthony F, Topart P, Martinez A et al (2005) Hypersensitive-like reaction conferred by the *Mex*-1 resistance gene against *Meloidogyne exigua* in coffee. Plant Pathol 54:476–482

Barbosa DHSG (2008) Manejo cultural, quimico e genético em áreas cafeeiras infestadas por *Meloidogyne exigua* Göldi, 1887 na região noroeste fluminense. Universidade Estadual do Norte Fluminense Darcy Ribeiro, DS thesis, Campos dos Goytacazes

Barbosa DHSG, Vieira HD, Souza RM et al (2004) Survey of root-knot nematode (*Meloidogyne* sp.) in coffee plantations in the state of Rio de Janeiro, Brazil. Nematol Bras 28:43–47

Barker KR (1985) Sampling nematode communities. In: Barker KR, Carter CC, Sasser JN (eds) An advanced treatise on *Meloidogyne*. Vol. 2, Methodology. North Carolina State University Department of Plant Pathology and USAID, Raleigh

Been TH, Schomaker CH (2006) Distribution patterns and sampling. In: Perry RN, Moens M (eds) Plant Nematology. CABI, Wallingford

Bertrand B, Cilas C, Herve G et al (1998) Relations entre les populations des nematodes *Meloidogyne exigua* et *Pratylenchus* sp., dans les racines de Coffea arabica au Costa Rica. Plante Rech Dev 5:279–286

Bertrand B, Nunez C, Sarah J-L (2000) Disease complex in coffee involving *Meloidogyne arabicida* and *Fusarium oxysporum*. Plant Pathol 49:383–388

Boneti JIS, Ferraz S, Braga JM et al (1982) Influência do parasitismo de *Meloidogyne exigua* sobre a absorção de micronutrientes (Zn, Cu, Fe, Mn e B) e sobre o vigor de mudas de cafeeiro. Fitopatol Bras 7:197–207

Campos PS, Quartin V, Ramalho JC et al (2003) Electrolyte leakage and lipid degradation account for cold sensitivity in leaves of *Coffea* sp plants. J Plant Physiol 160:283–292

Campos VP (1997) Café (*Coffea arabica* L.) Controle de doenças causadas por nematóides. In: Valle FXR, Zambolin L (eds) Controle de doenças de plantas. Universidade Federal de Viçosa, Viçosa

Campos VP (2007) Soil plowing and irrigation for *Meloidogyne* spp. control. Proceedings XXVII Congr Bras Nematol:18

Campos VP, Villain L (2005) Nematode parasites of coffee and cocoa. In: Luc M, Sikora RA, Bridge J (eds) Plant parasitic nematodes in subtropical and tropical agriculture, 2nd edition. CABI, Wallingford

Cannel MGR (1972) Photoperiodic response of mature trees of arabica coffee. Turrialba 22:198

Cardoso RML (1986) Ocorrência da murcha vascular do cafeeiro (*Coffea arabica*) no estado do Paraná – Brasil, induzida por *Fusarium oxysporum* f.sp. *coffeae*. Fitopatol Bras 11:753–760

Coste R (1992) Coffee – The plant and the product. MacMillan Press, London

DaMatta FM (2003) Drought as a multidimensional stress affecting photosynthesis in tropical tree crops. In: Hemantaranjan A (ed) Advances in Plant Physiology. Vol. 5. Scientific Publishers, Jodhpur

DaMatta FM (2004) Exploring drought tolerance in coffee: a physiological approach with some insights for plant breeding. Braz J Plant Physiol 16:1–6

DaMatta FM (2004a) Ecophysiological constraints on the production of shaded and unshaded coffee: a review. Field Crops Res. 86:99–114

DaMatta FM (2004b) Exploring drought tolerance in coffee: a physiological approach with some insights for plant breeding. Braz J Plant Physiol 16:1–6

DaMatta FM, Chaves ARM, Pinheiro HA et al (2003) Drought tolerance of two field-grown clones of *Coffea canephora*. Plant Sci 164:111–117

DaMatta FM, Loos RA, Rodrigues R et al (2001) Actual and potential photosynthetic rates of tropical crop species. Braz J Plant Physiol 13:24–32

DaMatta FM, Ramalho JDC (2006) Impacts of drought and temperature stress on coffee physiology and production: a review. Braz J Plant Physiol 18:55–81

De los Santos-Briones C, Hernández-Sotomayor SMT (2006) Coffee biotechnology. Braz J Plant Physiol 18:217–227

Di Vito M, Crozzoli R, Vovlas N (2000) Pathogenicity of *Meloidogyne exigua* on coffee (*Coffea arabica* L.) in pots. Nematropica 30:55–61
Dutra MR, Campos VP (2000) Patogenicidade de nematóides em cafeeiros submetidos a estresse. Proceedings XXII Congr Bras Nematol:107
Eisenback JD (1993) Interactions between nematodes in cohabitance. In: Khan MW (ed) Nematode Interactions. Chapman and Hall, London
Fahl JI, Carelli MLC (2007) Os estudos sobre a fisiologia do cafeeiro no Instituto Agronômico. O Agronômico 59:41–43
Fonseca HS, Ferraz LCCB, Machado SR (2003a) Caracterização do vacuoma de células gigantes induzidas por espécies de *Meloidogyne* em raizes de seringueira 'RRIM600'. Nematol Bras 27:193–198
Fonseca HS, Ferraz LCCB, Machado SR (2003b) Ultraestrutura comparada de raizes de seringueira parasitadas por *Meloidogyne exigua* e *M. javanica*. Nematol Bras 27:199–206
Gheysen G, Jones JT (2007) Molecular aspects of plant-nematode interactions. In: Perry RN, Moens M (eds) Plant Nematol CABI, Wallingford
Gilmore AM, Govindjee (1999) How higher plants respond to excess light: Energy dissipation in photosystem II. In: Singhal GS, Renger G, Irrgang H-D et al (eds) Concepts in Photobiology. Kluwer Academic Publishers, Dordrecht
Goldberg AD, Bierny O, Renard C (1984) Évolution compare des parameters hydriques chez *Coffea canephora* Pierre et l'hydride *Coffea arabusta* Capot et Aké Assi. Café Cacao Thé 28:257–266
Gonçalves W, Mazzafera P, Ferraz LCCB et al (1995) Biochemical basis of coffee tree resistance to *Meloidogyne incognita*. Plante Rech Dev 2:54–60
Gonçalves W, Ramiro DA, Gallo PB et al (2004) Manejo de nematóides na cultura do cafeeiro. Proceedings Reunião Itinerante Fitossanidade Inst Biol:48–66
Gonçalves W, Silvarolla MB (2007) A luta contra a doença causada pelos nematóides parasitos. Agronômico 59:54–56
Gonçalves W, Silvarolla MB, Lima MMA (1998) Estrategias visando a implementação do manejo integrado dos nematóides parasitos do cafeeiro. Inf Agropecuário 19:36–47
Haarer AE (1962) Modern Coffee Production. Leonard Hill, London
Herve G, Bertrand B, Villain L et al (2005) Distribution analyses of *Meloidogyne* spp. and *Pratylenchus coffeae sensu lato* in coffee plots in Costa Rica and Guatemala. Plant Pathol 54:471–475
Huang SP, Souza PE, Campos VP (1984) Seasonal variation of a *Meloidogyne exigua* population in a coffee plantation. J Nematol 16:115–117
Hurchanik DDP, Schmitt DP, Hue NV et al (2003) Relationship of *Meloidogyne konaensis* population densities to nutritional status of coffee roots and leaves. Nematropica 33:55–64
Hurchanik DDP, Schmitt DP, Hue NV et al (2004) Plant nutrient partitioning in coffee infected with *Meloidogyne konaensis*. J Nematol 36:76–84
Hussey RS (1985) Host-parasite relationships and associated physiological changes. In: Sasser JN, Carter CC (eds) An advanced treatise on *Meloidogyne*. Vol. 1 – Biology and control. North Carolina State University Department of Plant Pathology and USAID, Raleigh
Ikeda T, Suzaki T (1984) Influence of pine-wood nematodes on hydraulic conductivity and water status in *Pinus thunbergii*. J Jpn For Soc 66:412–420
Iwasaki Y, Yoshikawa K, Sakamoto K et al (1999) Changes in water relation parameters and photosynthetic rate of pine-wood nematode-infested *Pinus densiflora* Sieb. et Zucc. seedlings under several soil moisture conditions. J Jpn Soc Reveget Technol 24:186–191
Jaehn A (1990) Desenvolvimento de *Meloidogyne incognita* raça 2 em cafeeiro, afetado pela temperatura. Nematol Bras 14:89–102
Jaehn A (1991a) Determinação da constante térmica das raças 1, 2 e 4 de *Meloidogyne incognita* em cafeeiro. Nematol Bras 15:135–142
Jaehn A (1991b) Estimativa do número de gerações de três raças de *Meloidogyne incognita* em cafeeiro para o estado de São Paulo. Nematol Bras 15:143–151

Jaehn A, Monteiro AR, Lordello LGE et al (1983) Efeito de nitrogênio e de potássio em *Meloidogyne incognita* (Kofoid e White, 1919) Chitwood, 1949, como parasito do cafeeiro. Soc Bras Nematol 7:189–208

Jaehn A, Monteiro AR, Lordello LGE et al (1984) Efeitos na penetração de *Meloidogyne incognita* (Kofoid e White, 1919) Chitwood, 1949, em raizes de cafeeiros (*Coffea arabica* L.) do nitrogênio e do potássio. Nematol Bras 8:235–256

Jaehn A, Rebel EK (1984) Sobrevivência do nematóide de galhas *Meloidogyne incognita* em substrato infestado, para produção de mudas de cafeeiro sadias. Nematol Bras 8:319–324

Larcher W (1981) Effects of low temperature stress and frost injury on plant productivity. In: Johnson CB (ed) Physiological processes limiting plant productivity. Butterworths, London

Leguizamon-Caycedo J (1976) Relacion entre poblaciones de *Meloidogyne* spp. en el suelo y el daño causado em cafetales establecidos. Cenicafe 27:174–179

Leguizamon-Caycedo JE (1997) Efecto de *Meloidogyne* spp. en plantaciones establecidas de cafe variedad Caturra. Cenicafe, Informe Anual de Actividades 1998–1997, Chinchina

Lima RD, Ferraz S (1985a) Biologia de *Meloidogyne exigua*. 1. Desenvolvimento embriogênico e efeito da temperatura na embriogênese. Rev Ceres 32:339–347

Lima RD, Ferraz S (1985b) Biologia de *Meloidogyne exigua*. 2. Desenvolvimento pós-embriogênico em cafeeiro 'Mundo Novo'. Rev Ceres 32:349–361

Long SP, Humphries S, Falkowski PG (1994) Photoinhibition of photosynthesis in nature. Ann Rev Plant Physiol Plant Mol Biol 45:633–662

Lordello RRA, Lordello LGE (1983) Desenvolvimento de *Meloidogyne exigua* Göldi, 1887, em raizes de cafeeiros, em três ambientes. Ann Escola Superior de Agric Luiz de Queiroz 40:271–295

Macedo MCM, Haag HP, Lordello LGE (1974) Influência do nematóide *Meloidogyne exigua* na absorção de nutrientes em plantas jovens de cafeeiro. Ann Escola Superior de Agric Luiz de Queiroz 31:91–104

Maximiniano C, Campos VP, Souza RM et al (2001) Flutuação populacional de *Meloidogyne exigua* em cafezal naturalmente infestado por *Pasteuria penetrans*. Nematol Bras 25:63–69

Mazzafera P, Kubo RK, Inomoto M (2004) Carbon fixation and partitioning in coffee seedlings infested with *Pratylenchus coffeae*. Eur J of Plant Pathol 110:861–865

McSorley R (1987) Extraction of nematodes and sampling methods. In: Brown RH, Kerry BR (eds) Principles and practice of nematode control in crops. Academic Press, Sydney

Meinzer FC, Grantz DA, Goldstein G et al (1990b) Leaf water relations and maintenance of gas exchange in coffee cultivars grown in drying soil. Plant Physiol 94:1781–1787

Melakeberhan H, Webster JW (1993) The phenology of plant-nematode interaction and yield loss. In: Khan MW (ed) Nematode Interactions. Chapman and Hall, London

Mendes BV, Ferraz S, Shimoya C (1976) Histopatologia de raizes de cafeeiro parasitadas por *Meloidogyne exigua*. 1. Formação de células gigantes. Summa Phytopathol 2:97–102

Mendes BV, Ferraz S, Shimoya C (1977) Observações histopatológicas de raizes de cafeeiro parasitadas por *Meloidogyne exigua* Göldi, 1887. Nematol Bras 2:207–229

Moraes MV, Lordello LGE, Reis AJ et al (1977) Ensaio de rotação de culturas para reaproveitamento, com cafeeiro, de terras infestadas por *Meloidogyne exigua*. Nematol Bras 2:257–265

Nakasono K, Lordello RRA, Monteiro AR et al (1980) Desenvolvimento das raizes de cafeeiros novos transplantados e penetração por *Meloidogyne exigua*. Nematol Bras 4:33–46

Negron JA, Acosta N (1987) Studies on host-parasite relationships of *Meloidogyne incognita* and *Coffea arabica* cv Borbon. Nematropica 17:71–77

Negron JA, Acosta N (1989) The *Fusarium oxysporum* f.sp. *coffeae-Meloidogyne incognita* complex in 'Bourbon' coffee. Nematropica 19:161–168

Netto AT, Campostrini E, Oliveira JG et al (2005) Photosynthetic pigments, nitrogen, chlorophyll a fluorescence and SPAD-502 readings in coffee leaves. Scient Horticult 104:199–209

Nunes MA (1988) Environmental affects on the stomatal and mesophyll regulation of photosynthesis in coffee leaves. Photosynthetica 22:547–553

Nunes MA, Bierhuizen JF, Ploegman C (1968) Studies on the productivity of coffee. I. Effects of light, temperature and CO2 concentration on photosynthesis of *Coffea arabica*. Acta Bot Neerl 17:93–102

Nunes MA, Brumby D, Davies DD (1973) Estudo comparativo do metabolismo fotossintético em folhas de cafeeiro, beterraba e cana-de-açúcar. Ser Estud Agron 1:1–14

Nunes MA, Correia MM (1983) Regulação estomática da água disponivel no solo em *Coffea arabica* L. (cvs Caturra, Catuai e Harrar). Ser Estud Agron 10:83–90

Oliveira JG, Alves PLCA, Magalhães AC (2002) The effect of chilling on the photosynthetic activity in coffee (*Coffea arabica* L.) seedlings – the protective action of chloroplastid pigments. Braz J Plant Physiol 14:95–104

Pinheiro HA, DaMatta FM, Chaves ARM et al (2005) Drought tolerance is associated with rooting depth and stomatal control of water use in clones of *Coffea canephora*. Ann Bot 96:101–108

Pinochet J, Cordero D, Berrocal A (1986) Fluctuacion estacional de poblaciones de nematodos em dos cafetales em Panama. Turrialba 36:149–156

Rena AB, Barros RS, Maestri M et al (1994) Coffee. In: Schaffer B, Andersen PC (eds) Handbook of Environmental Physiology of Tropical Fruit Crops: Sub-Tropical and Crops. Vol. 2. CRC Press, Boca Raton

Rena AB, Guimarães PTG (2000) Sistema radicular do cafeeiro: estrutura, distribuição, atividade e fatores que o influenciam. EPAMIG, serie documentos # 37, Belo Horizonte

Rodrigues ACFO, Abrantes IMO, Melillo MT et al (2000) Ultrastructural response of coffee roots to root-knot nematodes, *Meloidogyne exigua* and *M. megadora*. Nematropica 30:201–210

Rodrigues IFD, Crozzoli R (1995) Efectos del nematode agallador *Meloidogyne exigua* sobre el crecimiento de plantas de cafe en vivero. Nematol Mediterr 23:325–328

Ronquim JC, Prado CHBA, Novaes P et al (2006) Carbon gain in *Coffea arabica* during clear and cloudy days in the wet season. Exper Agric 42:147–164

Santos JM, Ferraz S (1977) Efeito de exsudatos radiculares e temperatura sobre a eclosão de larvas de *Meloidogyne exigua* Göldi, 1887. Proceedings V Congr Bras Pesqu Cafeeiras, no pagination

Santos JM, Ferraz S, Oliveira LM (1981) Efeitos do parasitismo de *Meloidogyne exigua* sobre a absorção e translocação de nutrientes em mudas de cafeeiro. Fitopatol Bras 6:333–340

Schmitt DP, Riggs RD (1989) Population dynamics and management of *Heterodera glycines*. Agric Zool Rev 3:253–269

Serracin M, Schmitt DP (2000) *Meloidogyne konaensis* and coffee rootstock interactions at two moisture regimes in four soils. Nematropica 32:65–74

Sikora RA, Carter WW (1987) Nematode interactions with fungal and bacterial plant pathogens – fact or fantasy. In: Veech JA, Dickson DW (eds) Vistas on Nematology. Society of Nematologists, Hyattsville

Silva CP (2005) Efeito do nematóide *Meloidogyne exigua* no crescimento e fisiologia do cafeeiro (*Coffea arabica* L.). Universidade Estadual do Norte Fluminense Darcy Ribeiro, BS dissertation, Campos dos Goytacazes

Silva RV, Oliveira RDL, Pereira AA et al (2007) Respostas de genótipos de *Coffea* spp. a diferentes populações de *Meloidogyne exigua*. Fitopatol Bras 32:205–212

Souza MC (2006) Análise fisiológica dos danos causados ao cafeeiro pelo nematóide-das-galhas (*Meloidogyne* sp.). Universidade Estadual do Norte Fluminense Darcy Ribeiro, BS dissertation, Campos dos Goytacazes

Souza RM, Volpato AR, Viana AP (2008a) Epidemiology of *Meloidogyne exigua* in an upland coffee plantation in Brazil. Nematol Mediterr, *in press*

Souza RM, Volpato AR, Viana AP (2008b) Field assessment of different sampling strategies for coffee plantations parasitized by *Meloidogyne exigua*. Nematropica 37:345–355

Tronconi NM, Ferraz S (1985) Comportamento de *Meloidogyne exigua* em mudas de cafeeiro plantadas em diferentes tipos de solo. Fitopatol Bras 10:356

Tronconi NM, Ferraz S, Santos JM et al (1986). Influência da temperatura na patogenicidade e reprodução de *Meloidogyne exigua* em mudas de cafeeiro. Nematol Bras 10:69–83

Vaast Ph, Caswell-Chen EP, Zasoski RJ (1998) Effects of two endoparasitic nematodes (*Pratylenchus coffeae* and *Meloidogyne konaensis*) on ammonium and nitrate uptake by arabica coffee (*Coffea arabica* L.) App Soil Ecol 10:171–178

Villain L, Anzueto F, Hernandez A et al (1999) Los nematodos parasitos del cafeto. In: Bertrand B, Rapidel B (eds) Desafios de la caficultura em centroamerica. CIRAD/ICCA-PROMECAFE, San Jose

Villalba-Gault D, Fernandez-Borrero O, Baeza-Aragon CA (1983) Ciclo de vida de *Meloidogyne incognita* raza 5 (Kofoid and White 1919) Chitwood 1949, em *Coffea arabica* variedad Caturra 1. Cenicafé 34:16–33

Von Mende N (1997) Invasion and migration behaviour of sedentary nematodes. In: Fenoll C, Grundler FMW, Ohl SA (eds) Cellular and molecular aspects of plant-nematode interactions. Kluwer Academic Publishers, Dordrecht

Vovlas N, Di Vito M (1991) Effect of root-knot nematodes *Meloidogyne incognita* and *M. javanica* on the growth of coffee (*Coffea arabica* L.) in pots. Nematol Mediterr 19:253–258

Went FW (1957) The experimental control of plant growth. Chronica Botanica. An International Biological and Agricultural Series # 17. Ronald Press, New York

Wilcox-Lee D, Loria D (1987) Effects of nematode parasitism on plant-water relations. In: Veech JA, Dickson DW (eds) Vistas on Nematology. Society of Nematologists, Hyattsville

Wormer TM (1965) The effect of soil moisture, nitrogen fertilization and some meteorological factors on stomatal aperture of *Coffeae arabica* L. Ann Bot 29:523

Wyss U (2002) Feeding behaviour of plant-parasitic nematodes. In: Lee DL (ed) The biology of nematodes. Taylor and Francis, London

Zhang F, Schmitt DP (1995a) Embryogenesis and postinfection development of *Meloidogyne* konaensis. J. Nematol 27:103–108

Zhang F, Schmitt DP (1995b) Relationship of *Meloidogyne konaensis* population densities to coffee growth. Plant Dis 79:446–449

Zhang F, Schmitt DP (1995c) Spatial-temporal patterns of *Meloidogyne konaensis* on coffee in Hawaii. J Nematol 27:109–113

Chapter 8
Management of *Meloidogyne* spp. in Coffee Plantations

Vicente P. Campos and Juliana R.C. Silva

Abstract This chapter deals with management of coffee-parasitic root-knot nematodes (RKNs), *Meloidogyne* spp. Throughout the chapter, this is discussed according to the different situations that may be faced by coffee growers. For instance, certain procedures are recommended to avoid the introduction of RKNs into coffee fields. On the other hand, a field may be diagnosed as infested by these nematodes before or after the plantation has been established; these situations require distinct management approaches. Management is also discussed according to the *Meloidogyne* species involved; for instance, *M. exigua* can be eradicated from the soil by one-year rotation with non-host crops, and it can be profitably managed through nematicide or organic matter applications. On the other hand, *M. incognita* and *M. paranaensis* cannot be managed with those applications, and they cannot be eradicated from the soil. Coffee plantations infested by *M. incognita* or *M paranaensis* can be profitable if their soil population is decreased and nematode-resistant rootstocks are used. This chapter also discusses the prospects of controlling coffee-parasitic RKNs through naturally occurring nematicides, biological control and induced resistance.

Keywords Control · management · *Meloidogyne exigua* · *M. paranaensis* · *M. incognita*

8.1 Introduction

As a perennial crop, coffee (*Coffea* sp.) stays in the field for decades, subjected to nematode parasitism from the seedling stage through the economic time life of the plantation. Therefore, coffee plantations should not be established in areas infested with damaging nematodes, such as those of most concern in Brazil: the root-knot nematodes (RKNs) *Meloidogyne incognita* (Kofoid and White) Chitwood, *M. paranaensis* Carneiro, Carneiro, Abrantes, Santos and Almeida, *M. coffeicola*

V.P. Campos
Universidade Federal de Lavras, Lavras, Brazil
e-mail: vpcampos@ufla.br

R.M. Souza (ed.), *Plant-Parasitic Nematodes of Coffee*,

Lordello and Zamith and *M. exigua* Göldi (Campos and Villain, 2005). Indeed, *M. incognita* and *M. paranaensis*, which cause the greatest yield losses, are the limiting factor to coffee cultivation in certain areas in Brazil. These species destroy the plant's root system, are easily disseminated, persist for a long time in the soil in the absence of a host, and are not efficiently controlled by nematicides. In the case of *M. incognita*, the existence of physiological races complicates breeding for resistance and crop rotation, which led Gonçalves (1995) to advise growers not to grow susceptible coffee cultivars in areas infested by that species.

Even in nematode-free fields, RKNs may eventually be introduced during the long years of cultivation, especially in areas intensely cultivated with coffee and infested by RKNs. Therefore, it is extremely important that growers be advised on RKNs before planting coffee, and that they are made aware of the management strategies available.

This chapter begins by discussing each of these management strategies. In the following section, these strategies are discussed in an integrated manner, considering some specific situations that coffee growers may face in their farms. The management of coffee-parasitic RKNs is further discussed in the Part V of this book, in which crop, nematode and climate specificities of several countries are outlined, and the valuable experience of many nematologists, growers and extensionists is presented.

8.2 Management Strategies

8.2.1 Exclusion

The planting of nematode-free coffee seedlings avoids the introduction of RKNs into a new area. Therefore, any infected seedling should be discarded and not used in a nematode-free area.

In Brazil, the regulatory restrictions to avoid the introduction of infected seedlings into new coffee-growing areas were more effective in the past than today. In the past, the government subsidized new coffee plantations, but it imposed the use of modern agronomic practices and prohibited planting coffee (i) in areas previously cultivated with coffee or even close to them, (ii) from seedlings infected with nematodes, and (iii) in regions not recommended for coffee growing. Since 1980, the subsidies have no longer been available, and the government withdrew its control over the expansion of the crop. Nowadays, it is up to the coffee growers to obtain technical information from the extension service network, universities, research institutions or private sources, although the official inspection of commercial coffee nurseries is still in place. In Minas Gerais, Brazil's most important coffee-producing State, nurseries must have a certificate issued by an official nematology laboratory stating the absence of RKNs in their premises.

When coffee growers are to produce their own seedlings, care must be taken with the source of irrigation water and the planting soil. The use of water from dams filled with runoff water from hillsides cultivated with coffee should be avoided. Nematode-infested or -suspected water should be heated at temperatures above 70°C or be stored in containers and exposed to the sun, during the summer, for at least two weeks.

The planting soil should be treated by solarization for 30 days or be treated in solar collectors for two days for complete elimination of RKNs (Randig et al., 1998). The fumigant metan sodium can also be used to treat the planting soil. Alternatively, a site on the farm that has not been cultivated with any crop and that is not located downhill from any crop could provide planting soil with no need for anti-nematode treatments. Finally, coffee growers can use commercial, soil-less substrates to produce their own seedlings, but these can be expensive.

New coffee plantations should be located with special care, avoiding areas from which old coffee plants have been eradicated recently and those close to or downhill from RKN-infested fields. In certain circumstances, a furrow can be dug to prevent runoff waters from infested areas. Equipment and farm implements used in infested fields should be washed free of soil debris before being used in nematode-free plantations.

8.2.2 Containment of Focal Infestations

When an infestation by RKNs is found in a small number of coffee plants and/or in a restricted site in the field, one should try to contain the nematode and prevent it from spreading. In upland plantations, the lower limit of a focal infestation should be determined by downward samplings of coffee roots in all planting rows. A drainage furrow should be dug 10 planting rows below the last coffee plant to be found with RKN-induced root galls. This furrow should avoid downhill infestation by nematode-contaminated runoff waters, which should be diverted away from the plantation. New samplings should be performed every year.

8.2.3 Nematicides

The chemicals used today to control plant-parasitic nematodes on coffee and other crops are mostly restricted to granular products that act on the nematodes either by direct contact or systemically through the plant (Tables 8.1 and 8.2). In the group of fumigant chemicals used for controlling nematodes in the past (Anonymous, 1968), methyl bromide has been the most widely used to disinfest nursery soils, but international restrictions on its use have been in place for some years.

The organophosphate and organocarbamate systemic insecticides with potential for nematode control are rarely phytotoxic at the dosages recommended for field use. Their major disadvantage is being dispersed and lost through water. Their nematicidal activity is usually confined to a shallow root zone (the rhizosphere), and it

Table 8.1 Globally important nematicides

Active substance	Chemical group	LD_{50}[a]	Examples of trading names	Manufacturer
Aldicarb	Oxime carbamate	0.93	Temik 10G®, Temik 15 G®	Bayer CropScience
Carbofuran	Carbamate	8	Furadan 15G®, Furadan 4F®	FMC Corporation
Cadusafos	Organophosphorus	87	Rugby 200CS®, Rugby 10G®	FMC Corporation
Dazomet	Methyl isothiocyanate liberator	77–220[b]	Basamid®	BASF Corporation
1,3-Dichloropropene	Halogenated hydrocarbon	150	Telone II®, Telone EC®	Dow AgroScience
Ethoprophos	Organophosphorus	62	Mocap 10G®, Mocap EC®	Bayer CropScience
Fenamiphos	Organophosphorus	6	Nemacur 15G®, Nemacur 3®	Bayer CropScience
Fosthiazate	Organophosphorus	73	Nemathorin 10G®	Syngenta
Metam sodium (sodium N-methyldithiocarbamate)	Methyl isothiocyanate liberator	77–220	Vapam®, Vapam HL®	Amvac Chemical Corporation
Oxamyl	Oxime carbamate	3.1	Vydate 10G®, Vydate L®	Du Pont

[a] acute oral male rats.
[b] LD_{50} calculated for methyl isothiocyanate.
Adapted from Haydock et al. (2006).

Table 8.2 Nematicides registered for use in Brazil in 2005

Chemical group	Active substance	Trading name
Fumigant	(–)	(–)
Hidrocarbonate halogenate alifatic brometane	Methyl bromide	Bromex®, Bromo Fersol®, Bromo Flora®
Non-fumigants	(–)	(–)
Organophosphorus	Ethoprophos	Rhocap®
Oximecarbamate	Aldicarb	Temik 150®
Carbamate	Carbofuran	Furadan 50G®, Furadan 100G®, Furadan 350TS®, Furadan 350SC®, Diafuran 50®, Ralzer 350 SC®, Ralzer 50GR®
Organophosphorus	Terbuphos	Counter 150G®

Adapted from Anonymous (2005).

is often a result of narcotization of the nematodes, which suffer a disabling change in their behavior rather than death. By disrupting the eclosion of second-stage juveniles (J2) from the eggs, and subsequent root penetration, development and reproduction, nematicides can reduce or nearly cancel the rate of population increase in the field for a period of up to 90 days. These chemicals give little or no control of fungal or bacterial diseases, but do provide insecticidal protection depending upon the chemical involved (Van Gundy and McKenry, 1977). For example, aldicarb can control root-boring and leaf-mining insects at the end of the rainy season. On the other hand, parasitism by RKNs may reduce the root's uptake of systemic fungicides applied in the soil against 'leaf rust' caused by *Hemileia vastatrix* Berk and Br. (Otoboni et al., 2001, 2003).

In coffee, the effective dosage of aldicarb, carbofuran, phenamiphos and terbuphos are in the range of 1.6–6.0 g of active ingredient/plant, in one or two applications during the year. The first application should be made at the beginning of the rainy season, followed by the second three months later, because water availability is important for the release of the active ingredient. In Brazil, Campos et al. (2005) recommended the first application in November. Usually, a furrow is dug along both sides of the planting row, at the edge of the plant's canopy; the chemical is applied and incorporated into the soil, by an automated or manual application device.

The application of systemic or contact granular nematicides on coffee plants severely damaged by *M. incognita* or *M. paranaensis* has been considered ineffective due to the destruction of large portions of the plant's root system by the nematode (Curi et al., 1977). Accordingly, Jaehn (1984) has shown that although the rhizosphere population of *M. incognita* J2 decreases with the application of nematicides, with this effect lasting up to 60 days, the plants do not recover their vigor and the plantation's productivity is not recovered to a satisfactory level. Poor yield recovery was also observed by Gonçalves and Silvarola (2001) in *M. incognita*-infested plantations that had been treated with nematicides, in comparison to nematode-free plantations. Also, nematicides give poor protection to coffee seedlings parasitized by *M. incognita* (Jaehn et al., 1984).

Therefore, nematicides are not recommended for management of coffee plantations infested by *M. incognita, M. paranaensis, M. coffeicola* or other species causing similar symptoms.

For most *Meloidogyne* species that induce typical root galls, such as *M. exigua*, many granular nematicides are effective in decreasing nematode populations up to three months after application (Huang et al., 1983). After this period, the nematode population may increase on treated plants, but these usually have good foliage cover by this time in the rainy season. Apparently, the plants' vigor is achieved by other factors besides nematode control (Campos and Lima, 1986). Cadusaphos, carbofuran and carbosulfan have been tested for their efficacy against *M. exigua* (Volpato et al., 2001), and some of them have potential to control coffee-parasitic nematodes. Indeed, an *M. exigua*-infested coffee plantation treated with nematicide for five consecutive years yielded 30.9% more than a non-treated one. As expected, the nematicide did not eradicate the nematodes (Lordello et al., 1990).

8.2.4 Grafting

In Brazil, the widespread distribution and aggressiveness of *M. incognita* in the western region of São Paulo State led nematologists to seek alternatives to chemical control. An accession of *C. canephora* Pierre ex Froehner '2258' from CATIE's germplasm collection in Costa Rica showed high resistance to *M. exigua* and resistance and/or tolerance to several populations of *M. incognita* (Fazuoli, 1986). The same accession was later reported as resistant to races one, two and three of *M. incognita* (Gonçalves et al., 1996) and to *M. paranaensis* (Fazuoli et al., 2002). Initially, '2258' had a resistance level around 70%, but this level has been raised considerably by subsequent selection in fields highly infested by *M. incognita*. This improved line was later released as 'Apoatã', a rootstock resistant to *M. incognita* and *M. paranaensis* and immune to *M. exigua* (Fazuoli et al., 2002).

In fields infested with *M. incognita* race one, arabica coffee (*C. arabica* L.) 'Mundo Novo' grafted onto '2258' yielded 3.6 times as much as non-grafted plants (da Costa et al., 1991). In Brazil, the preventive planting of arabica coffee grafted onto 'Apoatã' is widespread in non-infested areas of São Paulo and Paraná States, which in the past suffered the most from *M. incognita* and *M. paranaensis*. In some municipalities, the planting of grafted coffee has revived the local coffee industry (Campos, 1997). Although using 'Apoatã' is the only feasible solution to growing arabica coffee in *M. incognita*- or *M. paranaensis*-infested fields, this rootstock showed intolerance to *Pratylenchus brachyurus* (Godfrey) Filipjev and S. Stekhoven in greenhouse tests (de Oliveira, 1996).

The same *C. canephora* line that originated 'Apoatã' was crossed with the RKN-resistant *C. canephora* line T3751, giving rise to a new rootstock cultivar named 'Nemaya', which shows resistance to a number of *Meloidogyne* species and populations (see Chapter 9).

The development of arabica coffee rootstock cultivars has become a possibility with the finding of *M. incognita*-resistance in *C. arabica* accessions (Anzueto et al., 2001). However, *C. canephora, C. congensis* A. Froehner and *C. dewevrei* De Wild. and T. Durand are the breeders' main focus to produce nematode-resistant rootstocks because these species present abundant root systems and resistance to other pathogens as well (Gonçalves and Silvarola, 2001). However, resistance genes found in wild or semi-wild lines of *C. arabica* from Ethiopia or Yemen could be used in interspecific hybridizations with resistant, diploid *Coffea* sp. lines. For example, the rootstock hybrid Arabusta (*C. canephora* x *C. arabica*) presents high vigor, nematode resistance and better adaptability to regions with mild climate, in comparison to *C. canephora* rootstocks (Capot, 1972; Berthaud, 1978a,b). Likewise, arabica coffee rootstocks should be more adapted to mild climates than 'Apoatã'.

In Brazil, non-grafted, nematode-resistant arabica cultivars have been released on the market, giving more options to coffee growers managing RKNs. For example, 'Iapar 59' and 'H 419-5-4-5-2 Paraiso' are resistant to *M. exigua*, although virulent populations have been reported by Barbosa et al. (2007). These cultivars are also susceptible to *M. incognita* populations from São Paulo and Parana states (Muniz, M.F.S., Embrapa/Cenargen, unpublished results).

8.2.5 Crop Rotation, Intercropping and Organic Matter Application, Alone or in Combination with Nematicides

In *M. exigua*-infested fields, de Moraes et al. (1977) observed that after eradicating the coffee plantation, a one year-rotation with cotton, soybean or maize drastically reduces the nematode population, allowing for a safe return to coffee cultivation. Almeida and Campos (1991a,b) studied rotation with common bean, soybean, sorghum, *Crotalaria spectabilis* Roth, *Stilozobium aterrinum* Piper and Tracy and *Panicum maximum* Jacq., and also concluded that a one-year rotation allows the replanting of *M. exigua*-susceptible cultivars. These authors monitored the field for 18 months after the replanting, and they did not observe nematode-induced root galls nor J2 in the soil. A one year-rotation is also effective against *M. coffeicola*. Nonetheless, reinfestation of coffee fields may occur through runoff waters from adjacent fields cultivated with coffee or other nematode hosts, as well as through soil debris carried by animals, implements or human traffic. Indeed, this chapter's first author witnessed an RKN-free field, previously cultivated with *Brachiaria decumbens* Stapf., be planted with RKN-free coffee seedlings; two years later, *M. exigua*-induced root galls could be seen in the new plantation because no preventive measure had been taken to avoid runoff waters from an uphill nematode-infested field.

As regards *M. incognita*, Carneiro and Carneiro (1982) screened 29 plant species as candidates for crop rotation, concluding that *Arachis hypogea* L. and *Ricinus communis* L. were immune to that nematode species, while *Styzolobium deeringianum* Bort. and *C. spectabilis* were resistant. Santiago et al. (2001) observed no root penetration by J2, root galls induced by, or egg masses produced by *M. incognita* races one, two, three or four or *M. paranaensis* when these species were inoculated on *Arachis pintoi* Krapov. and W.C. Gregory, which makes this plant a suitable candidate for crop rotation.

Unlike *M. exigua, M. incognita*- and *M. paranaensis*-infested fields must be crop rotated for more than one year due to these species' longer survival in the soil. In such fields crop rotation is recommended before replanting coffee using 'Apoata', because its resistance to those species is not complete.

As far as intercropping is concerned, not many studies have been carried out to examine the management of coffee-parasitic nematodes. Fazuoli et al. (2002) assessed the cultivation of velvet bean between the coffee rows, with the former being incorporated into the soil at flowering stage. The authors concluded that velvet been protected coffee from wind, and improved the sandy soil's texture, organic matter content and fertility, hence favoring the development of the coffee plants and minimizing the damage by *M. incognita* and *M. paranaensis*.

In greenhouse, the progressive incorporation of coffee bean husk in the soil reduced *M. exigua* population, with its total inhibition when the proportion husk/soil reached 3:1, and when husk only was used. On the other hand, layering the husk on the soil's surface had minimal effect on the nematode (Tronconi et al., 1986). In the field, adding organic matter to the soil around the edge of the plant canopy can temporarily reduce *M. exigua* population, induce new root flushes and improve productivity.

On the other hand, it is difficult to enhance productivity in *M. incognita*- and *M. paranaensis*-infested areas through incorporation of organic matter, combined or not with nematicides. Jaehn and Rebel (1984) were unable to enhance productivity or reduce *M. incognita* population to an acceptable level in an infested new plantation in which coffee husk alone, or combined with nematicides, was applied to the holes dug for planting the coffee seedlings. Likewise, castor bean bran cake or nematicide applied in the planting hole did not protect coffee seedlings from *M. incognita* nor did it provide satisfactory yield in a sandy area (Jaehn and Cataneo, 1986).

In mature plantations, no substantial gain in productivity was obtained in an *M. incognita*-infested field in which nematicides or castor bean cake were applied to the soil, or if *C. spectabilis* was cultivated between the coffee rows (Jaehn, (1984). The same poor results were obtained by application of Temik 100G®, Furadan 50G® or castor bean bran cake in an *M. incognita*-infested field, although the J2 soil population was reduced temporarily (Ferraz et al., 1983).

Although it is difficult to reduce *M. incognita* and *M. paranaensis* populations and increase coffee productivity through intercropping or application of nematicides and organic matter, it is advisable to employ the latter in coffee plantations established in depleted soils because this practice can be helpful in delaying eventual eradication of the plantation.

8.2.6 Naturally Occurring Nematicides and Inducers of Plant Resistance

Continuous research efforts worldwide seek new nematicidal compounds and chemical or biological agents that improve plant resistance to nematodes. Examples of such compounds are given in Table 8.3. Amaral et al. (2002) noticed *in vitro* and *in vivo* toxicity of extracts of onion and *Ruta graviolens* L. on *M. exigua*. Salgado et al. (2003) observed a high mortality of *M. exigua* J2 in *in vitro* tests with essential oils of *Eucalyptus camaldulensis* Dehn, *E. saligma* Smith, *E. urophylla* S. T. Blake, *Bixa orellana* L., *Xilopia brasiliensis* Sprengel and *Melia azidarach* L. A high mortality of *M. exigua* J2 also occurred in aqueous extracts of *Cinnamomum zeylanicum* Blume, yeast and solution of milk whey (Salgado and Campos, 2003a). In greenhouse, extracts of *C. zeylanicum or B. orellana* and a probiotic mix reduced the population of *M. exigua* in coffee roots (Salgado and Campos, 2003b). On-going studies are focused on reducing *M. exigua* population through the application of plant extracts in the rhizosphere of coffee plants. In the future, new nematicidal compounds may become available on the market to control plant-parasitic nematodes, including those parasitic on coffee.

Another promising strategy for managing *M. exigua*-infested areas is the use of biotic or abiotic agents to induce coffee resistance through the activation of the plant's latent defense mechanisms. Calcium and potassium silicates do not cause

Table 8.3 Examples of available non-synthetic nematicides

Trading name	Active substance(s) and mode of action (if known)	Source	Manufacturer	Comments
Dragonfire CPP Ontrol®	Aldehydes, ketones, linolenic acids	Sesame seed oil and seed meal	Poulenger.	Marketed for control of pathogenic nematodes in turf, horticultural and agricultural situations
Neo-trol®	Aldehydes, ketones, linolenic acids	Sesame stalk	Barmac Industries	A pelletized product for use in golf greens
CropGuard®	2-Furfuraldehyde, a pentose sugar derivative	Woody biomass	Illovo Sugar Ltd.	A solvent produced by acid hydrolysis of the pentosan contained in woody biomass such as maize cobs
Clandosan 618®	Chitin and urea increase chitin-feeding soil microorganisms	Crab and shrimp shells	Igene Biotechnology	By-product of seafood processing. The ground shells, along with urea, are formed into granules
DiTera®	Fermentation extracts from bacteria	*Myrothecium verrucaria*	Valent BioSciences	Approved by EPA for commercial use in the USA
Neem®, Nemate 10G® or liquid formulations	Azadirachtin	Neem plant extract or cake	Many suppliers, particularly in India	Residue from neem oil extraction process

Adapted from Haydock et al. (2006).

mortality of *M. exigua* J2 *in vitro* nor do they reduce their penetration into the coffee roots. However, these compounds reduce root galling induced by *M. exigua* and the nematode's reproduction (Dutra, 2004; Paiva et al., 2005; 2006). Also, the formation of giant cells is reduced or totally inhibited as the coffee plant absorbs silicon from the soil.

The compound acibenzolar-S-methyl is a plant resistance inducer recently registered in Brazil as Bion 500WG® for use against plant-pathogenic fungi. Nonetheless, it also works to reduce the reproduction of RKNs in some crops (Owen et al., 2002; Silva et al., 2004). Salicylic acid also enhances resistance in cowpea against *M. incognita* (Nandi et al., 2002).

8.2.7 Biological Control

Biological control is a promising strategy for managing coffee-parasitic nematodes, especially in the so-called 'organic' plantations where the use of synthetic chemicals is prohibited, and whose production sells for a higher market price. Among the many microorganisms reported as antagonistic to plant-parasitic nematodes, the bacterium *Pasteuria penetrans* (Thorne) Sayre and Starr has the advantage of presenting resistance to heat, soil drought and the pesticides commonly used in agriculture (Campos et al., 1998). *P. penetrans* was first observed in coffee fields by Baeza-Aragon (1978), and later by Sharma and Lordello (1992). In Brazil, up to 65% of *M. exigua* J2 have been found to be infested by *P. penetrans* throughout the year (Maximiniano et al., 2001), which suggests its importance to nematode control.

In Cuba, isolates of *Pochonia chlamydosporia* Zare, Gams and Evans (syn. *Verticillium chlamydosporium* Goddard) isolated from coffee plantations have potential for the biological control of coffee-parasitic RKNs (Hidalgo-Diaz et al., 2000). In Brazil, *P. chlamydosporia* has been found in an arabica plantation causing severe reduction in *M. exigua* J2 eclosion from the eggs (V.P. Campos, unpublished results). Other J2-predator and egg-parasitic fungi have been isolated from coffee fields (Naves and Campos, 1991; Ribeiro and Campos, 1993). The efficacy of *Arthrobotrys conoides* Drechsler, *A. musiformes* Drechsler, *Paecillomyces lilacinus* (Thom) Samson and *P. chlamydosporia* for the control of coffee-parasitic *M. exigua* was assessed by Campos (1997).

8.2.8 Fallowing, Plowing and Soil Irrigation

As cited above, the lack of suitable hosts leads to the decline of RKN-soil populations over time. However, maintaining a field free of hosts, including weeds, for many months is a difficult task, and it can be costly if herbicides or much labor are employed.

Furthermore, nematode survival in the soil varies with the *Meloidogyne* species involved. Following coffee eradication, *M. exigua* is no longer found in the soil after a six-month fallowing (Alvarenga, 1973; de Moraes and Lordello, 1977). *M. coffeicola* also presents a low persistence in the soil (Rebel et al., 1976; Carneiro Filho and Yamaguchi, 1995), and it seems to have reduced ability to infect coffee seedlings and young trees. On the other hand, *M. incognita* actively infects coffee plants after a six-month fallowing because it decreases only about a third of its original population during this period (Jaehn and Rebel, 1984).

To shorten the fallowing period, one can plow the soil to increase its water loss, which reduces the survival of nematode eggs and J2. Plowing followed by irrigation during hot days induces the eclosion of RKN J2, which, in the absence of suitable hosts, lose their infective ability and die in about 14 days at field temperatures ranging from 30 to 35°C (Dutra and Campos, 2003a; 2003b; Dutra et al., 2003). Nonetheless, fallowing may not result in a complete eradication of RKNs from an infested field, which can be achieved by crop rotation for some *Meloidogyne* species. Both procedures are recommended for *M. incognita*-infested fields in which 'Apoatã' will be used as a rootstock for arabica coffee.

8.2.9 Uprooting and Burning of Coffee Plants

Depending on the *Meloidogyne* species involved and the agronomic condition of the infested plantation, the plants' eradication may be the most suitable measure. In such cases, the plants should be pulled up (uprooted), gathered, left to dry and burned, because more than 80% of the nematode population lives in the roots. This procedure drastically reduces the population that will be combated by procedures such as soil plowing, fallowing and crop rotation.

8.3 The Timely Application of RKN-Management Strategies

Coffee growers should be advised to remain vigilant about RKNs at all times and during all farm practices. Furthermore, nematologists and extensionists should advise growers on the most important *Meloidogyne* species present in their region, and their corresponding management strategies. Three major issues should be considered for any coffee field: (i) the presence of RKNs and their identity, sometimes up to the level of physiological race, (ii) the nematode population level, and (iii) the choice of management strategies depending on whether the nematode was noticed before, at the time of, or after the establishment of the plantation.

Before the establishment of the plantation, if soil and/or root samplings reveal economically important RKNs in the field, the grower should employ strategies to eradicate the nematode, particularly if susceptible coffee cultivars are to be planted. If eradication is not possible or the prospective cultivar is highly susceptible to the particular nematode species found, the grower should be advised to employ strategies to reduce the nematode population and to use coffee seedlings grafted onto a resistant rootstock only. In regions where highly damaging *Meloidogyne* species are widespread, the grower should use grafted seedlings even if no RKNs were reported in the field. This preventive use of resistant rootstock has been proved fruitful in Brazil.

If infestation by RKNs is noticed after the establishment of the plantation, nematicides, intercropping and organic matter application can be applied, alongside strategies to contain the nematode's dispersal in the field.

8.4 The Agronomic Management of RKN-Infested Plantations

In addition to the strategies that aim to combat RKNs directly, coffee growers can employ agronomic practices to enhance the plantation's productivity. Such practices should nonetheless be tried with special care in *M. incognita*- or *M. paranaensis*-infested fields, since the economic return may be small in comparison to *M. exigua*-infested areas. Indeed, *M. incognita* and *M. paranaensis* destroy the plant's root system, especially the feeder roots responsible for nutrient uptake. In such cases, providing the plants with more fertilizers will not improve their vigor enough to substantially increase their yield.

Another agronomic approach would focus on alleviating all kinds of stress suffered by the plants. For example, Matiello et al. (2004) reported a worsening of *M. exigua*-related damage and an increase in the incidence of *Cercospora* sp. on coffee fruits six to eight months after drastically trimming the plant's plagiotrophic branches.

In most of the world's tropical coffee-producing regions it is common to have a dry season during the year. In such periods, nematode-infested plantations grown in clay soils are likely to suffer the least hydric stress, in comparison to those grown in sandy soils, because the former soil presents a higher capacity to hold water. Furthermore, high air and soil temperatures quicken the depletion of organic matter in tropical soils, a common phenomenon in sandy soils. Hence, coffee plantations grown in sandy soils are the most damaged by RKNs. In such areas, intercropping and application of organic matter to the soil may alleviate the nematode damage, as reported by Fazuoli et al. (2002).

Alleviating plant stress is likely to give better yield return in *M. exigua*-infested plantations. In Minas Gerais State, about 22% of the coffee plantations are infested by this species (Campos, 2002), which causes yield losses of up to 31% (Lordello et al., 1990). In this area, coffee growers have learned intuitively to manage the plantations. For example, to compensate for the stress suffered by the plants in the years of high yields, the growers step up the care given to the plants the year before. They apply higher fertilizer dosages and control appropriately 'leaf rust', leaf-mining and root-boring insects. The higher cost of systemic fungicides for proper control of 'leaf rust' is repaid by a better control of the fungus, which results in less defoliation of the plants, less damage by *M. exigua* and higher yields.

One or two months before harvesting, the soil underneath the coffee canopy is cleared of debris, which is moved together with some soil to the middle of the plantation's rows. Soon after harvest, the soil and debris should be moved back to the edge of, and under, the plant canopy. This material protects from drought the new feeder roots which are emitted in this area during the following dry season, during the plant's flowering stage. At this time, coffee husk and/or manure should be applied to the soil, which helps to maintain the soil humidity, provide nutrients to the plants, and release anti-nematode compounds produced during degradation of the organic matter. These processes prevent the plants suffering pronounced stress at the end of the dry season, which allows their fast vegetative growth once the first rains occur. In some regions, this may occur when the air and soil temperatures are still below the minimum required by the nematodes to eclode from the eggs, migrate and penetrate the coffee roots.

8.5 Concluding Remarks

In conclusion, coffee growers and nematologists stand a better chance against RKNs wherever regulatory restrictions are created and properly enforced to prevent the planting of non-certified, nematode-parasitized seedlings. This should prevent the relatively fast spread of damaging nematodes to new coffee-growing areas, as has occurred in many countries worldwide.

Furthermore, better nematode- and agronomic-management of the plantations should help coffee growers to maintain the crop's profitability wherever few destructive *Meloidogyne* species occur. In Brazil, this is possible in *M. exigua*-infested fields. Nonetheless, substantial yield improvements do not occur in *M. incognita*- or *M. paranaensis*-infested plantations.

In all coffee-producing regions, nematode management would certainly benefit from the selection of rootstocks resistant to the most important, if not all, *Meloidogyne* species parasitic to coffee. Nonetheless, coffee growers and all technical personnel should be aware of the climate adaptability of the rootstocks and cultivars available. For example, the rootstock 'Apoatã' is better adapted to hot, not mild, regions. This chapter's first author witnessed a grower using seedlings of 'Catuai' grafted onto 'Apoatã' to establish a plantation in a cold region of Minas Gerais. In this *M. paranaensis*-infested field, the plantation did not withstand two years because the cold climate inhibited rootstock growth, which then could not provide for the scion's growth. For that and other growers in a similar situation, not much is left besides giving up the coffee business. This emphasizes the urgency for the development of RKN-resistant arabica cultivars and rootstocks, with better adaptation to mild climates.

References

Almeida VF, Campos VP (1991a) Alternância de culturas e sobrevivência de *Meloidogyne exigua* em áreas de cafezais infestados e erradicados. Nematol Bras 15:30–42

Almeida VF, Campos VP (1991b) Reprodutividade de *Meloidogyne exigua* em plantas antagonistas e em culturas de interesse econômico. Nematol Bras 15:24–29

Alvarenga A (1973) Determinação preliminar da longevidade no solo do nematóide *Meloidogyne exigua*. Proceedings II Congr Bras de Pesquisas Cafeeiras:45

Amaral DF, Oliveira DF, Campos VP et al (2002) Efeito de alguns extratos vegetais na eclosão, mobilidade, mortalidade e patogenicidade de *Meloidogyne exigua* do cafeeiro. Nematol Bras 26:43–48

Anonymous (1968) Principles of plant and animal pest control. Vol. 4. Control of plant parasitic nematodes. National Academy of Sciences. Publication # 1966, Washington DC

Anonymous (2005) Compêndio de defensivos agricolas. 7th edition. Departamento de Defesa e Inspeção Vegetal do Brasil and Andrei Editora Ltda, São Paulo

Anzueto F, Bertrand B, Sarah JL et al (2001) Resistance to *Meloidogyne incognita* in Ethiopian *Coffea arabica* acessions. Euphytica 118:1–8

Baeza-Aragón CA (1978) Parasitismo de *Bacillus penetrans* en *Meloidogyne exigua* estabelecido en *Coffea arabica*. Cenicafé 29 94–97

Barbosa DHSG, Vieira HD, Souza RM et al (2007). Desenvolvimento vegetativo e reação de genótipos de *Coffea* spp. a uma população de *Meloidogyne exigua* virulenta a cultivares resistentes. Nematol Bras 31:1–6

Berthaud J (1978a) L'hybridation interspécifique entre *Coffea arabica* L. et *Coffea canephora* Pierre. Obtention et comparaison des hybrids triploids, Arabusta et hexaploides. Deuxième partie. Café Cacao Thé 22:87–112

Berthaud J (1978b) L'hybridation interspécifique entre *Coffea arabica* L. et *Coffea canephora* Pierre. Obtention et comparaison des hybrids triploids, Arabusta et hexaploides. Première partie. Café Cacao Thé 22:3–12

Campos HD, Campos VP (1997) Efeito da época e forma de aplicação dos fungos *Arthrobotrys conoides, A. musiformis, Paecilomyces lilacinus* e *Verticillium chlamydosporium* no controle de *Meloidogyne exigua* do cafeeiro. Fitopatol Bras 22:361–365

Campos VP (1997) Controle de doenças causadas por nematóides. In: Vale FXR, Zambolim L (eds) Controle de doenças de plantas. Vol. 1. Editora Universitária, Viçosa

Campos VP (2002) Coffee Nematode Survey in Minas Gerais State, Brazil. Research report to PNP & D/café. Embrapa, Brasilia

Campos VP, Lima RD (1986) Nematóides parasitas do cafeeiro. In: Rena AB, Malavolta E, Rocha M et al (eds) Cultura do cafeeiro, fatores que afetam a produtividade. Associação Brasileira para Pesquisa da Potassa e do Fosfato, Piracicaba

Campos VP, Lima RD, Almeida VF (1985) Nematóides parasitas do cafeeiro. Informe Agropecuário 11:50–58

Campos VP, Souza JT, Souza RM (1998) Controle de fitonematóides por meio de bactérias. Rev Ann Patol Plantas 6:285–327

Campos VP, Villain L (2005) Nematode parasites of coffee and cocoa. In: Luc M, Sikora RP, Bridge J (eds) Plant parasitic nematodes in subtropical and tropical agriculture. 2nd Edition. CABI, London

Capot J (1972) L'amélioration du caféier en Côte d'Ivoire. Les hybrides 'Arabusta'. Café Cacao Thé 16:3–18

Carneiro Filho F, Yamaguchi K (1995) Comportamento de progenies de *Coffea arabica* enxertados em *Coffea canephora* em área com nematóide *Meloidogyne coffeicola*, Lordello and Zamith, 1960, no Paraná. Proceedings XXI Congr Bras Pesquisas Cafeeiras:41–42

Carneiro RG, Carneiro RMDG (1982) Seleção preliminar de plantas para rotação de culturas em áreas infestadas por *Meloidogyne incognita* nos anos de 1979 e 1980. Nematol Bras 6:141–148

Curi SM, Silveira SGP, Elias EG Jr (1977) Resultados de produção e da proteção do sistema radicular de cafeeiros sob controle quimico do nematóide *Meloidogyne incognita* (Kofoid and White, 1969) Chitwood 1949, em condições de campo. Nematol Bras 2:93–99

da Costa WM, Gonçalves W, Fazuoli LC (1991) Produção de café Mundo Novo em porta enxertos de *Coffea canephora* em áreas infestadas por *Meloidogyne incognita* raça 1. Nematol Bras 15:43–50

de Moraes MV, Lordello LGE (1977) Estudo de três populações de nematóides nocivos do cafeeiro. Nematol Bras 2:249–255
de Moraes MV, Lordello LGE, Reis AJ et al (1977) Ensaio de rotação de culturas para reaproveitamento com cafeeiros de terras infestadas por *Meloidogyne exigua*. Nematol Bras 2:257–265
de Oliveira CMG (1996) Efeito de densidades populacionais de *Pratylenchus brachyurus* (Nemata: Pratylenchidae) no crescimento de plântulas de *Coffea arabica* cv. Mundo Novo e *Coffea canephora* cv. Apoatã. Escola Superior de Agricultura Luiz de Queiroz, MS dissertation, Piracicaba
Dutra MR (2004) Controle de nematóides de galhas, *Meloidogyne* spp com silicio. Universidade Federal de Lavras, DS thesis, Lavras
Dutra MR, Campos VP (2003a) Efeito do manejo do solo e da água na população de *Meloidogyne javanica* (Tremb, 1885) em quiabeiro no campo. Summa Phytopathol 29:249–254
Dutra MR, Campos VP (2003b) Manejo do solo e da irrigação como nova tática de controle de *Meloidogyne incognita* em feijoeiro. Fitopatol Bras 28:608–614
Dutra MR, Campos VP, Toyota M (2003) Manejo do solo e da irrigação para o controle de *Meloidogyne javanica* em alface. Nematol Bras 27:29–34
Fazuoli LC (1986) Genética e melhoramento do cafeeiro. In: Rena AB, Malavolta E, Rocha M et al (eds) Cultura do cafeeiro, fatores que afetam a produtividade. Associação Brasileira para Pesquisa da Potassa e do Fosfato, Piracicaba
Fazuoli LC, Medina Filho HP, Gonçalves W et al (2002) Melhoramento do cafeeiro: variedades tipo arabica obtidas no Instituto Agronômico de Campinas. In: Zambolim L (ed) O Estado da arte de tecnologias na produção de café. Editora UFV, Viçosa
Ferraz LCB, Rocha AD, Brancalion AM et al (1983) Considerações sobre a viabilidade do controle de *Meloidogyne incognita* visando a recuperação de cafezais infestados. Nematol Bras 6:117–123
Gonçalves W (1995) Problemas na produção brasileira de café devido a fitonematóides. Proceedings XIX Congr Int de Nematol Trop:216–223
Gonçalves W, Ferraz LCCB, Lima MMA et al (1996) Reações de cafeeiro às raças 1, 2 e 3 de *Meloidogyne incognita*. Summa Phytopathol 22:172–177
Gonçalves W, Silvarola MB (2001) Nematóides parasitos do cafeeiro. In: Zambolim L (ed) Tecnologias de produção de café com qualidade. Editora Universitária UFV, Viçosa
Haydock PPJ, Woods SR, Grove IG et al (2006) Chemical control of nematodes. In: Perry RN, Moens M (eds) Plant Nematology. CABI, London
Hidalgo-Diaz L, Bourne JM, Kerry BR et al (2000) Nematophagous *Verticillium* sp. in soils infested with *Meloidogyne* spp in Cuba: isolation and screening. Int J Pest Manag 46:277–284
Huang SP, Resende IC, Souza PE et al (1983) Effect of aldicarb, ethoprop and carbofuran on control of coffee root-knot nematode, *Meloidogyne exigua*. J Nematol 15:510–514
Jaehn A (1984) Recuperação da lavoura cafeeira recepada com utilização de *Crotalaria spectabilis*, torta de mamona e nematicidas, em área infestada por *Meloidogyne incognita*. Nematol Bras 8:257–264
Jaehn A, Cataneo A (1986) Uso de nematicida fumigante para expurgo de covas de cafeeiro de diferentes profundidades, com ou sem cobertura plástica, em área infestada por *Meloidogyne incognita*. Nematol Bras 10:191–206
Jaehn A, Rebel EK (1984) Sobrevivência do nematóide de galhas *Meloidogyne incognita* em substrato infestado, para produção de mudas de cafeeiro sadias. Nematol Bras 8:319–324
Jaehn A, Rebel EK (1984) Uso de palha de café, leguminosas e nematicida em mudas de cafeeiro, plantadas em área infestada por *Meloidogyne incognita*. Nematol Bras 8:308–317
Jaehn A, Rebel EK, Matiello JB (1984) Viabilidade de recuperação de mudas de cafeeiro infestadas por *Meloidogyne incognita* através de nematicidas. Nematol Bras 8:295–300
Lordello RRA, Lordello ALL, Martins ALM et al (1990) Plantio de cafezal em área infestada por *Meloidogyne exigua*. Nematol Bras 14:18–19
Matiello JB, Mendonça SM, Pimenta MLE et al (2004) Desfolha em cafeeiros recepados/esqueletados por efeito do complexo nematóide/cercospora na Zona da Mata de Minas Gerais. Proceedings XXX Congr Bras Pesquisas Cafeeiras: 3–4

Maximiniano C, Campos VP, Souza RM et al (2001) Flutuação populacional de *Meloidogyne exigua* em cafezal naturalmente infestado por *Pasteuria penetrans*. Nematol Bras 21:63–69

Nandi B, Sukul NC, Banerjee N et al (2002) Salicylic acid enhances resistance in cowpea against *Meloidogyne incognita*. Phytopathol Mediterr 41:39–44

Naves RL, Campos VP (1991) Ocorrência de fungos predadores de nematóides no Sul de Minas Gerais e estudo da capacidade predatória e crescimento in vitro de alguns de seus isolados. Nematol Bras 15:152–162

Otoboni CEM, Giroto F, Santos JM (2003) Efeito do tratamento quimico de solo em cafeeiros sobre o controle de nematóides, ferrugem e bicho-mineiro. Proceedings XXIX Congr Bras Pesquisas Cafeeiras:278–279

Otoboni CEM, Lemos EGM, Santos JM et al (2001) Influência do tratamento nematicida em cafeeiros de 8 anos de idade, infectados por *Meloidogyne* sp. sobre a eficácia do controle de ferrugem e bicho mineiro com aplicação via solo de tradimenol and dissulfoton. Proceedings XXIII Congr Bras Nematol:96

Owen KJ, Green CD, Deverall BJ (2002) A benzothiadiazole applied to foliage reduces development and egg deposition by *Meloidogyne* spp. in glasshouse-grown grapevine roots. Aust Plant Pathol 31:47–53

Paiva BRTL, Reis THP, Dutra MR et al (2006) Controle do nematóide *Meloidogyne exigua* com silicatos de cálcio e de potássio. Proceedings XXXII Congr Bras Pesquisas Cafeeiras:292

Paiva BRTL, Ribeiro Jr PM, Garcia ALA et al (2005) Controle do nematóide *Meloidogyne exigua* com silicato de cálcio. Proceedings XXXI Congr Bras Pesquisas Cafeeiras:383–384

Randig O, Medeiros CAB, Sperandio CA (1998) Efeito da desinfestação do solo pelo uso da energia solar sobre nematóides. Nematol Bras 22:1–11

Rebel EK, Gonçalves JC, Lordello LGE (1976) Considerações sobre o comportamento de *Meloidogyne coffeicola* em mudas, cafezais novos e em recepados. Proceedings IV Congr Bras Pesquisas Cafeeiras:11–12

Ribeiro RCF, Campos VP (1993) Isolamento e identificação e efeito de temperatura no crescimento 'in vitro' de fungos parasitos de ovos de *Meloidogyne* spp. no Sul de Minas Gerais. Nematol Bras 17:132–138

Salgado SML, Campos VP (2003a) Eclosão e mortalidade de *Meloidogyne exigua* em extratos e produtos naturais. Fitopatol Bras 28:166–170

Salgado SML, Campos VP (2003b) Extratos naturais na patogenicidade e reprodução de *Meloidogyne exigua* em cafeeiro e de *Meloidogyne incognita* raça 3 em feijoeiro. Nematol Bras 27:41–48

Salgado SML, Campos VP, Cardoso MG et al (2003) Eclosão e mortalidade de juvenis de segundo estádio de *Meloidogyne exigua* em óleos essenciais. Nematol Bras 27:17–22

Santiago DC, Homochin M, Krzyzanowski AA et al (2001) Efeito antagônico de *Arachis pintoi* sobre *Meloidogyne incognita* Kofoid and White, 1919 Chitwood, 1949, raças 1, 2, 3 e 4. Proceedings XXIII Congr Bras Nematol:150

Sharma RD, Lordello RRA (1992) Occurrence of *Pasteuria penetrans* in coffee plantations infested by *Meloidogyne exigua* in the State of São Paulo. Proceedings XXV Congr Bras Fitopatol:183

Silva LHCP, Campos JR, Dutra MR et al (2004) Aumento da resistência de cultivares de tomate a *Meloidogyne incognita* com aplicação de acibenzolar-S-metil. Nematol Bras 28:199–206

Tronconi NM, Ferraz S, Santos, JM et al (1986) Avaliação do efeito da palha de café, misturado ao solo, no desenvolvimento de *Meloidogyne exigua* Göldi 1887, em mudas do cafeeiro. Nematol Bras 10:85–102

Van Gundy SD, McKenry MV (1977) Action of nematocides. In: Horsfall JG, Cowling EB (eds) Plant disease – An advanced treatise, Volume 1. Academic Press, New York

Volpato AR, Otoboni CEM, Otoboni JAM et al (2001) Eficácia dos produtos casudafós, carbofuran e carbosulfan no controle de *Meloidogyne exigua* e *M. coffeicola* no cafeeiro. Proceedings XXIII Congr Bras Nematol:92

Chapter 9
Genetics of Resistance to Root-Knot Nematodes (*Meloidogyne* spp.) and Breeding

Benoît Bertrand and François Anthony

Abstract Genetic control of root-knot nematodes (RKNs) is an essential part of an integrated pest management as the use of resistant cultivars or rootstocks constitutes an easy, inexpensive, non-polluting method of control. This chapter presents the results achieved in understanding the genetic basis of coffee resistance to *Meloidogyne* spp. in Latin America, and in using genetic resistance in breeding programmes. The context of breeding for improving coffee resistance is firstly described. This is followed by an overview of works published on the identification of resistance sources among the genetic resources preserved in collections worldwide. The methods of resistance evaluation are discussed, and a standardized method is proposed in order to improve the reliability of resistance evaluation trials. The results obtained from studies on the genetics of resistance to RKNs are then given for the main species that parasitize coffee. So far, only one resistance gene has been identified and mapped in the coffee genome, the gene *Mex*-1 of resistance to *M. exigua*. The advent of large-scale molecular genomics will provide an access to previously inaccessible sources of genetic variation which could be exploited in breeding programmes. Strategies for using resistance sources are finally proposed in the context of coffee breeding.

Keywords Breeding · coffee · genetics · resistance gene · *Coffea*

9.1 Introduction

Arabica coffee (*Coffea arabica* L.) cultivation may have started in the species' centre of origin, in southwestern Ethiopia, around the fifth to eighth century. Modern coffee cultivars are derived from two base populations, known as Typica and Bourbon, which were disseminated worldwide in the eighteenth century (Anthony

B. Bertrand
Centre de Coopération Internationale en Recherche Agronomique pour le Développement, UMR RPB, TA A-98/IRD, Montpellier, France
e-mail: benoit.bertrand@cirad.fr

et al., 1999). Historical data indicate that these populations were composed of progenies of very few plants, i.e. only one for the Typica population. Breeders exploited these narrow genetic bases, resulting in Typica- and Bourbon-derived cultivars with a uniform agronomic performance and limited adaptability (Bertrand et al., 1999). In the twentieth century, the extension of coffee cultivation and the intensification of its production revealed their susceptibility to many pests [e.g. nematodes, the 'coffee berry borer' (*Hypothenemus hampei* Ferrari)] and diseases (e.g. 'leaf rust', caused by *Hemileia vastatrix* Berk and Br. and 'coffee berry disease', caused by Colletotrichum kahawae Waller and Bridge).

Natural interspecific hybrids between *C. arabica* and *C. canephora* Pierre ex A. Froehner (robusta coffee) (e.g. Timor Hybrid) or *C. liberica* W. Bull ex Hiern (e.g. S.26) were the first sources of resistance to 'leaf rust'. Other interspecific hybrids were created afterwards. Pedigree selection of those progenies led to the dissemination of introgressed lines resistant to leaf rust, called 'Catimor', 'Sarchimor', 'Icatu' and 'S.795', among others.

Root-knot nematodes (RKNs) (*Meloidogyne* spp.) are a major threat in the world's main coffee producing countries. The development of two international research projects funded by the European Commission (International Cooperation with Developing Countries), in 1997–2000 and 2002–2005, resulted in the definition of the natural diversity of *Meloidogyne* spp. parasitizing coffee in Latin America, and new species were collected (Carneiro et al., 2004; Hernandez et al., 2004a). In addition, specific enzymatic and molecular markers were used to complement taxonomic identification based on morphological traits (Carneiro et al., 2000; Carneiro et al., 2004). Confusion then appeared in the identification of certain isolates, which made it difficult to compare results between research centres. Seventeen species of *Meloidogyne* are now acknowledged as parasitic to coffee (see Chapter 6). Economic losses due to RKNs vary considerably depending on the species involved and its distribution. That information is essential for defining control policies and prioritizing targets.

Genetic control of diseases and pests is an essential part of integrated control, as it offers the advantage of being an easy, inexpensive, non-polluting control method, usually requiring no change in cultural practices (Luc and Reversat, 1985). Two strategies can be developed against coffee-parasitic nematodes: selection of resistant cultivars 'on their own roots' and/or resistant rootstocks. To achieve that, identification of molecular markers near resistance genes is a useful solution for controlling introgressions and thereby assisting the selection of improved cultivars (Lashermes and Anthony, 2007). This chapter describes the results achieved in understanding the genetic bases of coffee resistance to RKNs, and in using genetic resistance in breeding programmes. The first part is devoted to the context of breeding for improving coffee resistance to RKNs. The second part is an overview of works published on the identification of sources of resistance among the genetic resources preserved in collections worldwide. The third part describes the results obtained in the genetics of resistance to RKNs. The final part proposes strategies for using sources of resistance to improve cultivars.

9.2 Context of Breeding for Resistance Improvement

9.2.1 Origin of Cultivars

Coffee growing remained a monopoly of the Arabs on the shores of the Red Sea up to the fifteenth century, after a strong expansion in South Arabia (now Yemen) in the fourteenth century, and in the Middle East during the following century (see review by Anthony et al., 1999). Coffee trees were disseminated to the rest of the world from Yemen at the beginning of the eighteenth century. Two base populations are recognized for their strong impact on coffee growing, identified under the names Typica and Bourbon (Krug et al., 1939). The Typica population originated from a single plant from Indonesia that was subsequently cultivated in Amsterdam and Paris, whilst the Bourbon population came from several individuals introduced on the island of Bourbon (now Réunion). The individual from which the Typica population originated played an exceptional role in the history of varietal creation, as it gave birth to most of the world's cultivars up to the middle of the twentieth century (Carvalho, 1946). However, the cultivars derived from the Bourbon population proved to be more productive than those derived from Typica, which led to the latter being gradually less cultivated (see review by Bertrand et al., 1999). A molecular analysis of genetic diversity and polymorphism confirmed the low polymorphism in both populations, particularly in the Typica one (Anthony et al., 2002). The results also showed that there was little differentiation in the populations, which explains the genetic limitations encountered in traditional breeding programmes. Today, the world's most widely cultivated cultivars (i.e. 'Caturra', 'Catuai', 'Mundo Novo') are derived from those two populations, and their susceptibility to most parasites and diseases casts doubt on the sustainability of modern, pesticide-consuming coffee production systems.

Unlike the cultivars derived from Typica and Bourbon, wild coffee trees collected from the centre of diversity of *C. arabica* (e.g. Ethiopia), have been shown by molecular markers to have relatively high polymorphism (Anthony et al., 2001). Wild coffee tree accessions were recently used as parents to produce F1 hybrids by crossing them with cultivars (Bertrand et al., 2005). The F1 hybrid families produced between 20 and 50% more than the cultivars, which were used as the female parent in the crosses.

Breeding programmes based on the selection of F1 hybrids are an interesting alternative to traditional pedigree selection, by reducing the duration of selection to one generation and enabling multiple-trait selection. In particular, it is possible to accumulate the genetic resistance of both parents in an individual. However, Ethiopian wild *C. arabica* coffee trees have been found to have little resistance to biotic stresses, which explains their limited use for breeding purposes (see review by Anthony et al., 1999). That explains why breeders had to exploit resistance genes existing in other cultivated coffee species to control 'leaf rust'. Some progenies of a natural interspecific hybrid (*C. arabica* × *C. canephora*) known as the Timor Hybrid (Bettencourt, 1973) were selected and gave rise to introgressed lines in generation F5-F7, known under the generic names of Catimor and Sarchimor (see review by Bertrand et al., 1999). The progenies of other interspecific hybrids also underwent selection for resistance to 'leaf-rust', primarily the 'Icatu' hybrids (*C. arabica* × *C. canephora*) in Brazil and the 'S.795' hybrid (*C. arabica* × *C. liberica*) in India. Current resistance breeding programmes are geared against 'coffee berry disease', which causes serious damage in Africa, and to RKNs, which are the subject of this chapter.

9.2.2 Diversity of the Soil-borne Pathogen Complex on Coffee Trees

Coffee is a host for several nematode genera and species worldwide. In Latin America, root-lesion nematodes (*Pratylenchus* spp.) and RKNs are of major concern. In the tropics, these nematodes frequently occur associated in coffee roots (Luc and Reversat, 1985). Hence, when resistant cultivars are employed, the mixture of nematode species in the field has to be assessed. In addition, nematodes are often associated with other pathogens, fungi and/or bacteria, which may increase damage considerably (Powell, 1971). For example, Negrón and Acosta (1989) demonstrated the existence of a complex pathology involving *Fusarium oxysporum* (Schltdl.) Snyder and Hansen and *M. incognita* (Kofoid and White) Chitwood. Recently, Bertrand et al. (2000a) showed that *F. oxysporum* and *M. arabicida* López and Salazar lay behind an aetiology called 'corchosis'.

RKNs are sedentary endoparasites, i.e. they need to be harboured by a host plant to complete their biological cycle. They are also polyphagous, being able to parasitize several wild or cultivated plant species. Generally, the coffee-parasitic species present a parthenogenetic reproduction system (see Chapter 7), which in theory could limit the appearance of diversity in each generation. As nematode motility is highly limited, few exchanges occur between populations under natural conditions. Human activity would thus seem to be the main factor in disseminating these parasites.

Recent studies by Semblat et al. (2000) suggest that the parthenogenetic reproduction method does not prevent an evolution that results in a species's substantial genetic diversity, as it occurs with those with sexual (amphimitic) reproduction method. In parthenogenic species, the evolution is probably based on mutations that occur in line with the diversity of the environment (host plant, physical conditions) in which the parasite develops. This mechanism, which ensures the species's survival, results in substantial variability, as revealed by molecular markers (Semblat et al., 2000). The populations that specialize and develop on preferential hosts or environments are called 'pathotypes' (Dropkin, 1988) or 'biotypes' (Triantaphyllou, 1987).

9.2.3 Origin of the Coffee Tree/Nematode Pathosystem in Latin America

The origin of the coffee-*Meloidogyne* pathosystem in Latin America dates from less than 200 years ago, when large-scale coffee cultivation began. This pathosystem involves several nematode species but only a narrow genetic base of arabica coffee. Hence, three hypotheses can be put forward:

1) The base populations (Typica and Bourbon) were susceptible at the time of their introduction, as coffee is a host for other *Meloidogyne* species in its centre of origin (Whitehead, 1969).
2) Several indigenous species of *Meloidogyne* mutated and adapted to the new host.
3) RKN are highly polyphagous parasites and parasitized coffee without having to overcome any resistance mechanisms.

The third hypothesis has often been put forward. Yet it does not explain the existence of *M. incognita* populations (also called physiological races) with little or no ability to parasitize coffee, nor the existence of resistance sources to *M. incognita* in *Coffea* sp. The few examples where those same cultivars were inoculated with an RKN that was, in theory, not known to be parasitic to coffee resulted in incompatible interactions. Such is the case with *M. javanica* (Treub) Chitwood taken from tomato roots in the USA (Araya and Caswel-Chen, 1995) and with a *M. incognita* isolate also taken from tomato in Costa Rica (Hernández, 1997). It is therefore likely that compatible interactions result from RKN populations adapting to coffee as a new host. That adaptation would appear to occur all the more quickly the greater the environmental pressure.

Modern agriculture, which generally employs cultivars with substantial genetic homogeneity and cultural practices that reduce the diversity of the soil's microfauna, is conducive to the emergence of populations adapted to new hosts. The dispersal of virulent populations would then appear to be promoted by human activity. These two factors combine to make cultivars remarkably susceptible hosts (Trugdill and Blok, 2001). It is interesting to note that two 'biotypes' of the same species may acquire the same virulence in relation to the same host. Recent results obtained using AFLP analyses (Semblat et al., 2000) on tomato-parasitizing RKNs show that DNA polymorphism (such as it is detected) is independent of population virulence. In other words, it would be possible to find very close or even genotypically similar populations differing solely in their virulence.

9.2.4 Definition of Priorities in Nematode Control

Reliable inferences of the losses caused by RKNs are essential to guide coffee breeders in their choice of the resistances that need to be improved with higher priority. Figures for economic losses can be estimated in microplots, and projected for the total area (in hectares) infested by the nematode. In practice, it is often difficult to obtain a reliable estimate of the two terms of this equation. On the one hand, it is difficult to estimate average losses in the microplots, and it is complex to link economic losses to the extent of changes and/or damages caused in the roots by RKNs. On the other hand, the hectarage infested by RKNs is not always known with certainty. Sampling operations conducted in a few countries have shown that large areas are involved: at least 40% and 54% of the coffee growing areas are infested by

RKNs in Guatemala and El Salvador, respectively (A. Hernández, Procafe, personnal communication; Villain et al., 1999). In Guatemala, Alvarado (1997) suggested a 20% drop in production due to nematodes in one of the most important regions of the country. Infestation and yield loss figures for several other countries are given elsewhere in this book (Part V).

Given the high cost of breeding programmes for perennial plants such as coffee, the choice of resistances to be improved has to be reasoned from data on the distribution of the nematode species and on the damage caused in plantations (Table 9.1). In Latin America, *M. exigua* Göldi is the most widespread species but with low severity at plant level. The symptoms are limited to the development of numerous galls of various sizes (Fig. 9.1). In Costa Rica, a 10–20% drop in productivity was estimated by comparison of plantations with susceptible and resistant cultivars (Bertrand et al., 1997). However, in southeast Brazil the extent of productivity decrease was found to be variable according to the management level: low in poorly or just fairly managed plantations *vs.* high in the best managed plantations (Barbosa et al., 2004). In contrast, *M. arabicida* associated with *F. oxysporum*, causing 'corchosis' symptoms (Fig. 9.2), cause serious damage in plantations, leading to destruction of 40–80% of the root system of susceptible coffee trees five years after planting (Fig. 9.3) (Bertrand et al., 2000a). *M. paranaensis* Carneiro, Carneiro, Abrantes, Santos and Almeida and *M. incognita* are the species that cause most concern due to their vast distribution in Brazil, Central America and Hawaii (USA) (Table 9.1). Lastly, *M. arenaria* (Neal) Chitwood and *M. izalcoensis* Carneiro, Almeida, Gomes and Hernandez in El Salvador cause serious damage, but seem to be relatively limited in distribution (Carneiro et al., 2004; Hernandez et al., 2004b).

9.3 Identification of Sources of Resistance

The search for sources of resistance among the genetic resources available in genebanks is a prior step for studying the heritability of resistance and its use in breeding. The origin of resistance genes, their frequency and how they are transmitted are essential elements for defining a breeding programme. In fact, transferring resistance genes into cultivars may prove to be a relatively difficult task.

9.3.1 Genetic Resources Conserved in Coffee Genebanks

Coffee genebanks constitute a valuable source of resistance genes since approximately 120 species have been identified in the genera *Coffea* and *Psilanthus* (Bridson, 1987; Davis et al., 2005; 2006), and new species are still being discovered (Anthony et al., 2006). Although coffee species display considerable variation in morphology and ecological adaptation, they hybridize readily with one another and produce interspecific hybrids that are more or less fertile, even between species belonging to different genera (Couturon et al., 1998). Genetic material can thus

Table 9.1 Main species of root-knot nematodes observed on *Coffea arabica* in Latin America, classified according to the symptoms and the severity of damage they cause: + = root galls induced and decrease in production; ++ = root galls and necrosis, with generalized weakening of the tree, and some deaths; + + + = root galls and corky roots ('corchosis') induced, with generalized weakening of the tree, followed by death

Species	Damage severity	Association with other pathogens	Countries where reported	Distribution	References
M. exigua	+	Not observed	Brazil	Wide	Lordello (1972)
			Venezuela	Wide	Esbenshade and Triantaphyllou (1985)
			Colombia	Wide	
			Costa Rica	Wide	Flores and López (1989)
			Nicaragua	Wide	Hernandez et al. (2004a)
			Honduras	Wide	J.L. Sarah (not published)
			Dominican Rep.	Wide	Luc Villain (Cirad, pers. comm.)
M. hapla	+	Not observed	El Salvador	Limited	Hernández (2004)
M. arenaria	++	Not observed	El Salvador	Limited	Bertrand et al. (2000b)
			Guatemala	Limited	Carneiro and Villain (not published)
M. izalcoensis	++	Not reported	El Salvador	Limited	Hernandez et al. (2004a)
					Carneiro et al. (2005)
M. incognita	+ to + + +[a]	*Fusarium oxysporum*	Brazil	Wide	Esbenshade and Triantaphyllou (1985)
			Puerto Rico	Limited	Negrón et al. (1989)
					Carneiro et al. (2000)
M. arabicida	+ + +	*F. oxysporum*	Costa Rica	Limited	López and Salazar (1989)
					Bertrand et al. (2000a)
M. paranaensis	+ + +	*Fusarium* sp. suspected	Brazil	Wide	Carneiro et al. (1996)
			Guatemala[b]	Wide	Carneiro et al. (2004)

[a] Four races of *M. incognita* have been reported. It seems the severity varies according to the race.
[b] Isolate collected in the 1990 and identified as *Meloidogyne* sp., and later as *M. incognita*.

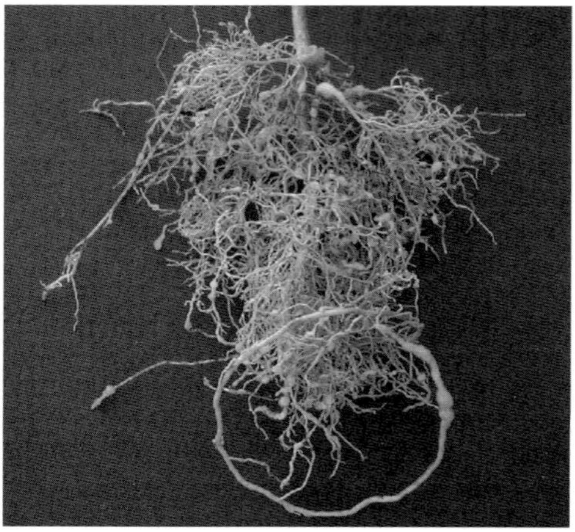

Fig. 9.1 Root system of a *C. arabica* 'Caturra' seedling susceptible to *Meloidogyne exigua*, showing numerous galls of different sizes (Photo by F. Anthony) (see color Plate 9, p. 323)

Fig. 9.2 Root system of a *C. arabica* 'Caturra' seedling susceptible to *Meloidogyne paranaensis*, showing symptoms of 'corchosis' on the main root (Photo by F. Anthony) (see color Plate 10, p. 323)

Fig. 9.3 Coffee plantation affected by *Meloidogyne arabicida*, with several dead trees in the foreground (Photo by F. Anthony) (see color Plate 11, p. 324)

be transferred from wild plants into cultivars, at either intraspecific or interspecific levels. Worldwide, the efforts of breeding programmes are now being turned to transference of resistance genes from wild *C. arabica* coffee trees or other species (see below).

Most coffee genebanks were established during the first half of the twentieth century, the oldest being the Indonesian Coffee and Cocoa Research Institute (1900), the Agronomic Institute of Campinas in Brazil (1924), and the Central Coffee Research Institute in India (1925) (van der Vossen, 2001). Coffee growers supplied genebanks with materials displaying good agronomic performance or specific traits. Many mutants were isolated from the Typica and Bourbon populations, as well as numerous cultivars and homozygous lines of *C. arabica* and clones of *C. canephora*. The interest in wild plants increased during the second half of the twentieth century, when breeders became aware that deforestation was causing destruction of coffee habitats, thereby threatening its genetic resources. Given the socio-economic importance of *C. arabica* cultivation, two large surveys were organized in the species' centre of diversity (Ethiopia) in 1964/65 (Fernie, 1968) and in 1966 (Guillaumet and Hallé, 1978). The collection of other species began at the same time in the Madagascar region, then followed in seven African countries between 1975 and 1987 (Anthony et al., 2007). At least 11,700 accessions representing 70 *Coffea* species were collected and conserved in only two field genebanks, namely in Madagascar for the *Mascarocoffea* species and in the Ivory Coast for the African mainland species. Only a few genotypes of those African and

Madagascan species have been spread worldwide and are available for breeding programmes.

Genetic diversity has been assessed in the species of agronomic interest, *C. arabica* and *C. canephora*. Using molecular markers, wild accessions of *C. arabica* were classified in four genetic groups that clearly differ from cultivars derived from the Typica and Bourbon populations (Anthony et al., 2001). One group included the accessions from southwest Ethiopia, while the other groups contained accessions from east and south Ethiopia. The genetic structure thus seems to be arranged in two large complexes separated by the tectonic rift that cuts through Ethiopia from the northeast to the southwest. Such a structure was also suggested on the basis of an agro-morphological study (Montagnon and Bouharmont, 1996).

In *C. canephora*, five genetic groups were identified using isozyme and molecular markers (Dussert et al., 2003). A differentiation was found between plants originating from West (the Guinean group) and Central Africa (the Congolese group), the latter being structured in several subgroups. The use of agro-morphological markers allows us to characterize accessions from two or three groups, depending on the material studied (Montagnon et al., 1992; Leroy et al., 1993; Dussert et al., 2003). However, only part of the known diversity has been conserved in each genebank, as recently shown in a coffee genebank in India (Prakash et al., 2005). This has dramatically limited the characterization and use of corresponding genetic resources in breeding programmes.

9.3.2 Methods of Resistance Evaluation

Reliable assessment methods are essential for studies of genetic factors of resistance, such as genes and Quantitative Trait Loci, among others. To obtain reliable data, one must control the conditions under which coffee plants grow and the nematode is inoculated. Once resistance has been confirmed under controlled conditions, the coffee trees should be evaluated in infested plots to assess the resistance efficacy in the field. Although field trials are necessary, their results can be misinterpreted when several *Meloidogyne* species or disease complexes are present in the soil.

9.3.2.1 Inoculum Preparation and Inoculation

For the last 10 years, most of the results published on the genetics of coffee resistance to RKNs were obtained using clonal nematode populations, which were established from single egg masses laid by single females. Using a clonal inoculum ensures good repeatability of experiment results.

Two methods have proved to be effective in extracting RKNs from coffee roots: centrifugation-flotation (Taylor and Sasser, 1978) and nebulization (Barker, 1985). The centrifugation-flotation method can be used to extract nematodes at all stages of development (eggs, juveniles and adults), whereas the nebulization method can only be used to extract young, second-stage juveniles (J2). To facilitate extraction by centrifugation, exposure to sodium hypoclorite is often used. However, it affects the physiological condition of the ecloded nematode stages, and only 20% of extracted eggs hatch into infectious J2 (Hussey and Barker, 1973). On the other hand, infectious J2 extracted by nebulization are very active and easy to count, allowing a precise quantification of the inoculum. The nebulization method is therefore recommended for inoculum preparation, but it requires greater investment in facilities (Fig. 9.4).

The number of eggs and/or juveniles present in the inoculum applied *per* plant varies considerably among published works, making it difficult to compare data. The inoculum most frequently applied contains two to three thousand nematodes (eggs and J2)/300 ml-pots. It was at this dose that the highest rate of *M. exigua* reproduction was observed 100 days after inoculation (Gonçalves, 1998), but doses below one thousand nematodes were not tested in the experiment. When the inoculum contains J2 only, it can be calibrated to 800–1,000 nematodes *per* pot.

9.3.2.2 Assessing Resistance

The resistance assessment method most frequently used for coffee plants is based on a visual estimation of the number of root galls on plants growing in greenhouse. The data are then grouped into five, six or 11 classes to form a gall index (GI); for example, in the five class-index proposed by Taylor and Sasser (1978), 0 = no galls, 1 = one or two galls, 2 = three to 10 galls, 3 = 11–30 galls, 4 = 31–100 galls, and 5 = more than 100 galls. A correspondence can be established between GI and the percentage of galled root system (Fig. 9.5). The six-class index offers the advantage of clearly distinguishing between highly resistant plants (GI $= 0-2$) and highly susceptible ones (GI = 4–5).

Fig. 9.4 Nebulization room for extraction of infectious *Meloidogyne* sp. juveniles. The infected roots are cut in 5 mm long segments and placed on a sieve nested onto a funnel, to facilitate nematode descent to the bottom of the white flasks (Photo by P. Topard, with permission) (see color Plate 12, p. 324)

For *M. exigua*, Anzueto et al. (2001b) found that susceptible plants displayed indexes three, four and five (on the five-class index), but not indexes zero, one or two. Therefore, the plants of indexes zero, one or two were considered to be resistant. Other GI shown in Fig. 9.5 enable a breakdown of plant resistance into 'highly resistant' (corresponding to less than 25% of galled root system), 'moderately resistant' (25–50%), 'moderately susceptible' (50–75%) and 'highly susceptible' (75–100%).

The assessment of resistance using GI takes from three to six months for experiments conducted in pots, and at least 15 months for assessments in the field. Extending these time frames facilitates gall observation, as they become more numerous and larger. The ease with which galls can be seen also depends on the nematode species considered. The number of galls induced by *M. exigua* can be estimated more easily than those induced by *M. arabicida*, since in the former the galls reach up to 7 mm in diameter, as opposed to 1–3 mm in the latter (Bertrand et al., 2000a). That problem can be overcome if one assesses the proportion of galled roots, rather than the gall number and/or size (Barker, 1985). For nematodes associated with bacteria or fungi, the resistance assessment can be based on the proportion of roots displaying necrosis – rather than the GI – which allows for the establishment of a damage index. Lastly, egg masses can be counted in root tissues under a dissecting microscope.

Although *M. exigua* induces large, easily seen galls in susceptible coffee, difficulties were faced in the evaluation of an F2 population based on a GI (Noir et al., 2003). In this work, four plants were classified ambiguously by the number of galls, in a five-class index. These were two plants with a GI of zero, which did not have any of the molecular markers linked to the resistance gene, and two plants with an index of four and five, both of which displayed the markers of the gene. In such cases when no reliable data can be obtained by repeated observations, it is advisable to combine a GI with another assessment criterion (e.g. nematode reproduction).

In coffee roots the nematode reproduction can be estimated by two methods, centrifugation-flotation (Taylor and Sasser, 1978) and nebulization (Barker, 1985). Nematode extraction by centrifugation-flotation is laborious due to successive handlings of centrifugation deposits and supernatants, and of filtrates recovered from sieves. The nebulization method is simpler to use, but two or even three counts,

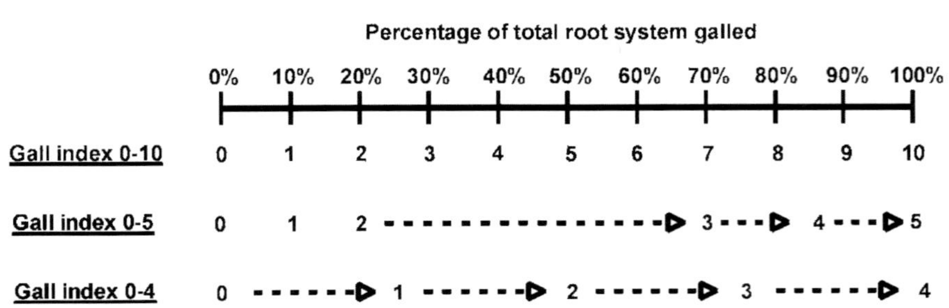

Fig. 9.5 Correspondence between the gall indexes mostly used to assess coffee tree resistance to *Meloidogyne* sp. (adapted from Barker, 1985)

one week apart, are necessary. Counting data can be converted into a reproduction index (Taylor, 1967), corresponding to the percentage of nematodes extracted from the plants being assessed in comparison to plants used as susceptible controls. Plants with a reproduction index below 10% can be considered as resistant. It is also possible to calculate a reproduction factor by the rate between the number of nematodes extracted and the number of those inoculated. Plants with a reproduction factor below 1 can be considered resistant. However, it is necessary to optimize inoculation conditions to reduce nematode losses.

In conclusion, it is difficult to define a standardized method of coffee resistance assessment for all *Meloidogyne* species because of the differences in their biology. For RKNs that induce large, easily seen galls (e.g. *M. exigua*), the best method would be a combination of gall number and reproduction factor estimation. This has proved to be useful to resolve certain ambiguities when classifying plants as resistant or susceptible (unpublished data). For the majority of RKNs which induce small galls (e.g. *M. paranaensis*), an estimation of the reproduction factor appears more pertinent than a GI since no clear relation has been established between both criteria (unpublished data). Regardless of the RKN considered, repeated experiments are required in order to estimate variations due to uncontrolled factors that could affect plant development and nematode inoculation in greenhouse. Data on experimental variations are indispensable for the development of genetic studies involving hybrid populations where each plant differs genetically from the others.

9.3.3 Resistance Identified in Genetic Resources

The screening results of the genetic resources available in field collections are described below for the different species of RKNs, beginning with those about which most studies have been published.

9.3.3.1 Resistance to *M. exigua*

Resistance to *M. exigua* has been evaluated in the coffee genebanks of several countries spread throughout the range of the species: Brazil, Colombia and Central America. No resistant accession has been found in *C. arabica*, in several cultivars ('Caturra', 'Catuai', 'Mundo Novo', among others), nor in wild coffee trees collected in Ethiopia (Curi et al., 1970; Fazuoli and Lordello, 1978; Arango et al., 1982; Bertrand et al., 1995). Research efforts have also drawn a blank in the little known species *C. pseudozanguebariae* Bridson and *C. sessiliflora* Bridson (Anthony et al., 2003). On the other hand, several resistant accessions have been identified in *C. canephora* and in some progenies of interspecific hybrids (*C. arabica* × *C. canephora*) (Curi et al., 1970; Bertrand et al., 1995; 1997; Gonçalves and Pereira, 1998; Silvarolla et al., 1998; Anthony et al., 2003). Some accessions resistant to *M. exigua* have also been identified in the species *C. racemosa* Ruiz and Pav. (Fazuoli, 1975; Anthony et al., 2003).

9.3.3.2 Resistance to *M. paranaensis*

Work on coffee resistance to *M. paranaensis* began in Guatemala before the description of this species from an isolate found in Brazil (Carneiro et al., 1996). Consequently, publications prior to that date, and a few after it, refer to *Meloidogyne* sp. or *M. incognita* for an isolate collected in Guatemala at the beginning of the 1990s (Carneiro et al., 2004). In Brazil, *M. paranaensis* was mistaken as *M. incognita* for more than 20 years (Carneiro et al., 1996). Two phenotypes of the esterase enzyme system and specific molecular markers allow now for distinction of *M. paranaensis* collected in Brazil from those of Guatemala (Carneiro et al., 2004). Once the taxonomy was clear, an analysis of the publications showed that accessions resistant to the isolate from Guatemala exist in the species *C. arabica* and *C. canephora* (Anzueto et al., 1993; 2001a; Bertrand et al., 2000b). In *C. arabica*, all cultivars have been considered susceptible to *M. paranaensis*, whereas numerous wild coffee trees from Ethiopia were considered resistant (Anzueto et al., 1991; 2001a). In *C. canephora*, it is not yet possible to state the proportion of resistant plants as only a small number of individuals have been assessed to date.

9.3.3.3 Resistance to *M. arabicida*

M. arabicida was described in Costa Rica by López and Salazar (1989). Since then few results have been published on this species due to its limited distribution. In plantations, this nematode is often associated with *F. oxysporum*, which causes a complex disease known as corky-root disease or 'corchosis' (Bertrand et al., 2000a). The symptoms of this disease are not found in assessments involving one or other of the two pathogens. Selection for resistance to *M. arabicida* appears to be an effective control strategy against 'corchosis' (Bertrand et al., 2002). In fields infested by both pathogens, resistant accessions (i.e. without root galls and 'corchosis' symptoms) have been found in wild *C. arabica* and *C. canephora* coffee trees. In another evaluation under controlled conditions, seven out of 16 accessions were considered resistant among wild *C. arabica* coffee trees from Ethiopia (Anthony et al., 2003). In *C. canephora*, resistance seems to exist at a high frequency in the species' main genetic groups (i.e. Guinean and Congolese) (Anthony et al., 2003). No accession resistant to *M. arabicida* has been identified in *C. pseudozanguebariae* and *C. sessiliflora* (Anthony et al., 2003).

9.3.3.4 Resistance to *M. incognita*

M. incognita has been reported in Brazil, El Salvador, Puerto Rico and Hawaii, but little work has been published on the search for resistance to this species. Four host races have been acknowledged in *M. incognita*, but only two phenotypes have been revealed in the esterase enzyme system, one for races one and four, and another for races two and three (Carneiro et al., 2000). These two phenotypes can also be distinguished by molecular markers (Carneiro et al., 2004). No *C. arabica* accession has proved to be resistant to *M. incognita* race three (Gonçalves and Ferraz, 1987), and a few 'Icatu' introgressed lines have been found to be tolerant to race two in the field (Carneiro, 1995). Accessions resistant to race one have been identified in *C. canephora* (Gonçalves et al., 1996). Lastly, a wild *C. arabica* tree and a line derived from the Timor Hybrid have been considered resistant to an *M. incognita* isolate in Brazil, for which the race was not specified (Hernandez et al., 2004b).

9.3.3.5 Resistance to Other RKNs

M. arenaria is easily identifiable by morphological traits, enzymatic and molecular markers (Carneiro et al., 2000; 2004). In El Salvador no resistance assessments have been published due to the limited distribution of this species. An assessment of *C. canephora* progenies derived from controlled crosses revealed variable degrees of resistance depending on the progenies, suggesting that it is possible to select resistant accessions to *M. arenaria* (Bertrand et al., 2000b). A resistant accession has been identified in wild *C. arabica* coffee trees (Hernandez et al., 2004b), showing that resistance is not limited to *C. canephora*.

Lastly, two RKN isolates from El Salvador were reported by Carneiro et al. (2004) and later described as *M. izalcoensis* (Carneiro et al., 2005). The search for resistance to this species remains to be done.

In conclusion, coffee genetic resources are a considerable reservoir of genes for breeding. Wild *C. arabica* and *C. canephora* coffee trees have expressed high levels of resistance to the main coffee-parasitic species of RKNs. Hence, a genetic solution can be applied to control these parasites by using resistant accessions as parents, or by creating rootstock cultivars. The successful transfer of resistance genes into cultivars will depend on how they are expressed (i.e. dominant, recessive or co-dominant), and the varietal creation scheme adopted (i.e. hybrids F1, backcrossing or successive selfing).

9.4 Genetics of Resistance

9.4.1 The Methodology for Revealing the Genetic and Molecular Bases of Resistance

Revealing the genetic bases of nematode resistance requires having an appropriate planting material. This can be obtained, for example, from a cross between a susceptible cultivar and a resistant accession identified in genebanks. The F1 hybrids can then be selfed (F2 population) or backcrossed with one of the parents (BC population), in order to produce a segregating population for the trait 'resistance to RKN' (Fig. 9.6). The frequency of resistant plants in such segregating populations will depend on its origin (selfing or backcrossing) and on the genetic determinism of the resistance trait. Once the F2 or BC plants have been evaluated, it is possible to compare DNA from resistant and susceptible plants in order to identify markers discriminating both statuses. The evaluation of at least 100 F2 or BC plants is

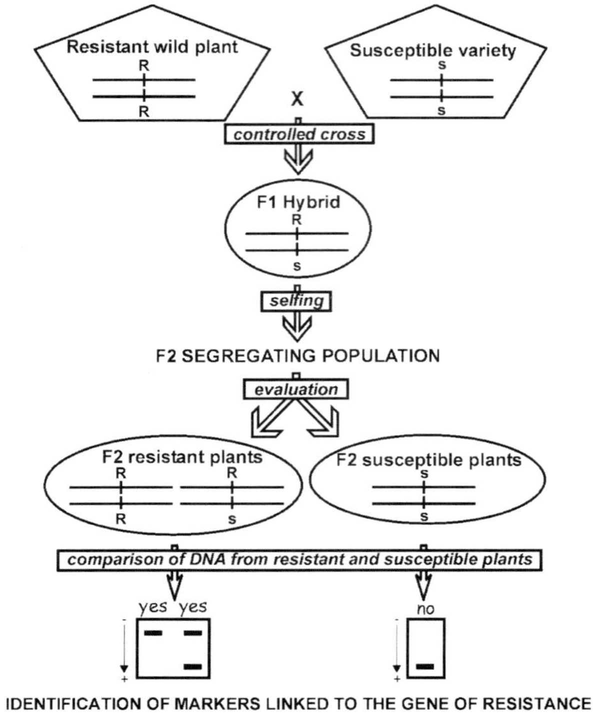

Fig. 9.6 Method for revealing the genetic and molecular basis of resistance. The sketch shows the case of a single dominant resistance gene (R = dominant allele, s = recessive allele) brought by a wild plant

necessary to map the chromosome region carrying the resistance gene. Increasing the number of evaluated plants allows a fine mapping of the targeted DNA region.

9.4.2 Resistance to M. exigua

9.4.2.1 Inheritance of Resistance in F1 Hybrids

Resistance to *M. exigua* appeared to be determined by a major dominant gene in the crosses between two *C. arabica* lines derived from the Timor Hybrid (Bettencourt, 1973), one of which was resistant ('Iapar 59') and the other susceptible ('Costa Rica 95') (Bertrand et al., 2001a). All but one of the F1 hybrids assessed (N = 274) were classified as resistant. The resistance gene, originated from the *C. canephora* parent of the Timor Hybrid, has been transmitted into certain progenies, as the one that gave rise to 'Iapar 59'. However, the F1 hybrids from crosses between 'Iapar 59' and susceptible wild coffee trees supported nematode reproduction rates higher than those estimated for the resistant 'Iapar 59'. This suggests the existence of intermediate resistance in F1 hybrids (Alpizar et al., 2007).

On tomato plants carrying the *Mi* resistance gene, Jaquet et al. (2005) showed that reproduction of *M. incognita* was greater on heterozygous genotypes than on homozygous resistant ones, what suggests a *Mi* gene dosage effect. These authors observed that intermediate resistance was associated with at least two heterozygous tomato genotypes. However, their experimental design was not appropriate for concluding whether that relation was consistent. Tzortzakakis et al. (1998) had suggested that further studies were needed on the influence of the number of *Mi* gene copies inserted into tomato hybrids after controlled hybridization.

9.4.2.2 Inheritance of Resistance in Segregating Populations

An analysis of a large F2 population (365 plants) segregating for resistance to *M. exigua* revealed that 76% of the plants were resistant and 24% were susceptible to this nematode (Bertrand et al., 2001a). These rates are close to the 3:1 proportion expected for a trait determined by a dominant gene. An analysis of another F2 population (96 plants) showed a similar proportion of resistant and susceptible plants, 70% and 30% respectively (Noir et al., 2003). The hypothesis that a major gene was involved in the resistance to *M. exigua* was further supported by these results.

9.4.2.3 Gene Mapping

The search for molecular markers associated with that resistance gene was undertaken by comparing DNA of well characterized resistant and susceptible plants (Noir et al., 2003). Discriminating markers were first sought by comparing the DNA of two resistant F2 plants and two susceptible ones. Out of the 564 polymorphic AFLP fragments identified, 33 appeared to be potentially linked to resistance or susceptibility. Their validation on a larger number of DNA samples (five from resistant plants and five from susceptible ones) allowed the selection of 14 markers associated with resistance, which were mapped over a distance of 8.2 cM. Cosegregation of the resistance gene with the marker Exi-11 was perfect, suggesting that the gene was located near to it.

9.4.3 Resistance to M. paranaensis

9.4.3.1 Inheritance of Resistance in F1 Hybrids

Sources of resistance to *M. paranaensis* exist in both cultivated coffee species. In *C. arabica*, an analysis of the families resulting from a factorial mating design (three cultivars × eight wild coffee trees) showed that three wild coffee trees produced resistant F1 hybrid families (Anzueto et al., 2001a). In *C. canephora*, the mating design involved four female parents and eight male parents (Bertrand et al., 2000b). Variable levels of resistance were found depending on the parents used, which highlights the merit of screening genetic resources prior to carrying out controlled crosses.

9.4.3.2 Inheritance of Resistance in Segregating Populations

The segregation of resistance to the Guatemala isolate of *M. paranaensis* was studied in small F2 populations (32 plants) (Anzueto et al., 2001a). The proportion of resistant to susceptible plants observed was 3:1 in two populations and 9:7 in a third. These results raised two hypotheses for the genetic determinism involved: a dominant major gene, as in the case of resistance to *M. exigua*, or two dominant complementary genes. Only an assessment of a large F2 population will enable researchers to determine the number and nature of the genetic factors controlling resistance to *M. paranaensis*.

9.4.4 Resistance to M. arabicida

As for *M. paranaensis*, accessions resistant to *M. arabicida* have been identified in both cultivated coffee species, but the transmission of the resistance trait has been studied in *C. arabica* only. Two out of five F1 hybrid families derived from crosses between wild coffee trees and cultivars displayed a high level of resistance to 'corchosis' five years after being planted in a plot infested by *M. arabicida* and *F. oxysporum* (Bertrand et al., 2002).

9.4.5 Resistance to M. arenaria

Transmission of resistance to *M. arenaria* has also been investigated in a limited study, in *C. canephora*. The mating design involved three female parents and four male parents (Bertrand et al., 2000b). One female parent and one male parent produced F1 hybrid families that displayed a high level of resistance. The family created by crossing those two plants displayed the highest level of resistance among the 12 families studied. The narrow sense heritability (h^2) of resistance to *M. arenaria* (0.308) appeared to be quite high.

9.5 Breeding Strategies for Resistance

Different breeding strategies can be employed to improve RKN-resistance according to goal of choice: developping resistant ungrafted cultivars or resistant rootstocks.

9.5.1 Selection of Ungrafted Cultivars

The transfer into *C. arabica* cultivars of resistance genes from another species has to be controlled to avoid the negative effects of other genes, which might affect the agronomic characteristics of introgressed cultivars, particularly beverage quality. Indeed, the amount of genetic material introgressed from *C. canephora* into 21 lines derived from the Timor Hybrid was substantial (8–20%, depending on the line) after at least four generations of selfing (Lashermes et al., 2000). All the molecular markers identified in that study corresponded to around 51% of the *C. canephora* genome, which shows the diversity of the fragments that were introgressed into those lines. A similar study conducted on introgressed lines of *C. liberica* led to the same conclusion (Prakash et al., 2002). Molecular markers are therefore valuable tools for identifying introgressions and monitoring their transmission over generations.

The effect of introgressions on the biochemical composition of coffee beans and the beverage's sensory value was assessed in lines derived from the Timor Hybrid (Bertrand et al., 2003). In comparison, the beverage produced from robusta coffee differs from the arabica one through its lower quality, higher caffeine and chlorogenic acid contents, and lower fat, sugar and trigonelline contents (Clifford, 1985). The study of 22 lines derived from the Timor Hybrid did not reveal any relation between the degree of *C. canephora* introgression and beverage quality or biochemical composition (Bertrand et al., 2003). Some lines displaying a large number of introgression markers produced a coffee of similar quality to that of the non-introgressed controls, and with a similar biochemical composition. When combined with data on resistance to 'leaf rust' and *M. exigua*, it appears possible to select lines with good cup quality and resistance traits introgressed from other species.

9.5.1.1 Monoresistant Cultivars

Where field populations are composed of a single *Meloidogyne* species, monoresistance cultivars may be a good choice, as some of the lines derived from the Timor Hybrid. Remarkable examples are 'Iapar 59' and the line 'T5296', both resistant to *M. exigua*. If the straight use of a resistant line derived from the Timor Hybrid is not possible, it may be sufficient to cross a resistant Catimor line with a susceptible or resistant parent. Depending on the resistant parents, this cross will give either totally immune or intermediate resistant plants. Several clones (hybrids) have been developed which displayed an intermediate resistance. A field trial revealed levels of 100–300 J2/gram of roots, as opposed to 1,000–2,000 J2/gram of roots for the susceptible control (Alpizar et al., 2007).

9.5.1.2 Multiresistant Cultivars

The creation of multiresistant cultivars is justified against polyspecific field populations of *Meloidogyne* or for economic reasons, to avoid the managing of a large catalogue of cultivars.

For example, for multiple resistance to *M. exigua* and *M. paranaensis* from Guatemala, resistance genes were provided in a complementary manner by two populations of Catimor and Ethiopian plants (Anzueto et al., 2001a). It was thus possible to create F1 hybrids that displayed both types of resistance. The hybrids 'T5296' × 'ET59' and 'T5296' × 'ET47' presented resistance to *M. exigua*, transmitted by the female parent 'T5296', and partial resistance to *M. paranaensis*, transmitted by the Ethiopian parent.

Certainly it will be difficult to find lines derived from the Timor Hybrid that display good agronomic and sensory traits as well as multiple nematode resistance. The best lines available are resistant to a single nematode species, such as 'T5296' and 'Iapar 59', which are resistant to *M. exigua* but susceptible to *M. paranaensis* and *M. arabicida*. Probably, the pyramiding of several resistance genes provided by different individuals would be lengthy and costly.

9.5.2 Selection of Rootstocks

9.5.2.1 Use of Pure Lines

A distinction needs to be made between the creation of monoresistant cultivars and multiresistant rootstocks. A pedigree selection scheme can be designed that would lead to creation of rootstock lines. Since this scheme would involve the selection for traits related to the root system, it should be possible to ensure varietal outputs as soon as the resistance genes have been fixed (in practice, in generations F3 or F4). Apart from the trait 'resistance to nematodes', it is also possible to select for resistance to other soil-borne pests or pathogens, such as cochineal insects (Garcia, 1991), as well as for vigour and adaptation to distinct agroecological conditions. For example, it has been found that lines derived from the Timor Hybrid display better resistance to drought and high aluminium contents (R. Santacreo, IHCAFE, personal communication). A selection scheme for rootstocks alongside a scheme to create pure lines or F1 hybrids allows for parallel selection for traits related to the shoot and the root system, thereby substantially reducing selection constraints. Simple, early selection tests based on vigour (collar diameter and plant height) can be used (Bertrand et al., 2001b). This strategy is currently being tested in a European INCO-DC project.

9.5.2.2 Use of *C. canephora* or other Species Close to *C. arabica*

This strategy consists of using species close to *C. arabica* as rootstocks, either directly or after selection. In Guatemala, grafting onto robusta rootstocks is very effective in the field against root-lesion nematodes, with the productivity of the grafted plants being four times that of the ungrafted ones (Villain et al., 2000). Against *M. exigua*, theoretically it is possible to use any robusta individual. Where *M. arabicida* is of concern, robusta plants develop the corky-root symptoms in the field (Table 9.2) whilst displaying good tolerance levels.

The multiresistant 'Nemaya' has been developed to collectively overcome the main problems associated with RKNs in Central America. 'Nemaya' is resistant to *M. exigua* and *M. paranaensis* from Guatemala and to *M. arenaria* from El Salvador (Bertrand et al., 2000b; Anzueto et al., 2001b). It also displays a good level of resistance to root-lesion nematodes (Villain, 2000). 'Nemaya' was derived from a cross between the *C. canephora* clones 'T3561' and 'T3751', being reproduced in the form of seeds produced in biclonal seed gardens. Somatic embryogenesis

Table 9.2 Incidence of symptoms caused by the complex *M. arabicida/F. oxysporum* on grafted and ungrafted *C. arabica* 'Caturra', under field conditions (from Bertrand et al., 2002). The mortality data are means of 40 replicates *per* genotype. Nematode development was rated on surviving plants, using the following classification system: class 0 = 0 galls, 1 = 1–10 galls, 2 = 11–30 galls, 3 = 31–100 galls, and 4 = more than 100 galls *per* root system. Corky-root development was recorded on plants showing symptoms, using a notation of the percentage of the entire root system affected by corky-root. n.a. = not applicable, P = probability level of the ANOVA

Accessions	Mortality after four years (%)	Gall index	Plants with corky-root symptoms (%)	% of the root system affected by corky-root
'Caturra' grafted on *C. canephora* rootstock	5.0 b	0.29 b	21.0	22.0 ± 18
Non-grafted 'Caturra'	30.0 a	2.3 a	42.8	29.0 ± 12
P	0.004	0.001	0.04	n.a.

had to be used to speed up propagation of the two mother plants (Bertrand et al., 2002).

Unfortunately, using *C. canephora* or close species as rootstocks will probably be limited by factors related to climate, primarily temperature. Most *Coffea* species originate from hot, tropical regions. At the altitudes and latitudes where arabica coffee is grown, *C. canephora* rootstocks encounter serious growth problems related to low temperatures. Bertrand et al. (2001b) have shown that this limitation leads to graft compatibility problems, which are reflected in substantial yield reductions in comparison to non-grafted arabica plants.

9.6 Concluding Remarks

During the last decade, extensive surveys and studies have shown that a wide range of *Meloidogyne* species parasitize coffee. The extent of this problem varies greatly from one country to another, according to the nematode species involved, their distribution and damage caused to plantations. In Latin America, RKNs are of concern in all coffee-producing countries.

The use of resistant cultivars or rootstocks constitutes an inexpensive, non-polluting and efficient control method. Compared to viruses, bacteria or fungi, RKNs are characterized by low natural dispersal, gene flow and genotype diversity between populations, which leads one to expect that durable resistance genes can be deployed to cultivated plants (McDonald and Linde, 2002). However, resorting to genetic control is not simple in the case of complex nematological situations. Resistance gene pyramiding is therefore the only breeding strategy for combining multiple resistances in an individual.

The strategy of using genetic resistance is now well established. A primary step consists in developing an efficient and repeatable assessment protocol to clearly distinguish between resistant and susceptible plants. A clear discrimination between these statuses is absolutely necessary before starting genetic studies. The next step – screening of genetic resources – provides sources of resistance to be used in breeding programmes. For example, resistant accessions can be used as progenitors in crosses with a susceptible cultivar, in order to produce segregating populations by selfing or backcrossing the F1 hybrids. The analysis of such populations is a key step in understanding the inheritance of resistance and mapping the region carrying the resistance gene(s).

The methodology described above has been adopted in the identification and mapping of *Mex*-1, which is the first resistance gene revealed in coffee (Noir et al., 2003). This gene induces a hypersensitive reaction in the roots of resistant plants, which reinforces the hypothesis of a gene-for-gene interaction between coffee and *M. exigua* (Anthony et al., 2005). Research efforts at IRD and CIRAD have now turned to the functional validation of the gene sequence. Similar studies should be extended to other coffee-parasitic RKNs, especially *M. incognita* and *M. paranaensis*. A reasonable short term objective could be the identification of molecular markers linked to resistance in order to assist genotype selection (Lashermes and Anthony, 2007). The data generated by these studies will be useful on understanding the distribution and organization of resistance genes in the coffee genome.

The recent development of high-output methods for analyzing the structure and function of genes – collectively termed 'genomics' – represents a new paradigm with broad implications, in particular for plant breeding. Although genomics are available for a few plant models only, it seems likely that such information will rapidly become available for most widely studied plant species, such as coffee.

The advent of large scale molecular genomics will provide a window to previously inaccessible sources of genetic variation, which will be exploited in breeding programmes. Anticipated outcomes in coffee breeding include i) rapid characterization and managing of germplasm resources, ii) enhanced understanding of the genetic control of priority traits, iii) identification of candidate genes or tightly linked genomic regions underlying important traits, and iv) identification of accessions in genetic collections with variants of genomic regions or alleles of candidate genes having a favorable impact on priority traits.

To fulfill this potential, there have been extremely encouraging recent efforts to set up an international commitment – the International Coffee Genome Network – to work jointly for the development of common sets of genomic tools, plant populations and concepts.

References

Alpizar E, Etienne H, Bertrand B (2007) Intermediate resistance to the *Meloidogyne exigua* root-knot nematode in *Coffea arabica*. Crop Prot 26: 903–910

Alvarado J (1997) Diagnostico sobre el parasitismo de los nematodos y cochinillas de la raiz en la zona cafetalera del Suroccidente de Guatemala. University of San Carlos, Dissertation of Agronomist Engineer, Quetzaltenango, Guatemala.

Anthony F, Astorga C, Berthaud J (1999) Los recursos genéticos: las bases de una solución genética a los problemas de la caficultura latinoamericana. In: Bertrand B and Rapidel B (eds) Desafios de la caficultura centroamericana. IICA/PROMECAFE-CIRAD-IRD-CCCR France, San José.

Anthony F, Bertrand B, Quiros O et al (2001) Genetic diversity of wild coffee (*Coffea arabica* L.) using molecular markers. Euphytica 118: 53–65

Anthony F, Combes M-C, Astorga C et al (2002) The origin of cultivated *Coffea arabica* L. varieties revealed by AFLP and SSR markers. Theor Appl Genet 104: 894–900

Anthony F, Dussert S, Dulloo E (2007) The coffee genetic resources. In: Engelmann F, Dulloo E, Astorga C et al (eds) Complementary strategies for ex situ conservation of coffee (*Coffea arabica* L.) genetic resources. A case study in CATIE, Costa Rica. Topical Reviews in Agricultural Biodiversity, Bioversity International, Rome.

Anthony F, Noirot M, Couturon E et al (2006) New coffee (*Coffea* L.) species from Cameroon bring original characters for breeding. Proceedings XXI International Scientific Colloquium on Coffee. Cd-Rom # B226.

Anthony F, Topart P, Anzueto F et al (2003) La resistencia genética de *Coffea* spp. a *Meloidogyne paranaensis*: identificación y utilización para la caficultura latinoamericana. Manejo Integrado de Plagas y Agroecologia 67: 4–11

Anthony F, Topart P, Martinez A et al (2005) Hypersensitive-like reactions conferred by the *Mex*-1 resistance gene against *Meloidogyne exigua* in coffee. Plant Pathol 54: 476–482

Anzueto F, Bertrand B, Sarah J-L et al (2001a) Resistance to Meloidogyne *incognita* in Ethiopian *Coffea arabica* accessions. Euphytica 118: 1–8

Anzueto F, Eskes AB, Sarah J-L et al (1991) Recherche de la résistance à *Meloidogyne* sp. dans une collection de *Coffea arabica*. Proceedings XIV Int Sci Colloq Coffee: 534–543

Anzueto F, Eskes AB, Sarah J-L et al (1993) Résistance de quelques descendances de *Coffea arabica* et C. *canephora* vis-à-vis de deux populations de *Meloidogyne* spp., originaires du Guatemala et du Brésil. Proceedings XV Int Scic Colloq Coffee: 338–349

Anzueto F, Molina A, Figueroa P et al (2001b) 'Nemaya', una variedad de Robusta resistente a nematodos. El Cafetal 3: 12–14

Arango LG, Baeza CA, Leguizamón JE (1982) Pruebas de resistencia a especies de *Meloidogyne* en plántulas de *Coffea* spp. Proceedings X Int Sci Colloq Coffee: 563–568

Araya M, Caswel-Chen EP (1995) *Coffea arabica* cvs Caturra and Catuai non hosts to California isolate of *Meloidogyne javanica*. Nematropica 25: 165–171

Barbosa DHSG, Vieira HD, Souza RM et al (2004) Field estimates of coffee yield losses and damage threshold by *Meloidogyne exigua*. Nematol Bras 28: 49–54

Barker KR (1985) Nematode extraction and bioassays. In: Barker KR, Carter CC and Sasser JN (eds) An advanced treatise on *Meloidogyne*, vol. 2 Methodology. North Carolina State University Graphics, Raleigh.

Bertrand B, Aguilar G, Bompard E et al (1997) Comportement agronomique et résistance aux principaux déprédateurs des lignées de Sarchimor et Catimor au Costa Rica. Plante Rech Dév 4: 312–321

Bertrand B, Aguilar G, Santacreo R et al (1999) El mejoramiento genético en América Central. In: Bertrand B and Rapidel B (eds) Desafios de la caficultura centroamericana. IICA/PROMECAFE-CIRAD-IRD-CCCR France, San José.

Bertrand B, Anthony F, Lashermes P (2001a) Breeding for resistance to *Meloidogyne exigua* of *Coffea arabica* by introgression of resistance genes of *C. canephora*. Plant Pathol 50: 637–643

Bertrand B, Anzueto A, Peña MX et al (1995) Genetic improvement of coffee for resistance to root-knot nematodes (*Meloidogyne* sp.) in Central America. Proceedings X Int Sci Colloq Coffee: 630–636

Bertrand B, Etienne H, Cilas C et al (2005) *Coffea arabica* hybrid performance for yield, fertility and bean weight. Euphytica 141:255–262.

Bertrand B, Etienne H, Eskes AB (2001b) Growth, production and bean quality of *Coffea arabica* as affected by interspecific grafting: consequences for rootstock breeding. Hort Sci 36: 269–273

Bertrand B, Guyot B, Anthony F et al (2003) Impact of *Coffea canephora* gene introgression on beverage quality of *C. arabica*. Theor Appl Genet 107: 387–39

Bertrand B, Nuñez C, Sarah J-L (2000a) Disease complex in coffee involving *Meloidogyne arabicida* and *Fusarium oxysporum*. Plant Pathol 49: 383–388

Bertrand B, Peña Durán MX, Anzueto F et al (2000b) Genetic study of *Coffea canephora* coffee tree resistance to *Meloidogyne incognita* nematodes in Guatemala and *Meloidogyne* sp. nematodes in El Salvador for selection of rootstock varieties in Central America. Euphytica 113: 79–86

Bertrand B, Ramirez G, Topart P et al (2002) Resistance of cultivated coffee (*Coffea arabica* and *C. canephora*) to corky-root caused by *Meloidogyne arabicida* and *Fusarium oxysporum*, under controlled and field conditions. Crop Prot 21: 713–719

Bettencourt AJ (1973) Considerações gerais sobre o 'Hibrido de Timor'. Circular n° 23. Instituto Agronômico de Campinas, Brazil.

Bridson DM (1987) Nomenclatural notes on *Psilanthus* including *Coffea* set. *ParaCoffea* (Rubiaceae tribe Coffeeae). Kew Bull 42: 453–460

Carneiro RG (1995) Reação de progenies de café 'Icatu' a *Meloidogyne incognita* raça 2 em condição de campo. Nematol Bras 19: 53–59

Carneiro RMDG, Almeida MRA, Gomes ACMM et al (2005) *Meloidogyne izalcoensis* n.sp. (Nematode: Meloidogynidae), a root-knot nematode parasiting coffee in El Salvador. Nematology 7: 819–832

Carneiro RMDG, Almeida MRA, Quénéhervé P (2000) Enzyme phenotypes of *Meloidogyne* spp. populations. Nematology 2: 645–654

Carneiro RMDG, Carneiro RG, Abrantes IMO et al (1996) *Meloidogyne paranaensis* n. sp. (Nemata: Meloidogynidae), a root-knot nematode paraziting coffee in Brazil. J Nematol 28: 177–189

Carneiro RMDG, Tigano MS, Randig O et al (2004) Identification and genetic diversity of *Meloidogyne* spp. (Tylenchida: Meloidogynidae) on coffee from Brazil, Central America and Hawaii. Nematology 6: 287–298

Carvalho A (1946) Distribução geográfica e classificação botânica do gênero *Coffea* com referência especial à espécie *arabica*. Boletim da superintendência dos serviços do café 21: 174–180

Clifford MN (1985) Chemical and physical aspects of green coffee and coffee products. In: Clifford MN and Wilson KC (eds) Coffee: botany, biochemistry and productions of beans and beverage. Croom Helm, London.

Couturon E, Lashermes P, Charrier A (1998) First intergeneric hybrids (*Psilanthus ebracteolatus* Hiern × *Coffea arabica* L.) in coffee trees. Can J Bot 76: 542–546

Curi SM, Carvalho A, Moraes FP et al (1970) Novas fontes de resistencia genética de *Coffea* no controle de nematóide do cafeeiro, *Meloidogyne exigua*. Biológico 36: 293–295

Davis A, Govaerts R, Stoffelen P (2006) An annotated taxonomic conspectus of the genus *Coffea* (Rubiaceae). Bot J Linn Soc 152: 465–512

Davis AP, Bridson D, Rakotonasolo F (2005) A reexamination of *Coffea* subgenus *BaraCoffea* and comments on the morphology and classification of *Coffea* and *Psilanthus* (Rubiaceae-Coffeeae) In: Keating RC, Hollowell VC and Croat T (eds) Festschrift for William G. D'Arcy: the Legacy of a Taxonomist. MBG Press, Missouri.

Dropkin, V H (1988) The concept of race in phytonematology. Annu Rev Phytopathol 26: 145–161

Dussert S, Lashermes P, Anthony F et al (2003) Coffee (*Coffea canephora*). In: Hamon P, Seguin M, Perrier X and Glaszmann JC (eds) Genetic diversity of cultivated tropical crops. Collection Repères, CIRAD, Montpellier.

Fazuoli LC (1975) Resistência de *Coffea racemosa* ao *Meloidogyne exigua*. Ciência e Cultura 27: 230

Fazuoli LC, Lordello RRA (1978) Fontes de resistência em espécies de cafeeiro a *Meloidogyne exigua*. Soc Bras Nematol 3: 49–52

Fernie LM (1968) FAO coffee mission to Ethiopia 1964–1965. FAO, Rome.

Garcia A (1991) Les Pseudococcidae déprédatrices du caféier (*Coffea arabica*) au Guatemala: cas particulier de *Dysmicocus cryptus* (Hempel, 1918). Université Paul Sabatier, PhD thesis, Toulouse.

Gonçalves W (1998) Efeito de diferentes niveis de inóculo na avaliação precoce da reação do cafeeiro a *Meloidogyne exigua*. Nematol Bras 22: 75–78

Gonçalves W, Barbosa LCC, Lima MMA et al (1996) Patogenicidade de *Meloidogyne exigua* e *M. incognita* raça 1 a mudas de cafeeiros. Bragantia 55: 89–93

Gonçalves W, Ferraz LCCB (1987) Resistência do cafeeiro a nematóides. II – Testes de progênies e hibridos para *Meloidogyne incognita* raça 3. Nematol Bras 11: 125–142

Gonçalves W, Pereira A (1998) Resistência do cafeeiro a nematóides. IV Reação de cafeeiros derivados do Hibrido de Timor a *Meloidogyne exigua*. Nematol Bras 22: 39–50

Guillaumet JL, Hallé F (1978) Echantillonnage du matériel récolté en Ethiopie. Bull IFCC 14: 13–18

Hernández A (1997) Etude de la variabilité intra et interspécifique des nématodes du genre *Meloidogyne* parasites des caféiers en Amérique Centrale. Université de Montpellier II, PhD thesis, Montpellier.

Hernandez A, Fargette M, Sarah J-L (2004a) Characterisation of *Meloidogyne* spp. (Tylenchida: Meloidogynidae) from coffee plantations in Central America and Brazil. Nematology 6: 193–204

Hernandez A, Fargette M, Sarah J-L (2004b) Pathogenicity of *Meloidogyne* spp. (Tylenchida: Meloidogynidae) isolates from Central America and Brazil on four genotypes of *Coffea arabica*. Nematology 6: 205–213

Hussey RS, Barker KR (1973) A comparison of methods of collecting inocula for *Meloidogyne* spp., including a new technique. Plant Dis Rep 57: 1025–1028

Jaquet M, Bongiovanni M, Martinez M et al (2005) Variation in resistance to the root-knot nematode *Meloidogyne incognita* in tomato genotypes bearing the Mi gene. Plant Pathol 54: 93–99

Krug CA, Mendes JET, Carvalho A (1939) Taxonomia de *Coffea arabica* L. Boletim Técnico n°62. Instituto Agronômico do Estado, Campinas.

Lashermes P, Andrzejewski S, Bertrand B et al (2000) Molecular analysis of introgressive breeding in coffee (*Coffea arabica* L.). Theor Appl Genet 100: 139–146

Lashermes P, Anthony F (2007) Gene mapping and molecular breeding in coffee. In: Kole CR (ed) Genome mapping and Molecular breeding. Volume 6: Technical Crops. Springer, Berlin.

Leroy T, Montagnon C, Charrier A et al (1993) Reciprocal recurrent selection applied to *Coffea canephora* Pierre. 1. Characterization and evaluation of breeding populations and value of intergroup hybrids. Euphytica 67: 113–125

López R, Salazar L (1989) *Meloidogyne arabicida* sp.n. (Nemata: Heteroderidae) nativo de Costa Rica: un nuevo y severo patógeno del cafeto. Turrialba 39: 313–323

Luc M, Reversat G (1985) Possibilités et limites des solutions génétiques aux affections provoquées par les nématodes sur les cultures tropicales. CR Acad Agri Paris 71: 781–791

McDonald BA, Linde C (2002) The population genetics of plant pathogens and breeding strategies for durable resistance. Euphytica 124: 163–180

Montagnon C, Bouharmont P (1996) Multivariate analysis of phenotic diversity of *Coffea arabica*. Genet Resour Crop Ev 43: 221–227

Montagnon C, Leroy T, Yapo A (1992) Etude complémentaire de la diversité génotypique et phénotypique des caféiers de l'espèce *C. canephora* en collection en Côte d'Ivoire. Proceedings XIV Int Sci Colloq Coffee: 444–450

Negrón JA, Acosta N (1989) The *Fusarium oxysporum* f.sp. *coffeae-Meloidogyne incognita* complex in 'Bourbon' coffee. Nematropica 19: 161–168

Noir S, Anthony F, Bertrand B et al (2003) Identification of a major gene (*Mex*-1) from *Coffea canephora* conferring resistance to *Meloidogyne exigua* in *Coffea arabica*. Plant Pathol 52: 97–103

Powell NT (1971) Interactions between nematodes and fungi in disease complexes. Annu Rev Phytopathol 9: 253–274

Prakash NS, Combes M-C, Dussert S et al (2005) Analysis of genetic diversity in Indian robusta genepool (*Coffea canephora*) in comparison with a representative core collection using SSRs and AFLPs. Genet Resour Crop Ev 52: 333–343

Prakash NS, Combes M-C, Somanna N et al (2002) AFLP analysis of introgression in coffee cultivars (*Coffea arabica* L.) derived from a natural inerterspecific hybrid. Euphytica 124: 265–271

Semblat JP, Bongiovanni M, Wajnberg E et al (2000) Virulence and molecular diversity of parthogenetic root-knot nematodes, *Meloidogyne* spp. Heredity 84: 81–89

Silvarolla MB, Gonçalves W, Lima MMA (1998) Resistência do cafeeiro a nematóides. V. Reprodução de *Meloidogyne exigua* em cafeeiros derivados da hibridação de *Coffea arabica* com *C. canephora*. Nematol Bras 22: 51–59

Taylor AL (1967) Introduction to research on plant nematology. N° PL/CP/5, FAO, Rome.

Taylor AL, Sasser JN (1978) Biology, identification and control of root-knot nematodes (*Meloidogyne* sp.). North Carolina State University Graphics, Raleigh.

Triantaphyllou AC (1987) Genetics of nematode parasitism on plants. In: Veech JA and Dickson DW (eds) Vistas on nematology. Society of Nematologists, Hyattsville.

Trugdill DL, Blok VC (2001) Apomictic, polyphagous root-knot nematodes: exceptionally successful and damaging biotrophic root pathogens. Annu Rev Phytopathol 39: 53–77

Tzortzakakis EA, Niebel A, Van Montagu M et al (1998) Evidence of a dosage effect of Mi gene on partially virulent isolates of *Meloidogyne javanica*. J Nematol 30: 76–80

Van der Vossen HAM (2001) Coffee breeding and selection: review of achievements and challenges. Proceedings XIX Int Sci Colloq Coffee, Cd-Rom # A001.

Villain L (2000) Caractérisation et bioécologie du complexe parasitaire du genre *Pratylenchus* (Nemata: Pratylenchidae) présent sur caféiers (*Coffea* sp.) au Guatemala. Ecole Nationale Supérieure Agronomique de Rennes, PhD thesis, Rennes.

Villain L, Anzueto F, Hernández A et al (1999) Los nematodos parasitos del cafeto. In: Bertrand B and Rapidel B (eds) Desafios de la caficultura centroamericana. IICA/PROMECAFE-CIRAD-IRD-CCCR France, San José.

Villain L, Molina A, Sierra S et al (2000) Effect of grafting and nematicide treatments on damage by root-lesions nematodes (*Pratylenchus* spp.) to *Coffea arabica* L. in Guatemala. Nematropica 30: 87–100

Whitehead AG (1969) The distribution of root-knot nematodes *Meloidogyne* spp. Tropical Africa. Nematologica 15: 315–333

Chapter 10
Genomic Tools for the Development of Engineered *Meloidogyne*-Resistant Coffee Cultivars

Mirian P. Maluf

Abstract This chapter discuss major issues related to the development of transgenic *Meloidogyne*-resistant coffee cultivars. Initially, the relevance of engineering cultivars *in vitro* is highlighted in relation to the limitations found in traditional coffee-breeding programs. Given that potential approaches to develop transgenic cultivars are transferring genes that confer traditional plant resistance or anti-nematode results, this chapter discuss the selection process of genes candidates for transference, including resistance and general-defense genes, and proteinase inhibitors. The use of gene-silencing as an approach to modify gene expression during plant-nematode interaction, resulting in plant resistance, is also discussed. A review is presented on recent progresses on the functional characterization of coffee genes and nematode-responsive promoters. Finally, this chapter presents successful examples of engineered nematode-resistant cultivars in other plant species, which support the feasibility of these strategies for the development of transgenic coffee cultivars.

Keywords Transgenic cultivars · transgenic coffee · defense genes · resistance genes · nematode-responsive promoters

10.1 Introduction

Plant genetic resistance is one of the key strategies to sustain the commercial production of coffee (*Coffea* sp.) in nematode-infested areas. Breeding programs aim to develop cultivars with durable nematode resistance, to be planted in infested fields to decrease yield losses and to reduce production costs considerably. To achieve this goal, basic breeding methods include the identification of resistance genes (RGs), either in the species undergoing breeding or in related ones, followed by an efficient transfer of these genes to susceptible cultivars and selection of resistant lines. Coffee breeding strategies for nematode resistance are discussed in detail in Chapter 9.

M.P. Maluf
Embrapa Café, Campinas, Brazil
e-mail: maluf@iac.sp.gov.br

R.M. Souza (ed.), *Plant-Parasitic Nematodes of Coffee*,

In recent years, breeding programs have been increasingly adopting genomic tools to improve the development of cultivars. This progress became possible for several reasons, such as the development of technology for efficient *in vitro* gene transfer and selection. Breeding is also benefiting from the vast amount of information available on every molecular aspect of plant genes, including sequence information, regulation of gene expression and genome interaction. This wealth of knowledge has been generated by large-scale sequencing projects.

A number of biotechnology approaches could be employed for the development of nematode-resistant coffee cultivars. Among them, the most promising are the production of transgenic plants bearing RGs or bearing mechanisms for silencing specific genes through RNA interference (RNAi). However, these approaches rely largely on the identification of plant candidate genes that arrest nematode invasion and/or development, and of nematode candidate genes that would be targeted for gene silencing. This chapter focuses on recent progress in these approaches for the development of coffee cultivars resistant to the root-knot nematode (RKN), *Meloidogyne* sp.

10.2 Difficulties Associated with Breeding for Nematode Resistance

Arabica coffee (*C. arabica* L.) is successfully parasitized by several *Meloidogyne* species. For example, cytological studies demonstrated that *M. incognita* (Kofoid and White) Chitwood and *M. exigua* Göldi infect and reproduce in the roots of arabica coffee, indicating that a compatible interaction is associated with susceptibility to these nematodes (Anthony et al., 2005; Barros et al., 2006).

The development of RKN-resistant arabica cultivars through traditional breeding is impaired by the fact that no reliable source of resistance has been identified in cultivated arabica genotypes. Therefore, a strategy commonly adopted in coffee breeding programs is the transference of RGs from other *Coffea* species, such as *C. canephora* Pierre ex Froehner and *C. racemosa* Ruiz and Pav. to arabica cultivars. One limitation of this approach is that crosses between *Coffea* species are not always efficient, and only a low number of viable hybrids are normally produced. Other aspects that interfere with breeding are the long life cycle of coffee plants and the complex trials required for resistance evaluation of segregating genotypes. These processes are normally expensive and time-consuming.

As an example, thousands of hybrid genotypes are currently under selection as part of the Agronomic Institute of Campinas' breeding program in Brazil, both in greenhouse and in RKN-infested fields. So far, no cultivar has been released for commercial use. The most promising inbred lines, derived from 'Icatu', are under field evaluation for resistance to *M. exigua, M. incognita* and *M. paranaensis* Carneiro, Carneiro, Abrantes, Santos and Almeida (W. Gonçalves, personal communication). Also, the *M. exigua*-resistant 'Tupi RN IAC-1669-13', to be released soon, will be an alternative for infested areas (Fazuoli et al., 2006). Thus far, only a handful of *M. exigua*-resistant arabica cultivars have been released, such as 'Acauã', 'Catucai' and 'Iapar-59'. However, these cultivars do not present complete resistance, since there have been reports of RKN populations capable of reproducing on these cultivars (Salgado et al., 2002; Matiello et al., 2004; Barbosa et al., 2006).

10.3 Transgenic Cultivars

Considering the limitations imposed by the inherent biology of coffee plants, the development of faster and more accurate methods of RG transference would result in considerable gain in time, resources and overall efficiency. Also, as host-pathogen interactions are often dynamic, new nematode races may arise and challenge plant resistance. Hence, development of new resistant cultivars should be a dynamic process as well, offering to the farmers novel choices of cultivars once nematodes become a problem. In this view, biotechnology and genomic tools are valuable alternatives since they can overcome breeding limitations such as barriers for interspecific crossings and the long time-span of back and self-crosses. An example of successful use of *in vitro* transformation technology as a tool for breeding was the development of nematode-resistant solanaceous cultivars (Milligan et al., 1998; Goggin et al., 2006). An allele conferring resistance to RKNs was isolated from tomato, and transferred to susceptible tomato and eggplant lines.

A major limitation for the development of coffee cultivars through bioengineering methods is that very little information is available regarding the genetic control of defense mechanisms in coffee plants. Indeed, the resistance of *Coffea* species to RKNs is not well characterized at the molecular level, and only a few gene loci related to nematode resistance have been identified (Noir et al., 2003). Limiting factors for the identification of resistance-related loci include the lack of either genetic or molecular maps, and the reduced number of genes identified so far. This last constraint should be soon overcome since valuable genomic information is being released by genome projects, such as the Coffee Genome Project in Brazil (see www.cenargen.embrapa.br/biotec/genomacafe) and the Solanaceae Genomics Network (see http://www.sgn.cornell.edu/).

Transgenic coffee plants are not available for commercial cultivation yet. However, two important pioneer works have provided the proof-of-principle. Transgenic coffee plants were developed bearing resistance to 'leaf-miner' (*Leucoptera coffeella* Guérin-Mèneville and Perrottet), one of the most important coffee pests in Brazil (Perthuis et al., 2005). In this case, genes encoding the Bt toxin *cry1Ac* were transferred from *Bacillus thuringiensis* Berliner to *C. canephora* plants, which are under field trial for the assessment of the toxin's effectiveness in 'leaf-miner'-infested areas. Using a different approach, transgenic *C. canephora* plants exhibiting low caffeine content in leaves were developed by Ogita et al. (2003). These authors silenced the gene expression of theobromine synthase, one of the caffeine biosynthetic enzymes, through the RNAi technique.

10.4 Candidate Genes for *In Vitro* Transfer

In theory, the transference of nematode RG from any plant species into a susceptible one could result in transgenic, resistant plants. Transformation techniques and selection methods are well established for most cultivated plant species, including coffee (Van Boxtel, 1995; Ribas et al., 2005). Hence, the major concerns regarding the development of engineered resistant crops are associated with taking a decision on which genes would be desirable for obtaining reliable, durable nematode resistance.

Nematodes are provided with a wide selection of tools that guarantee successful plant infection and parasitism. These tools include the synthesis of anti-defense proteins, such as superoxide dismutase, thierodoixin peroxidases and lipoxygenase-inhibiting proteins, cell-wall modifying proteins, and unknown proteins capable of controlling plant gene expression for the reprogramming of cell structure and metabolism. These concerted events lead to the establishment of nematode feeding sites (Davis et al., 2004; Lilley et al., 2005). All resistant cultivars, either transgenic or bred, must contain defense-related genes associated with mechanisms capable of overcoming the nematode's sophisticated capability to parasitize plant roots.

Several aspects should be considered for the selection of candidate defense genes. The gene to be delivered into the plant should be well characterized regarding its ability to promote resistance by preventing nematode penetration or survival in the root. Also, it is desirable that the expression of the introduced gene be induced either at the time of, or in response to nematode infection. The stability of the introduced gene as part of the genome, and the likelihood of chromosome rearrangements, should be assessed. Finally, there must be an assessment of the introduced gene's effects on the control of the plant's constitutive defense mechanisms, since modifications in the overall defense mechanisms and responses could affect the interactions of the plant with other pathogens.

Studies on candidate genes for plant transformation have focused on two major classes of RGs: those involved directly with nematode recognition during root penetration, which trigger plant defense responses, and those associated with the synthesis of anti-nematode compounds, such as proteinase-inhibitors and phytoecdysteroids. Recent advances in these areas will be outlined in the following sections, as well as the available nematode-responsive promoters and their potential use for developing coffee transgenic cultivars.

10.5 Resistance Genes

In resistant plants, pathogen recognition leads to a cascade of gene expressions, which result in specific intracellular reactions that lead to localized cell death. This hypersensitive response (HR) results from the action of several proteins that are specifically induced after a gene-for-gene interaction. Several plant genes have been identified as responsive to pathogen invasion, and they are related to the initiation of defense mechanisms in plants (see reviews by Lamb and Dixon, 1997; Innes, 2004).

Among these, RG or resistance gene analogs have been identified in several plant species, and they are apparently related to specific recognition of elicitors produced by pathogen avirulent genes (avr). This recognition may trigger the activation of several proteins in cascade, which results in HR.

Molecular analysis of RGs of diverse origins and pathogen-specificities revealed that they all share highly conserved amino acid domains (Bent et al. 1994; Lawrence et al. 1995). These domains include a nucleotide-binding site (NBS) and a leucine-rich repeat (LRR), which may respond to protein interactions during signal transduction and pathogen-specific recognitions. Several RGs have been identified in soybean, wheat, rice and maize, among others (Innes, 2004). Several studies using genetics and molecular mapping allowed physical localization and the association of resistance sequences with previously assessed RG loci (Kretschemer et al., 1997; Pflieger et al., 1999; Graham et al., 2000). However, very few of these genes have been associated with specific pathogen resistance mechanisms; thus, there has been no definite proof that these are not just pseudogenes.

One of the RGs identified is the *Mi* locus in tomato, responsible for resistance to RKNs (Gilbert and McGuire, 1956). A functional *Mi* allele was cloned from a BAC library containing the entire region to which *Mi* was localized (Milligan et al. 1998). Moreover, in transient expression analysis the functional *Mi* allele conferred *M. javanica* (Treub) Chitwood-resistance to a previously susceptible tomato line (Milligan et al., 1998). This was the first study demonstrating that a family of RGs is actually involved with a specific defense response, and that a transgenic cultivar could express nematode resistance. Also, pyramiding of RGs has been successful in the development of potato cultivars resistant to *Globodera* sp. (Dale and Scurrah, 1998), indicating that these genes are potential candidates for transference to other susceptible plant species.

Resistance genes have been also identified in coffee plants. Noir et al. (2001) identified several sequences using heterologous primers, corresponding to NBS domains. However, amplified fragments included only three amino acid sub-domains of the NBS conserved region. These authors identified 19 different RG sequences, which could be grouped into nine distinct classes.

In a similar approach, Orsi (2003) also identified and cloned homologous of RGs in *C. arabica, C. canephora* and *C. racemosa*. The amplified region included several motifs of the NBS-LRR region. However, the amplified sequences presented only a moderate variability, and only two families of RGs with four different sequences were observed. These authors also investigated the pattern of RG expression during the infection of susceptible and resistant coffee roots by *M. exigua*, in a time-course analysis. They demonstrated that 10 days after nematode infection resistance transcripts had accumulated in resistant roots, but not in susceptible ones. According to cytological analysis by Rodrigues et al. (2000), this time-period corresponds to the establishment of nematode feeding sites (NFSs).

During the establishment of NFSs the expression of several plant genes is altered, and the nematode uses plant cells as the source of nutrients required for their life cycle completion (see review by Gheysen and Fenoll, 2002). Indeed, several studies have suggested that plant genes that are not essential for the establishment of NFSs are preferentially turned off (Goddijn et al., 1993; Bar-Or et al., 2005). Moreover, several studies have demonstrated that nematodes induce the expression of plant genes involved in cell wall degradation, such as pectin acetylesterase and α-1, 4 glucanase, since cell wall degradation is an essential process for nematode development (Goellner et al., 2001; Vercauteren et al., 2002). Cytological analysis of coffee roots parasitized by *M. incognita* has demonstrated that giant cells are underdeveloped in resistant plants (Anzueto et al., 2001).

Collectively, these analyses suggest that in coffee plants defense mechanisms against nematodes are probably related to inhibition of NFS development, rather than obstruction of nematode infection and migration through the roots. This reinforces the role of specific RGs in triggering resistance response. Such genes are, therefore, the best candidates for transformation experiments.

The only RG identified in coffee is *Mex*-1, which confers resistance to *M. exigua* (Noir et al., 2003). Histological analysis of roots infected by that nematode have demonstrated that coffee plants bearing *Mex*-1 exhibit an HR-like reaction, indicating that this gene could be involved in triggering a resistance response (Anthony et al., 2005). The cloning of *Mex*-1 should demonstrate whether it is a member of a RG family or not. Also, the transference of *Mex*-1 to susceptible coffee cultivars is essential for evaluation of its potential for the development of transgenic resistant cultivars.

To confirm that *Mex*-1 or other RGs can be used to transfer nematode resistance to susceptible cultivars, specific complete gene sequences must be identified. Since RGs are members of multiallelic families, each one responsible for the recognition of a particular pathogen, future research efforts must concentrate on the identification of full-length genes involved with nematode recognition. To accomplish this task, complete gene sequences could be mined in coffee genome databases, which contain valuable sequence information for identification of functional alleles associated with nematode resistance. This strategy currently faces limitations since nematode resistance in *Coffea* sp. is poorly characterized, and resistant germplasm bearing RGs is not yet accessible for gene 'hunting'.

In the future, the life-span of specific RGs may turn out to be limited because most virulent pathogens evolve rapidly, with new genotypes arising with alleles not recognized by the plant's RGs. Another important issue is the influence of the number of RG copies on the interaction between the nematode and the plant. Jacquet et al. (2005) investigated whether tomato resistance to RKNs could be influenced by the plant's genetic background and the state of the allele in the *Mi* locus. Apparently, heterozygous plants exhibited higher nematode reproduction rates than homozygous ones, suggesting a possible dosage effect of the *Mi* gene. Finally, yet another important issue is how the introduction of one or more foreign genes would affect the plant's overall defense response. In an interesting study, Goggin et al. (2006) demonstrated that the resistance induced by the *Mi* gene can be extended to other Solanaceae species. Transgenic eggplants bearing the *Mi* gene exhibited resistance to *M. javanica*. However, these plants were susceptible to the potato aphid *Macrosiphum euphorbiae* Thomas, a phenomenon not observed in tomato plants bearing the *Mi* gene.

Altogether, these results indicate that care must be taken in the transference of genes between different plant species, since unbalanced copies of a foreign gene can negatively affect a plant's resistance response.

10.6 Anti-Nematode Compounds

Proteinase inhibitors (PIs) are major components of the plants' defenses, being synthesized normally in response to wounding or herbivory (Haq et al., 2004). Proteinase inhibitors accumulate also in seeds, since they play a role in their germination. The classification of PIs is based on their proteolitic capability, which is determined by the aminoacid that is active in their reaction center. The four known classes of PIs, cysteine, serine, aspartyl and metallo, form stable complexes with targeted proteases, thus inhibiting their action.

PIs are found in most plant tissues, although at higher concentrations in aerial tissues than in roots. Although this may represent a limitation for the use of PIs against root-feeding nematodes, some studies have established their potential for controlling these parasites. Indeed, PIs are strong candidate genes for the development of engineered nematode-resistant crops because they are present in several plant species, they act on different types of pathogens, and they share common biological mechanisms. Hence, a combination of distinct genes, targeting more than one pathogen, could be transferred to a susceptible plant cultivar. In addition, there has been no report of deleterious effects of PIs on mammals. Transgenic cultivars carrying PI genes have already been released on the market, with resistance to a broad range of pests. This reinforces the feasibility of this strategy for the development of nematode-resistant cultivars (Haq et al., 2004).

Transgenic crops bearing PI genes expressed in response to parasitism are an interesting alternative for nematode control, with several studies having reported enhancement of nematode resistance in different plant species (Atkinson et al., 2003). The most significant results were achieved by Urwin et al. (1995; 1997) using cysteine PIs, also called cystatins. In these studies, transgenic tomato and arabidopsis [(*Arabidopsis thaliana* (L.) Heynh.] plants were developed carrying previously cloned cystatin genes from rice, namely *Oc*-I and *Oc*-II, which were under the control of the constitutive promoter CaMV35S. Parasitism by the sugarbeet cyst nematode *Heterodera schachtii* Schmidt and *M. incognita* was suppressed in the transgenic lines, which harbored fewer nematode mature females, in comparison to the control, non-transformed lines. In the former lines, the females were also smaller and less fecund. Besides resistance to *H. schachtii* and *M. incognita*, transgenic arabidopsis lines also exhibited resistance to *Rotylenchulus reniformis* Lindford and Oliveira. In another interesting study, transgenic potato expressing cystatin exhibited resistance to *Globodera pallida* (Stone) Behrens and *G. rostochiensis* (Wolleweber) Behrens, but had no negative effect on the non-target herbivorous insect *Eupteryx aurata* (L.) Curtis (Atkinson et al., 2003). These studies have suggested that PI genes could effectively lead to the development of nematode-resistant crops without major risks to non-target species.

The development of PI-transformed coffee plants will depend largely on the identification of PIs with negative effects on RKNs parasitic to coffee. Since proteases are essential for nematodes during root penetration and establishment of NFSs, the characterization of all proteins synthesized by the nematodes during these stages would offer insights into which types of proteases are most important for nematode parasitism. The first proteinase cloned from *M. incognita* parasitic on coffee was a serine proteinase, apparently encoded by a single gene, *Mi-serI* (Fragoso et al., 2005). This putative protein exhibits a single chymotrypsin-like catalytic domain, what suggests a digestive role for it. As a potential target for inhibition, further studies will be necessary to identify its corresponding PI.

Ongoing studies are focused on the characterization of protein profiles of *M. paranaensis*-resistant and -susceptible coffee cultivars during nematode parasitism (Andrade et al., 2005). In the resistant plants, time-course experiments have revealed the differential expression of several proteins. Some of these may turn out to be PIs with potential for transgenic transformation of susceptible cultivars.

Alternatively, PI genes isolated from other plant species and already characterized can be used for transformation of coffee plants. Cabos et al. (2006) transferred cysteine and serine PI genes from rice and cowpea to *C. arabica* lines. These authors detected transcripts of PI genes in the coffee roots, but further assays involving nematode parasitism are necessary to certify that these genes are active in the plants, and that they result in nematode resistance.

Phytoecdysteroids are another group of molecules that act directly on nematodes. These are analogs of steroid hormones with a defensive role during pathogen attacks (Schmelz et al., 1999; 2000). Recent reports indicated that these compounds have anti-nematode effects, including immobility and death of *M. javanica* (Soriano et al., 2004). Also, the nematode's capability for root infection was reduced in spinach plants in which the synthesis of phytoecdysteroids had been over-induced with the use of methyl-jasmonate. These results suggest that nematode resistance could be achieved by enhancing the plant's synthesis of phytoecdysteroids through over-expression of the genes involved in the biosynthesis of methyl-jasmonate. It should be considered, however, that methyl-jasmonate is an intermediate compound of the ethylene biosynthetic pathway, and that other pathways regulated by these compounds, such as pollinator signaling and fruit development, could be affected in such transformed plants.

10.7 General Defense-Related Genes

An ideal candidate RG should encode nematode-specific *avr* proteins. However, since such genes have not been identified in plants (Williamson and Gleason, 2003), other pathogenesis-related genes could be used to improve the overall defense response of host plants. Studies on the expression profile of plant genes regulated during nematode infection have shown that pathogenesis-related genes are up- or down-regulated during this process (Bar-Or et al., 2005). These authors used microarray analysis to examine gene expression in tomato roots during the different stages of parasitism by *M. javanica*. Genes associated with hormone biosynthesis and signaling pathway, the antimicrobial protein defensin and transcriptional activator factors, such as two members of the *wrky* family, were regulated in a compatible reaction, indicating that these pathways could represent potential targets for expression control in transgenic nematode-resistant plants.

Recently, several studies have been set up with the aim of characterizing potential defense-genes in coffee. Functional analysis of genes expressed during nematode parasitism was performed in susceptible and resistant *C. arabica* plants inoculated with *M. exigua* (Silvestrini et al., 2005). The expression of six different classes of genes was evaluated through the RT-PCR technique. These genes included transcription factors, oxidative stress-related proteins, resistance proteins and proteins with unknown function. The analyses demonstrated an active expression of defense-related genes during nematode parasitism. However, no significant differential expression of these genes were observed between roots of susceptible and resistant plants.

Using a different approach, based on the construction of subtractive cDNA libraries enriched with genes induced during the early stages of HR, Lecouls et al. (2006) identified coffee genes expressed during both compatible and incompatible responses to *M. exigua* infection. According to their analysis, only 4% of the identified expressed sequence tags were common to both kinds of interaction, indicating that a large number of genes are specifically expressed during compatible and incompatible interactions. A thoroughly functional analysis of these differentially expressed genes may result in the identification of several nematode-responsive candidates for *in vitro* transfer.

It is important to note that all these studies represent preliminary reports only; hence, more functional analyses of coffee-nematode interactions are necessary for the identification of strong candidate defense-genes for transference to new cultivars.

10.8 RNAi/Gene Silencing

A new procedure, originally described in the nematode *Caernorhabditis elegans* Maupas, is post-transcriptional gene silencing (PTGS) through RNAi. This highly conserved mechanism, also present in plants, is gene-specific and results in a sequence-specific degradation of selected RNA. Transgenes have been demonstrated to trigger PTGS in plants (Napoli et al., 1990), suggesting that this process could play a role in the plants' defense strategy. Also, there have been reports that PTGS can be redirected to silence endogenously expressed genes in plants, thus representing an alternative for knock-down of specific pathways. Indeed, PTGS has been used in tomato, tobacco and arabidopsis plants to silence specific genes in pathways such as carotenoid biosynthesis, flowering, and meristem maintenance (Peele et al., 2001; Ratcliff et al., 2001).

The use of PTGS could represent an alternative for the development of nematode-resistant cultivars, since a PTGS vector has been developed that knocks-down genes specifically expressed in roots. This vector was constructed by modification of the tobacco rattle virus (TRV), which is transmitted from plant to plant via nematodes, and it can efficiently regulate foreign gene expression in roots. In a pioneer study associating vector control of a gene related to nematode infection, the expression of *Mi* was repressed in transgenic plants bearing the TRV-modified vector (Valentine et al., 2004). As a consequence *Mi*-bearing resistant tomato cultivars were successfully parasitized by *M. javanica*, demonstrating that this system can be used to modulate expression of defense-related genes, and consequently to control nematode resistance.

To effectively use this system to control plant-parasitic nematodes, the best candidates for silencing would be those genes involved with nematode invasion and/or development, such as those associated with the recruitment of plant cell metabolism in NFSs. Several recent studies aimed at the characterization of genes expressed during nematode infection and development improved our knowledge of the molecular mechanisms involved during nematode-plant interactions (see review by Bird, 2004). Several of these genes show homology to plant genes involved in meristem growth and differentiation, such as orthologous of *Phantastica* (PHA), *Clavata* and *Knotted1* (KNOX) (Koltai and Bird, 2000; Olsen and Skriver, 2003; Wang et al., 2005). Analyses by *in situ* RT-PCR localization indeed demonstrated that expression of PHA and KNOX is up-regulated in giant cells (Koltai et al., 2001). Future studies could verify whether the silencing of these genes in transgenic plants could impair nematode development, resulting in nematode resistance.

Another pathway candidate for gene silencing is the synthesis of plant hormones. Since the levels of auxin and cytokinin increase significantly during plant-nematode interaction (Bird, 2004), down-regulation of the genes involved in the biosynthesis of these hormones could be targeted through RNAi. Transgenic *Lotus japonicus* (Regel) Larsen plants expressing low levels of cytokinin exhibited reduced number of NFSs upon infection with RKNs (Lohar et al., 2004). These findings suggest that this approach could be feasible for nematode control. However, knocking-down genes from biosynthetic hormone pathways would certainly affect several other aspects of plant development. Thus, the use of this approach should be associated with an effective control of gene expression so that it will be activated in localized root tissues upon nematode infection only.

The first successful attempt to interfere with nematode development in plants through RNAi was achieved through silencing of the parasitism gene *16D10* (Huang et al., 2006a). Apparently this gene is associated with early signaling events during RKN-plant interactions (Huang et al., 2006b). When arabidopsis plants transformed with *16D10* dsRNA were infected with *M. javanica*, *M. incognita*, *M. arenaria* (Neal) Chitwood and *M. hapla* Chitwood, the reproduction of these nematodes was significantly reduced in comparison to non-transformed control plants. This result is very interesting, since it suggests that RNAi can be used to transform crops for a broad nematode resistance.

10.9 Nematode-Responsive Promoters

One of the major criticisms of transgenic crops is that upon introduction, exogenous genes are normally expressed in most plant tissues and organs, where they are not needed. This occurs when one uses constitutive promoters, such as CaMV35S from the Cauliflower mosaic virus, which directs the expression of fused genes in all plant tissues and organs. Also, the expression of transgenes using virus promoters is neither efficient nor guaranteed (Zheng and Murai, 1997; Green et al., 2002; Neuteboom et al., 2002). To avoid these problems it is advisable to use promoters capable of controlling transgene expression in specific organs or tissues, and upon specific inducible stimuli. Therefore, transgenic cultivars resistant to root-feeding nematodes should use promoters induced specifically by nematodes and expressed in root tissues only.

Several promoters have been associated with gene expression in roots and in response to nematode infection (see review by Gheysen and Fenoll, 2002). Most of these promoters are associated with defense genes that are either up- or down-regulated in NFSs. These promoters are particularly interesting for use in transgenic nematode-resistant plants since they will drive expression of transgenes upon nematode infection. On the other hand, most of those defense genes are not exclusively responsive to nematodes, since they are general-defense genes regulated during plant response to biotic and abiotic stresses. Therefore, transgenes using these promoters would be expressed upon several kinds of stresses, not exclusively by nematode infection.

As an example, Mazarei et al. (2004) isolated and functionally characterized the promoter of the arabidopsis gene *At17.1*, whose function is unknown. *At17.1* is homologous to the soybean gene *GM17.1* and it is up-regulated by the soybean cyst nematode *H. glycines* Ichinohe. Transient expression analyses using the *At17.1* promoter region fused to the reporter gene GUS demonstrated that this promoter could induce gene expression upon nematode infection in both soybean and arabidopsis transgenic plants.

Another important study aimed to identify the regulatory region of the nematode-responsive promoter of the arabidopsis endoglucanase gene *Atcel1* (Sukno et al., 2006). These authors developed transgenic tobacco and arabidopsis plants bearing deletions of that promoter region, and the expression of the reporter gene GUS was monitored upon nematode infection. The analysis allowed the identification of a regulatory fragment that is essential for the activity of the promoter. The characterization of a promoter's element that specifically regulates the expression of a gene upon nematode infection is an important step towards minimizing pleiotropic effects of the promoter. In a preliminary study, Bertioli et al. (2001) evaluated the effect of nematode-responsive promoters on the expression of the tomato *Cf* genes and of their counterpart *avr* genes. The results showed that some *Cf/avr* combinations can activate a HR in tobacco plants, even in the absence of nematodes, indicating that different regulation features may be associated with nematode-responsive promoters.

In coffee, searches for nematode-responsive promoters are underway. In a recent study, several root-specific genes were evaluated regarding their responsiveness to nematode infection (Brandalise et al., 2005). A transcript was identified that exhibits expression on *C. arabica* roots upon *M. exigua* infection, and the corresponding promoter region was isolated and cloned from 'Mundo Novo'. Further transient expression analyses confirmed that this promoter controls gene expression in the roots of coffee seedlings (Brandalise et al., unpublished results). These authors also examined transgenic tobacco plants bearing that promoter fused to the reporter gene GUS upon infection by *M. javanica*. The results showed GUS expression in nematode-infected roots, where it was localized preferentially in the root's cortex and nematode-induced galls. This result indicates that the promoter can control gene expression in response to nematode infection in coffee and in other plant species as well. This is the first coffee tissue-specific promoter to be identified that potentially regulates nematode-responsive gene expression. Therefore, that promoter would be a suitable choice for transferring genes to develop transgenic RKN-resistant coffee cultivars. However, before that and other promoters can be used for developing transgenic cultivars it is necessary to identify specific nematode-responsive elements in the promoter sequence to avoid undesired biological responses.

10.10 Concluding Remarks

Transgenic plants represent a promising alternative for the development of new coffee cultivars, including nematode-resistant ones. However, the expectations should be kept in perspective regarding these cultivars as a definitive solution for RKN-infested areas. Since nematodes are extremely sophisticated parasites, which are capable of controlling plant metabolism through a cascade of events not yet completely understood, it is reasonable to expect that single transgenes will not be capable of sustaining complete, durable nematode resistance.

Hence, a long-term strategy is likely to require a combination of features to arrest nematode infection and/or development. Therefore, the goal should be the transference of a number of foreign genes, which would be accurately expressed in the transgenic plants, and aimed at disrupting distinct aspects of the nematode biology. Also, since gene identification through genomic analysis relies heavily on thoroughly characterized germplasm resources, a concerted effort involving nematologists, coffee breeders and molecular biologists is essential for this task. Furthermore, it is mandatory that transgenic approaches be associated with classical breeding methods for successful plant selection for the character being improved.

These recommendations may seem idealistic at this point, when knowledge of nematode-plant interactions is just starting to reach its molecular events. However, in the years to come we should expect ever more information to become available regarding the expression of nematode- and plant-genes during all phases of the interaction between these organisms. Therefore, the development of transgenic nematode-resistant coffee cultivars and their availability to growers should be just a matter of time.

References

Andrade AE, Albuquerque EVS, Grossi de Sá MF et al (2005) Expressão diferencial de proteinas em raizes de *Coffea canephora* infectadas com o nematóide endoparasita *Meloidogyne paranaensis*. Proceedings IV Simp de Pesqu Cafés Bras:22

Anthony F, Topart P, Martinez A et al (2005) Hypersensitive-like reaction conferred by the *Mex-1* resistance gene against *Meloidogyne exigua* in coffee. Plant Pathol 54:476–482

Anzueto F, Bertrand B, Sarah JL et al (2001) Resistance to *Meloidogyne incognita* in Ethiopian *Coffea arabica* accessions. Euphytica 118:1–8

Atkinson HJ, Urwin PE, McPherson MJ (2003) Engineering plants for nematode resistance. Ann Rev Phytopathol 41:615–639

Bar-Or C, Kapulnik Y, Koltail H (2005) A broad characterization of the transcriptional profile of the compatible tomato response to the plant parasitic root-knot nematode *Meloidogyne javanica*. Europ J Plant Pathol 111:181–192

Barbosa DHSG, Vieira HD, Souza RM et al (2006) Avaliação de genótipos de *Coffea arabica* em áreas isenta ou infestada com *Meloidogyne exigua* no noroeste fluminense. Proceedings XXXII Congr Bras de Pesqu Cafeeiras:236–237

Barros EVSA, Costa PM, Gomes ACMM et al (2006) Characterization of histological effects of *Meloidogyne incognita* infection in resistant and susceptible *Coffea arabica* genotypes. Proceedings XXI Int Conf Coffee Sci:PA191

Bent AF, Kunkel BN, Dahlbeck D et al (1994) RPS2 of *Arabidopsis thaliana*: a leucine rich repeat class of plant disease resistance genes. Science 265:1856–1860

Bertioli DJ, Guimarães PM, Jones JDG et al (2001) Expression of tomato *Cf* genes and their corresponding avirulence genes in transgenic tobacco plants using nematode responsive promoters. Ann Appl Biol 138:333–342

Bird DM (2004) Signaling between nematodes and plants. Curr Opinion Plant Biol 7:372–376

Brandalise M, Maluf MP, Guerreiro-Filho O et al (2005) Analysis of tissue-specific genes from leaf and root of *Coffea arabica*. Proceedings IV Simp Pesqu Cafés Bras:143

Cabos R, Sipes B, Schmitt D et al (2006) Transformation of *Coffea arabica* for root-knot nematode resistance using cysteine and serine proteinase inhibitors. Proceedings XXI International Conference on Coffee Science:PB280

Dale MFB, Scurrah MM (1998) Breeding for resistance to the potato cyst nematodes *Globodera rostochiensis* and *Globodera pallida*: strategies mechanisms and genetic resources. In: Marks RJ, Brodie BB (eds) Potato cyst nematodes – Biology, distribution and control. CABI Publishing, Oxford.

Davis EL, Hussey RS, Baum TJ (2004) Getting to the roots of parasitism by nematodes. Trends in Parasitol 20:134–141

Fazuoli LC, Gonçalves W, Braghini MT et al (2006) Tupi RN 1669-13: a coffee cultivar resistant to *Hemileia vastatrix* and *Meloidogyne exigua*. Proceedings XXI Intl Conf Coffee Sci:PB305

Fragoso RR, Batista JAN, Neto OBO et al (2005) Isolation and characterization of a cDNA encoding a serine proteinase from the root-knot nematode *Meloidogyne incognita*. Exp Parasitol 110:123–133

Gilbert JC, McGuire DC (1956) Inheritance of resistance to severe root-knot from *Meloidogyne incognita* in commercial type tomatoes. Proc Am Hort Soc 68:437–442

Gheysen G, Fenoll C (2002) Gene expression in nematode feeding sites. Ann Rev Phytopathol 40:191–219

Goddijn OJM, Lindsey K, van der Lee FM et al (1993) Differential gene expression in nematode induced feeding structures of transgenic plants harbouring promoter-gusA fusion constructs. Plant J 4:863–73

Goggin FL, Jia LL, Shah G et al (2006) Heterologous expression of the *Mi-12* gene from tomato confers resistance against nematodes but not aphids in eggplant. Mol Plant-Microbe Inter 19:383–388

Goellner M, Wang X, Davis EL (2001) Endo-β-14-glucanase expression in compatible plant-nematode interactions. Plant Cell 13:2241–2255

Graham MA, Marek LF, Lohnes D et al (2000) Expression and genome organization of resistance genes analogs in soybean. Genome 43:86–93

Green J, Vain P, Fearnehough MT et al (2002) Analysis of the expression patterns of the *Arabidopsis thaliana* tubulin-1 and *Zea mays* ubiquitin-1 promoters in rice plants in association with nematode infection. Physiol Mol Plant Pathol 60:197–205

Haq SK, Atif SM, Khan RH (2004) Protein proteinase inhibitor genes in combat against insects pests and pathogens: natural and engineered phytoprotection. Arch Bioch and Bioph 431:145–159

Huang GZ, Allen R, Davis EL et al (2006a) Engineering broad root-knot resistance in transgenic plants by RNAi silencing of a conserved and essential root-knot nematode parasitism gene. Proc Nat Acad Sci USA 103:14302–14306

Huang GZ, Dong R, Allen R et al (2006b) A root-knot nematode secretory peptide functions as a ligand for a plant transcription factor. Mol Plant-Microbe Interact 19:463–470

Innes RW (2004) Guarding the goods: new insights into the central alarm system of plants. Plant Physiol 135:695–701

Jacquet M, Bongiovanni M, Martinez M et al (2005) Variation in resistance to root-knot nematode *Meloidogyne incognita* in tomato genotypes bearing the *Mi* gene. Plant Pathol 54:93–99

Koltai H, Bird DMK (2000) Epistatic repression of PHANTASTICA and class 1 KNOTTED genes is uncoupled in tomato. Plant J 22:455–459

Koltai H, Dhandaydham M, Opperman C et al (2001) Overlapping plant signal transduction pathways induced by a parasitic nematode and a rhizobial endosymbiont. Mol Plant-Microbe Interact:14 1168–1177

Kretschemer JM, Chalmers KJ, Manning S et al (1997) RFLP mapping of the Ha2 cereal cyst nematode resistance gene in barley. Theor Appl Genet 94:1060–1064

Lamb C, Dixon RA (1997) The oxidative burst in plant disease resistance. Annu Rev Plant Physiol 48:251–275

Lawrence GJ, Finnegan EJ, Ayliffe MA et al (1995) The *L6* gene for flax rust resistance is related to the *Arabidopsis* bacterial resistance gene *RPS2* and the tobacco viral resistance gene *N*. Plant Cell 7:1195–1206

Lecouls AC, Petitot AS, Fernandez D (2006) Early expressed genes in the coffee resistance response to root-knot nematodes (*Meloidogyne* sp) infection. Proceedings XXI Int Conf Coffee Sci:PB265

Lilley CJ, Atkinson HJ, Urwin PE (2005) Molecular aspects of cyst nematodes. Mol Plant Pathol 6:577–588

Lohar DP, Schaff JE, Laskey JG et al (2004) Cytokinins play opposite roles in lateral root formation and nematode and rhizobial symbioses. Plant J 38:203–214

Matiello JB, Queiroz AR, Almeida SR et al (2004) Competição de cafeeiros hibridos com resistência à ferrugem e nematóide *M exigua* na Zona da Mata em Minas Gerais. Coffea – Rev Bras Tecnol Cafeeira 1:16–17

Mazarei M, Lennon KA, Puthoff DP et al (2004) Homologous soybean and *Arabidopsis* genes share responsiveness to cyst nematode infection. Mol Plant Pathol 5:409–423

Milligan SB, Bodeau J, Yaghoobi J et al (1998) The root knot nematode resistance gene *Mi* from tomato is a member of the leucine zipper nucleotide binding leucine-rich repeat family of plant genes. Plant Cell 10:1307–1319

Napoli C, Lemieux C, Jorgensen R (1990) Introduction of a chimeric chalcone synthase gene into petunia results in reversible co-suppression of homologous genes in trans. Plant Cell 2:279–289

Neuteboom LW, Kunimitsu WY, Webb D et al (2002) Characterization and tissue-regulated expression of genes involved in pineapple (*Ananas comosus* L) root development. Plant Sci 163:1021–1035

Noir S, Combes M-C, Anthony F et al (2001) Origin diversity and evolution of NBS-type disease-resistance gene homologues in coffee trees (*Coffea* L). Mol Genet Genomics 265:654–662

Noir S, Anthony F, Bertrand B et al (2003) Identification of a major gene (*Mex-1*) from *Coffea canephora* conferring resistance to *Meloidogyne exigua* in *Coffea arabica*. Plant Pathol 52:97–103

Ogita S, Uefuji H, Yamaguchi Y et al (2003) Producing decaffeinated coffee plants. Nature 423:823

Olsen AN, Skriver K (2003) Ligand mimicry? Plant-parasitic nematode polypeptide with similarity to CLAVATA3. Trends Plant Sci 8:55–57

Orsi CH (2003) Identification and analysis of RGA expressions in species of *Coffea* resistant and susceptible to the nematode *Meloidogyne exigua* State University of Campinas, MS Dissertation, Campinas.

Peele C, Jordan CV, Muangsan N et al (2001) Silencing of a meristematic gene using geminivirus-derived vectors. Plant J 27:357–366

Perthuis B, Pradon JL, Montagnon C et al (2005) Stable resistance against the leaf miner *Leucoptera coffeella* expressed by genetically transformed *Coffea canephora* in a pluriannual field experiment in French Guiana. Euphytica 144:321–329

Pflieger S, Lefebvre V, Caranta C et al (1999) Disease resistance gene analogs as cadidates for QTLs involved in pepper-pathogen interactions. Genome 42:1100–1110

Ratcliff F, Martin-Hernandez AM, Baulcombe DC (2001) Tobacco rattle virus as a vector for analysis of gene function by silencing. Plant J 25:237–245

Ribas AF, Kobayashi AK, Pereira LFP et al (2005) Genetic transformation of *Coffea canephora* by particle bombardment. Biol Plantarum 49:493–497

Rodrigues ACFO, Abrantes IM, Melillo MT et al (2000) Ultrastructural response of coffee roots to root-knot nematodes *Meloidogyne exigua* and *M megadora*. Nematropica 30:201–210

Salgado SL, Campos VP, Resende MLV et al (2002) Reprodução de *Meloidogyne exigua* em cafeeiros Iapar-59 e Catuai. Nematol Bras 26:205–207

Schmelz EA, Grebenok RJ, Galbraith DW et al (1999) Insect-induced synthesis of phytoecdysteroids in spinach *Spinacia oleracea*. J Chem Ecol 25:1739–1757

Schmelz EA, Grebenok RJ, Ohnmeiss TE et al (2000) Phytoecdysteroid turnover in spinach: long-term stability supports a plant defense hypothesis. J Chem Ecol 26:2883–2896

Silvestrini M, Maluf MP, Guerreiro-Filho O et al (2005) Defense gene expression in response to nematode infection. Proceedings IV Simp Pesqu Cafés Bras:133

Soriano IR, Riley IT, Potter MJ et al (2004) Phytoecdysteroids: a novel defense against plant-parasitic nematodes. J Chem Ecol 10:1885–1899

Sukno S, Shimerling O, McCuiston J et al (2006) Expression and regulation of the *Arabidopsis thaliana* cell *endo 14 beta glucanase* gene during compatible plant-nematode interactions. J Nematol 38:354–361

Urwin PE, Atkinson HJ, Waller DA et al (1995) Engineered *oryzacystatin-I* expressed in transgenic hairy roots confers resistance to *Globodera pallida*. Plant J 8:121–131

Urwin PE, Lilley CJ, McPherson MJ et al (1997) Resistance to both cyst- and root-knot nematodes conferred by transgenic *Arabidopsis* expressing a modified plant cystatin. Plant J 12:455–461

Valentine T, Shaw J, Blok VC et al (2004) Efficient virus-induced gene silencing in roots using a modified tobacco rattle virus vector. Plant Physiol 136:3999–4009

Van Boxtel J, Berthouly M, Carasco C et al (1995) Transient expression of β-glucuronidase following biolistic delivery of foreign DNA into coffee tissues. Plant Cell Rep 14:748–752

Vercauteren I, Engler JD, De Groodt R et al (2002) An *Arabidopsis thaliana* pectin acetylesterase gene is upregulated in nematode feeding sites induced by root-knot and cyst nematodes. Mol Plant Micr Inter 15:404–407.

Wang X, Mitchum MG, Gao B et al (2005) A parasitism gene from a plant-parasitic nematode with function similar to CLAVATA3/ESR (CLE) of *Arabidopsis thaliana*. Mol Plant Pathol 6:187–191

Williamson VM, Gleason CA (2003) Plant–nematode interactions. Curr Opin Plant Biol 6:327–333

Zheng Z, Murai N (1997) A distal promoter region of the rice seed storage protein glutelin gene enhanced quantitative gene expression. Plant Sci 128:59–65

Part IV
Other Coffee-Associated Nematodes

Chapter 11
Other Coffee-Associated Nematodes

Ricardo M. Souza

Abstract This chapter reviews the information available on the many nematode genera and species that have been associated with coffee, with the exception of *Meloidogyne* spp. and *Pratylenchus* spp., which are dealt with in other chapters. For many of these species, their coffee-parasitic status cannot be asserted beyond doubt, because they have been found in soil samples collected around coffee plants; this sampling method does not preclude the possibility of reporting nematodes that were actually parasitizing weeds or intercropped plants, or even plants that had grown in the field previously to coffee. On the other hand, coffee-parasitic status can be assigned to many species from the proper sampling and extraction methods used in the surveys, or from laboratory or greenhouse studies. For a subset of the parasitic species, there have been reports of damage to coffee, particularly by *Xiphinema* spp., *Hemicriconemoides* spp., *Radopholus* spp., *Rotylenchulus reniformis* and *Helicotylenchus* spp.; in some cases, controlled studies have confirmed the pathogenicity to coffee. This review examines critically all these reports, and outlines initiatives that could contribute to assessing the real importance of these species to coffee production worldwide.

Keywords Minor coffee-parasitic nematodes · coffee-associated nematodes

11.1 Introduction

In addition to *Meloidogyne* sp. and *Pratylenchus* sp., a plethora of plant-parasitic nematode genera and species has been reported from surveys in coffee (*Coffea* sp.) plantations and nurseries throughout the world. In several reports, some of them published as conference proceedings, it is difficult to apprehend from the methodology described whether the nematodes reported were actually parasitizing coffee plants or doing so in weeds or intercrops. From a scientific standpoint, even if a

R.M. Souza
Universidade Estadual do Norte Fluminense Darcy Ribeiro, Lab. Entomologia e Fitopatologia, Campos dos Goytacazes, Brazil
e-mail: ricmsouza@censanet.com.br

R.M. Souza (ed.), *Plant-Parasitic Nematodes of Coffee*,

nematode taxon has repeatedly been found in samples collected around coffee plants this does not constitute proof of its coffee-parasitic status. In line with this scientific strictness, this review considers that soil samples collected around plants with shovel or auger do not constitute samples from the rhizosphere, which has been defined as a narrow region of soil that is directly influenced by root secretions and associated soil microorganisms. Naturally, finding a nematode in a true coffee rhizosphere sample does constitute strong evidence of its parasitic status.

For many nematode taxa, the coffee-parasitic status has been confirmed through appropriate methods for nematode extraction from plant roots, controlled seedling inoculations or histological studies. For some taxa, a pathogen status has been suggested by the association between high soil population and damage to coffee seedlings or plantations; for a subset of those, this presumed status has been confirmed through assessments under controlled conditions.

This chapter reviews the information available on those nematodes considered of minor importance to coffee cultivation. The list of coffee-associated nematodes presented here is possibly not all-inclusive, since many local or regional surveys have been published in sources not easily retrievable. On the other hand, as discussed in Chapters 3 and 6 for *Pratylenchus* sp. and *Meloidogyne* sp., taxonomy to species level requires expertise, proper methodology and often access to previous publications and types; classification instability itself complicates matters. The same difficulties apply to nematode genera discussed in this chapter. Therefore, producing an accurate worldwide list of coffee-parasitic nematode species would be a daunting task which would require an international effort involving experts in several nematode groups to conduct resamplings, examine long-stored glass slides and survey notes and carry out proper host status assessments.

By critically examining the information available on 'minor' coffee-associated nematodes, this review aims to stimulate nematologists to examine such associations more accurately, performing basic and applied studies to confirm their parasitic status and to assess their real importance to coffee production worldwide. As has occurred many times in plant pathology, systematic studies have, on the one hand, not confirmed presumed parasitic associations; on the other hand, many important plant-pathogens have been unveiled from the realm of the 'little-known' or 'not important' organisms.

In this chapter, the taxonomic identification given to coffee-associated nematodes in the original publications has been verified against the classification reviews by Sturhan and Brzeski (1991), Jairajpuri and Ahmad (1992) and Siddiqi (2000) and updated accordingly.

11.2 Field Surveys and Species Descriptions

Many nematode surveys have been carried out in coffee plantations and nurseries throughout the world. In several of these the sampling and/or nematode extraction method employed apparently did not preclude the possibility that the taxa reported had been parasitizing weeds or intercropped plants. As early as 1960, Luc and de Guiran stressed the importance of confirming the parasitic status of any nematode taxon found in coffee plantation surveys. In studies conducted in the Ivory Coast, Guinea, Togo, Senegal and Cameroon, those authors have found the following nematodes in soil samples collected around plants of robusta coffee (*C. canephora* Pierre ex A. Froehner): *Criconemoides limitaneus* [= *Discocriconemella limitanea* (Luc) De Grisse and Loof *sensu* Siddiqi, 2000], *Helicotylenchus* spp. [including *H. erythrinae* (Zimmermann) Golden], *Hemicycliophora paradoxa* [= *Hemicaloosia* paradoxa (Luc) Ray and Das], *Rotylenchoides affinis* Luc and *Tylenchorhynchus* sp. Around plants of arabica coffee (*C. arabica* L.) they have found *Criconemoides onoensis* [= *Macroposthonia onoensis* (Luc) De Grisse and Loof], *Helicotylenchus* spp. (including *H. erythrinae*), *Scutellonema bradys* (Steiner and LeHew) Andrássy and *Xiphinema* spp. (including *X. ebriense* Luc). Those authors stressed that coffee-parasitism by these species was probable, but had not been proved. As seen below, later surveys or controlled experiments confirmed the parasitic status of several of those taxa.

In 1969, Whitehead listed several ectoparasitic nematodes that had been reported associated with coffee worldwide, but with no certainty as to their parasitic status. Those included *Ditylenchus procerus* (Bally and Reydon) Filipjev (species inquirenda to Sturhan and Brzeski, 1991), *Paratylenchus besoekianus* Bally and Reydon, *P. macrophallus* (de Man) Goodey (species inquirenda to Siddiqi, 2000), *Trichodorus christiei* [= *Paratrichodorus christiei* (Allen) Siddiqi *sensu* Jairajpuri and Ahmad, 1992], *T. monohystera* [= *Monotrichodorus monohystera* (Allen) Andrássy], *Xiphinema insigne* Loos and *X. radicicola* Goodey.

In an excellent taxonomic study on dorylaimid nematodes associated with coffee plantations in the State of São Paulo, Brazil, Monteiro (1970a; b) has listed 43 species, including 11 new species, *Xiphinema brevicolle* Lordello and Costa and *X. krugi* Lordello, which are recognized as parasitic to coffee. *X. krugi* has also been reported from south Brazil (Lordello et al., 1974).

Surveys from which the coffee-parasitic status cannot be apprehended with certainty from the methodology employed include those by Garcia et al. (1988), Lima and Almeida (1989), Dias et al. (1996) and Lordello and Lordello (2001) in Brazil. In these surveys, the nematodes found associated with arabica and/or robusta coffees have not been identified beyond the generic level; these were *Helicotylenchus* sp., *Trichodorus* sp., *Ditylenchus* sp. and *Rotylenchulus* sp., among others. Early in 1928, Rahm reported *Tylenchorhynchus robustus* Cobb from coffee roots, but this species has not been listed by Siddiqi (2000). Instead, this author has considered *T. robustus* var *pseudorobustus brasiliensis* Rahm a *nomina nuda*.

Also in Brazil, Ferraz (1980) has sampled soil around coffee plants and found *Aphelenchus avenae* Bastian, *Helicotylenchus dihystera* (Cobb) Sher, *H. erythrinae, H. pseudorobustus* (Steiner) Golden, *Macroposthonia curvata* (Raski) De Grisse and Loof, *M. onoensis, M. sphaerocephalus* (Taylor) De Grisse and Loof, *Rotylenchulus reniformis* Linford and Oliveira, *Xiphinema brevicolle, X. krugi* and *X. surinamense* Loof and Maas. In addition, several nematodes were identified at the genus level only, such as *Filenchus* sp., *Nothocriconema* sp. (= *Criconema* sp.), *Pseudohalenchus* sp., *Tylenchulus* sp. and *Tylenchus* sp.

In an extensive and well-conducted survey carried out in Brazil, Castro et al. (2008) remained uncertain about coffee-parasitism by *Discocriconemella degrissei* Loof and Sharma, *D. limitanea, D. repleta* (= *D. limitanea*), *Aphelenchoides bicaudatus* (Imamura) Filipjev and Schuurmans Stekhoven, *A. coffeae* (Zimmermann) Filipjev, *A. tagetae* Steiner, *Tylenchus hamatus* (Thorne and Malek) Raski and Geraert, *T. sandneri* (Wasilewska) Raski and Geraert, *Criconema* sp., *Gracilacus* sp., *Hoplotylus* sp., *Merlinius* sp., *Ogma* sp., *Polenchus* sp., *Diphtherophora* sp., *Rotylenchus* sp. and *Tetylenchus* sp., among other genera.

In India, Kumar (1981; 1983) has found *Discocriconemella pannosa* Sauer and Winoto, *D. retroversa* Sauer and Winoto, *Tylenchorhynchus ewingi* Hopper, *Quinisulcius acti* [= *Q. capitatus* (Allen) Siddiqi], *Trichotylenchus astriatus* [= *Uliginotylenchus astriatus* (Khan and Nanjappa) Siddiqi, 1986] and *Trophurus similis* Khan and Nanjappa in soil collected around coffee plants. Using the same sampling approach, Giribabu and Saha (2003) have found *Aphelenchoides asterocaudatus* Das and *Aphelenchus avenae*.

In a review on management of coffee-parasitic nematodes, Kumar (1988) has listed additional nematodes that had been found associated with coffee in India: *Boleodorus thylactus* Thorne, *Gracilacus aculenta* (= *Paratylenchus aculentus* Brown), *G. mutabilis* (= *P. mutabilis* Colbran), *G. peperpotti* [= *P. peperpotti* (Shoemaker) Siddiqi and Goodey], *Hemicriconemoides chitwoodi* Esser, *H. cocophillus* (Loos) Chitwood and Birchfield, *Helicotylenchus erythrinae, H. dihystera, Hemicycliophora penetrans* [= *Aulosphora penetrans* (Thorne) Siddiqi], *H. typica* (= *H. thornei* Goodey), *Hoplolaimus coronatus* [= *H. galeatus* (Cobb) Thorne], *Paratylenchus coronatus* Colbran, *P. goodeyi* Oostenbrink, *P. vandenbrandei* de Grisse, *Rotylenchus robustus* (de Man) Filipjev, *Rotylenchulus reniformis, Scutellonema bradys, Trophorus imperialis* Loof, *Tylenchorhynchus dubius* [= *Bitylenchus dubius* (Butschli) Filipjev], *Xiphinema chambersi* Thorne, *X. index* Thorne and Allen, *X. ornatum* Loos [Jairajpuri and Ahmad (1992) have made no reference to this species], *Heterodera* sp. and *Longidorus* sp. Kumar (1988) has also listed five species about which Siddiqi (2000) has made no reference: *Hemicriconemoides cassiae* Kumar, *Macroposthonia grissei* Kumar, *Nothocriconema indicum* Kumar, *Radopholus colbrani* Kumar and *Scutellonema conlcaphalum siva* Kumar and Selvasekaran.

In a classification review, Dasgupta et al. (1969) have stated that previous authors had found *Hemicriconemoides gaddi* (Loos) Chitwood and Birchfield associated with coffee in India. Germani and Anderson (1991) have reported *H. mangiferae* Siddiqi associated with this crop in Vietnam.

In Chapter 15, Loang K. Tran (WASI) reports several nematodes associated with coffee in Vietnam, such as *Radopholus* sp., *Rotylenchus* sp., *Rotylenchulus reniformis, Tylenchorhynchus brassicae* Siddiqi, *Hoplolaimus seinhorsti* (Luc) Shamsi, *Helicotylenchus coffeae* Eroshenko and Nguen Vu Thanh, *H. concavus* Roman, *H. crassatus* Anderson, *H. crenacauda* Sher, *H. dihystera, H. digonicus* Perry in Perry, Darling and Thorne, *H. dignus* Eroshenko and Nguen Vu Thanh,

H. erythrinae, H. exallus Sher, *H. paraconcavus* Rashid and Khan, *H. pseudorobustus, Criconemoides goodeyi* de Guiran, *Macroposthonia onoensis, M. magnifica* Eroshenko and Tkhan, *Crossonema fimbriatum* (Cobb in Taylor) Mehta and Raski and *Xiphinema insigne*.

From São Tomé and Principe, Arias et al. (1995) have reported *Longidorus laevicapitatus* Williams associated with arabica and robusta coffees, *Xiphinema setariae* Luc and *X. vulgare* Tarjan (species inquirenda to Jairajpuri and Ahmad, 1992) associated with arabica, *X. dihysterum* Lamberti, Arias, Agostinelli and Santo with robusta and *X. longicaudatum* Luc with arabica and *C. liberica* W. Bull ex Hiern.

From the Ivory Coast, van Doorsselaere and Samsoen (1982) have reported *Aphelenchoides bicaudatus, Aphelenchus avenae, Criconemella onoensis* (= *Macroposthonia onoensis*), *Malenchus cognatus* (= *M. acarayensis* Andrássy), *Scutellonema bradys, Tylenchus clarki* [= *Filenchus clarki* (Egunjobi) Siddiqi] and *Tylenchus discrepans* [= *Ottolenchus discrepans* (Andrássy) Siddiqi and Hawksworth], which were found in soil collected around coffee plants. In Chapter 17, A. Adiko (CNRA) also reports *Helicotylenchus* sp. and *Paratylenchus* sp. from the Ivory Coast.

In Hawaii (USA), Schenck and Schmitt (1992) have reported that *Criconemella* sp. (= *Criconemoides* sp.) was common in coffee plantations following sugarcane, but much rarer in older plantations. They often found *Helicotylenchus* spp. (including *H. dihystera*) and *Paratylenchus minutus* Linford in Linford, Oliveira and Ishii.

In Guatemala, Herrera and Marban-Mendoza (1999) have reported a conspicuous presence of *Rotylenchulus reniformis* in coffee plantations, although they have not assessed whether this nematode was actually parasitizing coffee plants or causing yield losses.

A number of nematode species have been described from soil samples collected around coffee plants, although no further studies seem to have been conducted to assess their parasitic status. From India, these include *Scutellonema coffeae, Quinisulcius seshadrii, Xiphinema arubreviensis* and *Helicotylenchus shervarayensis* (Giribabu and Saha, 2002; 2006), *Caloosia loofi* and *Trophonema coffeae* (Kumar, 1979) [Siddiqi (2000) has made no reference to these species and has synonymized *Trophonema* to *Trophotylenchulus*], *Radopholus colbrani, Hemicriconemoides coffeae, H. cassiae, Nothocriconema indicum, Discocriconemella andrassyi* and *D. cardamomi* (Kumar, 1980; 1982; 1983) [Siddiqi (2000) has made no reference to this species], *Tylenchorhynchus amgi* (Kumar, 1981), *Rotylenchoides desouzai* [= *Orientylus desouzai* (Kumar and Rao) Orton Williams] and *Scutellonema conicephalum* (Sivakumar and Selvasekaran, 1982).

Paratylenchus holdemani has been described from El Salvador (Raski, 1975), *Dolichorhynchus prophasmis* [= *Neodolichorhynchus prophasmis* (Jairajpuri and Hunt) Talavera and Tobar] from Zimbabwe, *Hemicriconemoides snoecki* from the Ivory Coast (van Doorsselaere and Samsoen, 1982) and *Allotrichodorus loofi* from Brazil (Rashid et al., 1985).

It is clear that there is limited usefulness to surveys in which the sampling and processing strategies employed and the time and expertise required are not arranged in such a manner as to allow taxonomic identifications at the species level and confirmation of coffee-parasitic status. Even for nematodes that have been recognized as parasitic to coffee, such as *R. reniformis*, their presence in the soil sample does not constitute useful information because the nematode's host range may include weeds or cultivated plants common to coffee plantations. Geographically-distant populations of the same species may differ in their ability to parasitize coffee, as seems to be the case for *R. reniformis* (see below). Furthermore, because of the genetic diversity among coffee cultivars and varieties, any report on coffee-parasitism must specify the genotype involved.

Much more informative are those surveys in which the sampling and nematode extraction methods employed have allowed the authors to assert the taxa as parasitic to coffee. In Brazil, this parasitic status has been given by Prates et al. (1985), Campos et al. (1987), Souza et al. (1999), Kubo et al. (2001) and Castro et al. (2008) to *Macroposthonia* spp. [including *M. xenoplax* (Raski) De Grisse and Loof, *M. sphaerocephalus, M. ornata* (Raski) De Grisse and Loof, *M. onoensis, M. curvata, M. palustris* (Luc) Loof and de Grisse, *M. discus* (Thorne and Malek) Loof and de Grisse, *M. humilis* (= *Criconemoides humilis* Raski and Riffle) and *M. inusitata* (= *Criconemoides inusitatus* Hoffmann), *Xiphinema brevicolle, Paratrichodorus minor* (Colbran) Siddiqi, *Helicotylenchus* spp. (including *H. dihystera*) and *Rotylenchulus reniformis*. Those authors have also listed coffee-parasitic nematodes that were not identified at the species level, such as *Criconemella* sp. (= *Criconemoides* sp.), *Trichodorus* sp., *Discocriconemella* sp., *Criconema* sp., *Scutellonema* sp., *Rotylenchus* sp., *Xiphinema* sp., *Ditylenchus* sp., *Tylenchus* sp., *Nothotylenchus* sp., *Aphelenchus* sp., *Aphelenchoides* sp., *Ecphyadophora* sp., *Hemicycliophora* sp., *Paratylenchus* sp. and *Tylenchorhynchus* sp.

In a soil and root sampling of declining coffee plantations in the State of Bahia, Brazil, Sharma and Sher (1973) have found *Helicotylenchus dihystera, Xiphinema* spp. (including *X. basiri* Siddiqi and *X. brasiliense* Lordello), *Rotylenchulus reniformis, Criconemoides onoensis* (= *Macroposthonia onoensis*), *Dolichodorus* sp. and *Trichodorus* sp. in most of the samples. *Criconema decalineatum* [= *Ogma decalineatum* (Chitwood) Andrássy], *Paratylenchus minutus* and *Tylenchus* sp. were found less frequently. In coffee nurseries, Lordello (1980) and Santos and Silva (1984) have reported coffee seedlings infected with *R. reniformis*. Apparently, these authors' concerns that this species might become a serious problem for coffee production in Brazil have not materialized.

In Tanzania, Bridge (1984) found the following species to be parasitic to coffee: *Criconemella sphaerocephala* (= *M. sphaerocephalus*), *Hemicriconemoides cocophilus, Quinisulcius capitatus, Scutellonema africanum* Smit, *S. magniphasmum* (= *S. magniphasma* Sher), *Xiphinema elongatum* Schuurmanns Stekhoven and Teunissen, *Helicotylenchus mucronatus* Siddiqi, *Discocriconemella limitanea, Aphelenchus avenae, Hoplolaimus indicus* Sher, *Rotylenchoides brevis* Whitehead and *Tylenchorhynchus mashhoodi* Siddiqi and Basir. Several coffee-parasitic nematodes were identified at the genus level only, such as *Hemicycliophora* sp., *Gracilacus* sp., *Nothotylenchus* sp., *Hoplolaimus* sp. and *Ditylenchus* sp., among a few others.

From Uganda, J. Namaganda (NARO) reports in Chapter 17 that *Rotylenchulus reniformis, Helicotylenchus dihystera* and *Tylenchus* sp. have been found in roots of robusta coffee, while *Aphelenchus* sp., *Trichodorus* sp., *Xiphinema* sp. and *Paralongidorus* sp. have been found in the soil around the plants.

In Panama and El Salvador, Tarté (1970), Pinochet (1987) and Pinochet and Guzman (1987) have listed *Xiphinema americanum* Cobb, *Radopholus similis* (Cobb) Thorne, *Aphelenchoides* sp., *Ditylenchus* sp., *Paratylenchus* sp. and *Criconemella* spp. (= Criconemoides sp.) as parasitic to arabica coffee. In Cuba, Rodriguez et al. (2000) have found in coffee roots several nematodes identified at the genus level only, such as *Pratylenchus* sp., *Radopholus* sp., *Rotylenchulus* sp. and *Xiphinema* sp.

In India, a field survey conducted by Sekhar (1963) has reported *Xiphinema americanum, Tylenchorhynchus* sp., *Hoplolaimus* sp., *Hemicriconemoides* sp., *Rotylenchulus* sp., *Helicotylenchus* sp. and *Aphelenchoides* sp. from arabica and robusta coffee roots. In Vietnam, *Radopholus duriophilus* Nguyen, Subbotin, Madani, Trinh and Moens has been reported as parasitic to robusta coffee and *R. arabocoffeae* has been described from coffee roots in Vietnam (Trinh et al., 2004).

In Indonesia, S. Wiryadiputra (ICCRI) has found several nematode species associated with coffee (see Chapter 15). These include *Aphelenchus avenae*, *Criconemoides morgensis* (Hofmanner in Hofmanner and Menzel) Taylor, *Ditylenchus dipsaci* (Kühn) Filipjev, *Helicotylenchus dihystera, Hemicriconemoides chitwoodi, Hemicycliophora arenaria* Raski, *Paratylenchus besoekianus, Radopholus similis, Rotylenchulus reniformis, Rotylenchus robustus, Tylenchorhynchus dubius* (= *Bitylenchus dubius*) and *Tylenchus davainei* Bastian.

For most of the coffee-parasitic nematode species listed above, no studies have been conducted on their feeding behavior on coffee roots, their potential damage to root tissues and plant physiological processes, their population fluctuation or epidemiology (if pathogenic to coffee) or their economic importance to coffee production. As remarked by De Waele and Elsen (2007), in tropical countries plant-parasitic nematodes receive attention from nematologists and funding agencies only if the nematode's economic importance has been established, which generally occurs through preliminary surveys or reports from growers or extension personnel. Without human and funding resources available for 'exploratory' research on 'non-important' nematodes, only a few species have been studied in any detail. Apparently, with the exception of *Hemicriconemoides* spp. in India and *Radopholus* spp. in Vietnam, the so-called 'minor' nematodes have never been the subject of a long-term research program; studies and publications have been scattered in time and space during the last decades.

As stated by De Waele and Elsen (2007), surveys conceived only to list coffee-parasitic nematodes are of limited utility because they do not inform which nematode species are predominant and potentially damaging; they only rule out those which are not present in the region surveyed. Surveys should also bring additional information on the species' frequency and abundance, which could be correlated to observations on plant damage and plantation yield to identify potential nematode problems. Possibly, such preliminary data could at least support applications to funding agencies for 'exploratory' studies.

11.3 Pathogenicity Reports from Field Observations

As regards *Helicotylenchus* sp., D'Souza and Sreenivasan (1965) have stated that in India its parasitism invariably caused poor growth of coffee plants. To Sekhar (1963), no damage was observed if *Helicotylenchus* sp. only or *Rotylenchulus* sp. only were involved. To this author instead, decline and death of arabica coffee plants – robusta ones seemed tolerant – only occurred when *Helicotylenchus* sp. and *Rotylenchulus* sp. were associated with *Pratylenchus* sp. [mainly *P. coffeae* (Zimmermann) Filipjev and Schuurmans Stekhoven]. In Colombia, *H. erythrinae* has been associated with lesions in coffee secondary and tertiary roots, leading to invasion and destruction of the root system by *Fusarium* sp. and *Rosellinia* sp. (see Chapter 13).

In India, D'Souza and Sreenivasan (1965) have stated that *R. reniformis* was a serious problem in coffee plantations, invariably damaging the plants. In heavily infested areas (~10 nematodes/50 cc of soil) arabica coffee plants failed to grow despite all the good agronomic practices applied. The parasitized plants presented almost no feeder roots, poorly developed tap-root, yellowing and wilting of above-ground parts. In the Pacific Islands, Bridge and Page (1984) (cited by Bridge, 1988) have reported *R. reniformis* associated with leaf chlorosis and wilting of coffee plants.

According to Zimmermann (1898) and Bally and Reydon (1931) (both cited by Kumar and Samuel, 1990), *Radopholus similis* was highly pathogenic to arabica and robusta coffees in Java, causing root rotting. In India, D'Souza and Sreenivasan (1965) have stated that *R. similis* invariably caused poor growth of coffee plants. In Indonesia, this nematode has become a major concern in several coffee-producing provinces, although there has been no assessment of its damage.

In some locations in Guatemala, Thorne and Schieber (1962) have reported a high incidence of *Xiphinema americanum* in coffee plantations, often in high populations. According to these authors, at least in one location the combined parasitism by *Pratylenchus* sp., *Meloidogyne* sp. and *X. americanum* have produced a pathology that practically destroyed the plants' root system and caused the plantation's decline. In other locations, parasitism by *X. americanum* seemed to be manageable by proper agronomic practices.

In India, Kumar and Samuel (1990) have given an account of the widespread incidence and pathogenicity of *Hemicriconemoides* spp. to robusta and arabica coffees, causing 'crinkle-leaf' disorder. On the other hand, Kumar (1988) had stated that 'crinkle-leaf' was caused by concomitant high populations of *Hemicriconemoides* sp., *Nothocriconema* sp. (= *Criconema* sp.) and *Helicotylenchus* sp., among others. Nematicides have been considered inefficient for controlling these nematodes; instead, Kumar recommended eradicating the declining coffee plants, fallowing and replanting with robusta coffee or arabica grafted onto a robusta rootstock.

Ideally, such scattered and sometimes contradictory reports should be assessed through controlled studies under greenhouse and/or field conditions. Under the latter, special attention should be given to abiotic and/or biotic factors that could interplay with nematodes to cause or enhance plant damage. Therefore, such studies should necessarily monitor edaphic and climatic conditions, as well as investigate whether other soil-borne organisms, such as fungi and bacteria, could be involved to create a complex pathosystem.

11.4 Some Studies Under Controlled Conditions

According to Kumar and Samuel (1990) young arabica coffee plants parasitized by *Radopholus similis* exhibit retarded growth, undersized chlorotic leaves and enhanced susceptibility to drought. The tap- and secondary roots may be entirely destroyed and the plants tend to emit adventitious roots at the collar region. Accordingly, Milne and Keetch (1976) found arabica plants to be highly susceptible to *R. similis*' 'banana race'; the inoculated seedlings suffered severe growth retardation. Likewise, Zem and Lordello (1983) have attested the susceptibility of arabica coffee 'Mundo Novo' to a Brazilian population isolated from banana 'Nanicão'. Nonetheless, the ability of this nematode to reproduce on and damage coffee probably varies according to the plant and nematode genetic 'make-ups'. Indeed, Kumar and D'Souza (1969) and Kumar (1980) were unable to reproduce on coffee *R. similis* populations isolated from black pepper and banana.

Recent studies conducted in Vietnam under controlled conditions have revealed *R. arabocoffeae* Trinh, Nguyen, Waeyenberge, Subbotin, Karssen and Moens as more prolific on and pathogenic to seedlings of arabica coffee 'Catimor' than *Pratylenchus coffeae and R. duriophilus* (Trinh et al., 2004).

As regards *Rotylenchulus reniformis*, Ayala (1962) (cited by Macedo, 1974) has demonstrated its pathogenicity to arabica coffee 'Puerto Rico' under greenhouse conditions, although with restricted nematode reproduction. In the Philippines, Valdez (1968) reported *R. reniformis* as the causal agent of 'stubby root' disease of arabica and robusta coffee seedlings, as well as seedlings of *C. excelsa* (= *C. liberica* var *dewevrei*). Valdez observed a severe reduction in the plants' root system coupled with delay in their development and abundant nematode reproduction. In Brazil, Macedo (1974) has observed under greenhouse conditions a limited reproduction of *R. reniformis* in the arabica coffees 'Mundo Novo' and 'Catuai'; no reproduction was observed on the robusta coffee 'Guarini'. In India, Kumar and Samuel (1990) have considered erroneous previous reports that *R. reniformis* would be parasitic to coffee, although Vovlas and Lamberti (1990) have characterized the histological alterations caused by this nematode on arabica coffee roots.

Schenck and Schmitt (1992) have concluded that coffee is a poor host for *R. reniformis* from Hawaii (USA), although this nematode has often been found in soil samples collected in coffee plantations; they concluded that *R. reniformis* reproduces mostly in weeds and grasses intercropped to function as windbreaks. Through controlled inoculations, Schenck and Schneck (1994) have assessed the host status of several coffee genotypes for a population of *R. reniformis* from Hawaii. These authors observed that the nematode did infect the seedlings, but its population remained low and the plants remained not visibly damaged.

In conclusion, it seems that *R. reniformis* populations from Southeast Asia, e.g. the Philippines, are capable of reproducing abundantly on and being pathogenic to coffee, while most populations from other world regions seem marginally capable of reproduction, although capable of delaying plant development. Furthermore, coffee-parasitic populations seem to remain restricted geographically by mechanisms that are not understood. For example, in a survey covering plantations and nurseries located in 119 municipalities in the State of Minas Gerais, Brazil (total of 2,266 samples), *R. reniformis* has been found parasitizing coffee in just one location (Souza et al., 1999).

Certainly both nematode and plant genetic 'make-ups' interplay in this pathosystem in ways that have not been addressed by nematologists. Indisputably, the interactions between *Coffea* sp. and *R. reniformis* hold a great deal of exciting aspects for future studies.

Apparently, *Hemicriconemoides* sp. has been associated with damage to coffee plants in India only. Its pathogenicity to arabica and robusta coffee seedlings has been characterized under controlled conditions (Kumar and D'Souza, 1969). The nematode reduced the growth and weight of the seedlings' shoot and root system; their leaves did not fully develop and turned dull-green. The nematode successfully reproduced on the plants; those authors observed a reproduction factor varying from four to six, eight months after inoculation. In Anonymous (1986), it is said that in controlled inoculations of arabica and robusta coffee seedlings, *H. coffeae, H. cocophilus* and *H. gaddi* have significantly reduced the plants' stem height and root weight of both types of coffee. Nine months after the inoculations, the reproduction factor varied from three to nine depending on the nematode species and coffee type involved. The 'crinkle leaf' symptoms developed predominantly on arabica coffee.

It has been reported that in the field the population fluctuation of *H. gaddi* appears to be related to rainfall pattern: the soil population decreases during the winter and early summer, during which the soil is mostly dry; the new root flushes during April, May and June allow the nematode population to increase, while the heavy rainfall in July and August appears to be adverse to the nematode (Anonymous, 1986).

As for other nematodes, Vovlas (1987) has studied the histopathology of coffee roots parasitized by *Trophotylenchulus obscurus* (Colbran) Cohn and Kaplan. In 1985, Vovlas and Lamberti had done the same with coffee roots parasitized by a population of *Hoplolaimus pararobustus* from São Tomé. On this island Vovlas and Lamberti have observed a widespread incidence of this nematode on coffee plantations, but they have not assessed its possible damage to the plants. Through controlled inoculations, Lamberti et al. (1992) have concluded that robusta coffee seedlings are intolerant poor hosts to populations of *Xiphinema ifacolum* Luc and *X. longicaudatum* from Liberia.

11.5 Concluding Remarks

It is clear that a great many studies remain to be done on coffee-parasitic nematodes other than *Meloidogyne* sp. and *Pratylenchus* sp. As stated by Luc et al. (2005), establishing the pathogenicity of nematodes involved in subtropical and tropical agriculture should be a main priority. Careful laboratory, greenhouse and field studies should be conducted on those nematodes regardless of their presumed low importance to coffee production. Such investment of human and monetary resources may be considered unjustifiable in commodity-driven research institutions such as Cenicafe, Embrapa and CCRI in Colombia, Brazil and India, respectively. On the other hand, 'exploratory' multidisciplinary studies on 'minor' coffee-parasitic nematodes are easily justifiable in the university academic environment, which should favor all scientific enterprise regardless of its economic relevance.

For example, an effort to better understand the interactions between coffee and *R. reniformis* or *Radopholus* spp. may have all the ingredients for the scientific training of future nematologists, such as setting up collaboration to establish a collection of isolates, putting forward hypotheses, developing proper methodology, determining whether a pathosystem is involved and if applicable, gauging the nematode-induced yield loss and economic relevance.

Unquestionably, studies such as these can be all the more scientifically stimulating, easily granted and far-reaching within the framework of a national or international collaboration. For nematologists working in tropical countries, establishing collaborations with committed scientists abroad is not a condition for developing top-ranking research, but it may help guarantee a continuous flow of resources and may make several initiatives easier, such as obtaining nematode isolates and coffee genotypes or having molecular tasks performed if proper facilities or expertise are not readily available.

As has happened in many areas of plant pathology, taking off blindfolds, conducting sound research and keeping a constant flow of ideas through collaborations is a sure recipe for insights about new pathogens and into new areas of science. In tropical countries, this could result in more sustainable production systems to be delivered to growers, which in turn could enhance the role of science in these societies, thus creating a virtuous cycle that would please any agricultural scientist.

Acknowledgments The author is in debt to several nematologists from Brazil, USA, UK, India and Colombia who speedily provided copies of many publications cited in this review.

References

Anonymous (1986) Investigations on the 'crinkle leaf' disorder of arabica coffee caused by the three species of *Hemicriconemoides*. Coffee Board, 39th Annual Report. Central Coffee Research Institute, Karnataka.

Arias M, Lamberti F, Bello A et al (1995) Estudio agroecologico de los nematodos de la familia Longidoridae en São Tomé e Principe. Nematol Mediterr 23:167–175

Bridge J (1984) Coffee nematode survey of Tanzania – Report on a visit to examine plant parasitic nematodes of coffee in Tanzania. International Institute of Parasitology, St Albans.

Bridge J (1988) Plant-parasitic nematode problems in the Pacific Islands. J. Nematol 20:173–183

Campos VP, Lima RD, Almeida VF (1987) Nematóides parasitos de grandes culturas identificados em localidades de Minas Gerais e São Paulo. Nematol Bras 11:226–232

Castro JMC, Campos VP, Pozza EA et al (2008) Levantamento de fitonematóides em cafezais do sul de Minas Gerais. Nematol Bras 32:56–64

D'Souza GI, Sreenivasan CS (1965) A note on the reniform nematode, *Rotylenchulus reniformis* L and O (Pratylenchinae:Nemata) on arabica coffee in south India. Indian Coffee 29:11–13

Dasgupta DR, Raski DJ, Van Gundy SD (1969) Revision of the genus *Hemicriconemoides* Chitwood and Birchfield, 1957 (Nematoda: Criconematidae) J. Nematol 1:126–145

De Waele D, Elsen A (2007) Challenges in tropical plant nematology. Ann Rev. Phytopathol 45:457–485

Dias WP, Liberato JR, Fonseca AFA (1996) Nematóides associados ao cafeeiro no Estado do Espirito Santo. Ceres 43:808–812

Ferraz S (1980) Reconhecimento das espécies de fitonematóides presentes nos solos do Estado de Minas Gerais. Experientiae 26:256–328

Garcia A, Tihohod D, Caetano MF et al (1988) Nota sobre a ocorrência de fitonematóide em cafezais da região de Marilia. Nematol Bras 12:151–152

Germani G, Anderson RV (1991) Taxonomic notes on some *Hemicriconemoides* species and description of a new species. J Nematol 23:502–510

Giribabu P, Saha M (2002) Three new tylenchid nematode species associated with coffee plantations at Salem, Tamil Nadu, India. Int J Nematol 12:209–214

Giribabu P, Saha M (2003) Studies on nematode communities at different depths, altitudes and periods of year in coffee plantation. Ann Plant Prot Sci 11:360–363

Giribabu P, Saha M (2006) A new dorylaimid nematode species associated with coffee plantations at Salem, Tamil Nadu, India. Ann Plant Prot Sci 14:184–187

Herrera SIC, Marban-Mendoza N (1999) Efecto de coberturas vivas de leguminosas en el control de algunos fitonematodos del café en Nicaragua. Nematropica 29:223–232

Jairajpuri MS, Ahmad W (1992) Dorylaimida – free-living, predaceous and plant-parasitic nematodes. E.J. Brill, Leiden

Kubo RK, Inomoto MM, Oliveira CMG et al (2001). Nematóides associados a cafeeiros do Estado de São Paulo. Proceedings XXIII Braz Cong Nematol:91

Kumar AC (1979) Studies on nematodes in coffee soils of south India. 1. Three new species of Criconematoidea. J. Coffee Res 9:102–111

Kumar AC (1980) Studies on nematodes in coffee soils of South India. 3. A report on *Radopholus similis* and description of *R. colbrani* n.sp. J Coffee Res 10:43–46

Kumar AC (1981) Studies on nematodes in coffee soils of south India. 5. Descriptions of three new species of *Tylenchorhynchus* and occurrence of four other tylenchid species. J Coffee Res 11:89–99

Kumar AC (1982) Studies on nematodes in coffee soils of South India. 6. *Hemicriconemoides cassiae* n.sp. J Coffee Res 12:14–17

Kumar AC (1983) Studies on nematodes in coffee soils of south India. 8. Five new species of Criconematoidea. J Coffee Res 13:14–26

Kumar AC (1988) Nematode problem of coffee and its management. Indian Coffee 52:12–19

Kumar AC, D'Souza GI (1969) Pathogenicity of *Hemicriconemoides* species to coffee. Plant Dis Reporter 53:15–16

Kumar AC, Samuel SD (1990) Nematodes attacking coffee and its management – a review. J Coffee Res 20:1–27

Lamberti F, Boiboi JB, Ciancio A et al (1992) Plant parasitic nematodes associated with tree crops in Liberia. Nematol Mediterr 20:79–85

Lima RD, Almeida VF (1989) Ocorrência e distribuição de fitonematóides em cafeeiros do Triângulo Mineiro e Alto Paranaiba. Nematol Bras 13:6

Lordello AIL, Lordello RRA (2001) Nematóides encontrados em cafezais do Estado de São Paulo. Proceedings XXIII Braz Cong Nematol:85

Lordello LGE (1980) Estado atual do nematóide reniforme como parasito do cafeeiro. Rev Agric 55:62

Lordello LGE, Carneiro Filho F, Rebel EK et al (1974) Identificação de nematóides em cafezais do Estado do Paraná. Nematol Bras 1:16–24

Luc M, Bridge J, Sikora RA (2005) Reflections on nematology in subtropical and tropical agriculture. In: Luc M, Sikora RA, Bridge J (eds) Plant parasitic nematodos in subtropical and tropical agricultura, 2nd edn. CABI Publishing, Wallingford

Luc M, de Guiran G (1960) Les nématodes associés aux plantes de l'ouest african. L'agronomie tropicale 15:434–449

Macedo MCM (1974) Suscetibilidade de cafeeiros ao nematóide reniforme. Solo 66:15–16

Milne DL, Keetch DP (1976) Some observations on the host plant relationships of *Radopholus similis* in Natal. Nematropica 6:13–17

Monteiro AR (1970a) Acêrca de alguns Dorylaimodea (Nemata, Dorylaimida). Anais Escola Superior Agric Luiz de Queiroz 27:239–279

Monteiro AR (1970b) Dorylaimoidea de cafezais paulistas (Nemata, Dorylaimida). Universidade de São Paulo, DS thesis, Piracibaba

Pinochet J (1987) Management of plant parasitic nematodes in Central America: the Panama experience. In: Veech JA, Dickson DW (eds) Vistas on nematology. Society of Nematology, Hyattsville.

Pinochet J, Guzman R (1987) Nematodos asociados a cultivos agricolas en El Salvador: su importancia y manejo. Turrialba 37:137–145

Prates HS, Curi SM, Silveira SGP (1985) Levantamento de nematóides na cafeicultura paulista. Nematol Bras 9:24

Rahm G (1928) Alguns nematodes parasitas e semi-parasitas das plantas culturaes do Brasil. Arch Inst Biol 1:239–252

Rashid F, de Waele D, Coomans A (1985) Trichodoridae (Nematoda) from Brazil. Nematologica 31:289–320

Raski DJ (1975) Revision of the genus *Paratylenchus* Micoletzky, 1922, and descriptions of new species. Part II of three parts. J Nematol 7:274–295

Rodriguez MG, Sánchez L, Rodriguez ME (2000) Plant parasitic nematodes associated with coffee crop (*Coffea arabica*) in Cajálbana, Cuba. Rev Prot Veg 15:38–42

Santos BB, Silva, LAT (1984) Ocorrência de *Rotylenchulus reniformis* Linford and Oliveira, 1940 em mudas da cafeeiro no estado do Paraná. Rev Agric 59:27–28

Schenck S, Schmitt DP (1992) Survey of nematodes on coffee in Hawaii. J. Nematol 24: 771–775

Schenck S, Schneck, D (1994) Determination of a management strategy for nematode pests of Hawaiian coffee. Int J. Pest Manag 40:283–285

Sekhar OS (1963) Apuntes sobre los nematodos en plantaciones de cafe em sur India. Turrialba 5:1–5

Sharma RD, Sher SA (1973) Nematodes associated with coffee in Bahia, Brazil. Arq. Inst. Biol 40:131–135

Siddiqi MR (2000) Tylenchida parasites of plants and insects, 2nd edn. CABI Publishing, Wallingford

Sivakumar CV, Selvasekaran, E (1982) Description of two new species of *Scutellonema* Andrassy 1958 (Hoplolaimoidea:Nematoda). Indian J. Nematol 12:118–123

Souza JT, Maximiano C, Campos VP (1999) Nematóides parasitos encontrados em cafeeiros em campo e viveiros de mudas do Estado de Minas Gerais. Summa Phytopathol 25:180–183

Sturhan D, Brzeski, MW (1991) Stem and bulb nematodes, *Ditylenchus* spp. In: Nickle WR (ed) Manual of Agricultural Nematology. Marcel Dekker, New York

Tarté R (1970) Reconocimiento de nematodos asociados con diversos cultivos en Panama. Turrialba 20:401–406

Thorne G, Schieber E (1962) American dagger nematode (*Xiphinema americanum*) on coffee in Guatemala, with suggestions for nematode control in nurseries. Plant Dis Reporter 46:857

Trinh PQ, Nguyen CN, Waeyenberge L et al (2004) *Radopholus arabocoffeae* sp. n. (Nematoda:Pratylenchidae), a nematode pathogenic to *Coffea arabica* in Vietnam, and additional data on *R. duriophilus*. Nematology 6:681–693

Valdez RB (1968) Stubby roots of coffee seedlings caused by *Rotylenchulus reniformis*. The Philippine Agric 51:672–679

van Doorsselaere R, Samsoen L (1982) Some tylenchids from coffee fields of Ivory Coast, with the description of *Hemicriconemoides snoecki* n. sp. (Nematoda:Tylenchida) Rev Nematol 5:51–63
Vovlas N (1987) Parasitism of *Trophotylenchulus obscurus* on coffee roots. Rev Nematol 10: 337–342
Vovlas N, Lamberti F (1985) Observations on the morphology and histopathology of *Hoplolaimus pararobusts* attacking coffee in São Tomé. Nematol Mediterr 13:73–80
Vovlas N, Lamberti F (1990) Histological alterations induced by *Rotylenchulus reniformis* on *Coffea arabica* roots. Nematol Mediterr 18:77–81
Whitehead AG (1969) Nematodes attacking coffee, tea and cocoa. In: Peachey JE (ed) Nematodes of tropical crops. CAB, St Albans
Zem AC, Lordello LGE (1983) Estudo sobre hospedeiros de *Radopholus similis* e *Helicotylenchus multicinctus*. Nematol Bras 7:175–187

Part V
World Reports

Abstract In this book section, nematologists from Brazil, Colombia, India, Indonesia, the Ivory Coast, Uganda and Vietnam present reports on their coffee-parasitic nematodes. A region report – Central America – was included to represent Costa Rica, El Salvador, Guatemala, Honduras and Nicaragua. In these accounts, the authors present a brief outline of the crop in their countries, followed by historical landmarks in coffee-related nematology. They also present results from regional or national surveys, assessments of damage caused by their main nematode species, and from assays using biological, cultural, chemical and genetic control approaches. They conclude their reports by outlining their country's infrastructure and personnel dedicated to research and extension on coffee-parasitic nematodes, as well as their prospects for the upcoming years.

Keywords Brazil · Colombia · Honduras · Costa Rica · Nicaragua · El Salvador · Guatemala · Vietnam · Indonesia · India · Uganda · The Ivory Coast · Historical Accounts · Surveys · Nematode Management · Chemical Control · Biological Control · Cultural Control · Genetic Control · Nematology Extension · Nematology Research

Chapter 12
Brazil

Luiz Carlos C. B. Ferraz

12.1 Brief Outline of the Crop

Brazil is a major producer of agricultural products. In 2004, this country exported US$ 30.9 billion worth of food and other agricultural products, which makes it the world's third largest exporter of agricultural goods, after the United States and the European Union. Coffee is one of Brazil's major agricultural exports, besides sugar, soybeans, cotton and orange juice (Council and Hanrahan, 2006). This country ranks first in coffee production, with a yield of around 44 million 60 Kg-bags in 2006; this represents 30% of the world's coffee production and US$ 5.1 billion on the international commodity market. It is worth noting that predictions suggested this output would be achieved only by the year 2010 (Anonymous, 2001).

Although Brazil is currently the world's largest coffee exporter (27.2 million bags in 2006, corresponding to US$ 3.3 billion), this crop represented only 2.5% of the country's exports in that year. Germany, the United States, Italy and France are the most important importers, but Japan, China and some Arabian countries are becoming important too (Council and Hanrahan, 2006).

It is widely known that coffee-producing countries have large populations involved with this crop, even when these countries have a diversified export portfolio. Mexico and Indonesia are good examples, with three and five million people, respectively, working in the coffee industry. In Brazil, some 3.5 million people, mostly in rural areas, are involved with this crop, which generates around seven million direct and indirect jobs (Rice, 2003; Anonymous, 2004b).

In Brazil, coffee (*Coffea* sp.) plantations are spread over 2.7 million hectares (ha), corresponding to approximately six billion trees, of which 74% is comprised of the arabica type (*C. arabica* L.) and 26% of the robusta one (*C. canephora* Pierre ex A. Froehner). Traditional varieties and cultivars are the most cultivated, but these have been progressively replaced by modern cultivars which are resistant to

L.C.C.B. Ferraz
Escola Superior de Agricultura Luiz de Queiroz/USP, Piracicaba, Brazil
e-mail: lccbferr@carpa.ciagri.usp.br

R.M. Souza (ed.), *Plant-Parasitic Nematodes of Coffee*,

pests and/or diseases and are recommended for planting under high density system (Anonymous, 2004b; Mattielo, 2004a).

In Brazil, coffee is cultivated in different geographic regions and under different edaphic and climatic conditions, from the south up to the Amazon basin (Fig. 12.1), mostly at altitudes ranging from 350 to 1,000 masl. Nowadays, the most important producing areas are situated in the States of Minas Gerais, Espirito Santo, São Paulo, Paraná, Bahia and Rondônia. In this list, the first three States rank as the top three producers. The States of Rio de Janeiro, Goiás, Mato Grosso, Pará and Acre are of minor importance, producing from 100 to 500 thousand bags (Mattielo, 2004b). Coffee production remains vulnerable to both frost and drought. These factors combined reduced the 1994/1995 and 1995/1996 production yields by about 40% (Anonymous, 2001).

Due to the influence of several factors, such as climatic and edaphic conditions, cultivar or variety planted, planting and harvesting system adopted and the incidence of pests, diseases and plant-parasitic nematodes, the average productivity varies greatly between and even within Brazilian States. In 2006/2007, the national production was 42.5 million bags, with a mean productivity of 19.75 bags/ha and a range between 7.7 and 23 in Mato Grosso and Bahia, respectively. In Minas Gerais, the average productivity was 21.7 and the range between 16.1 and 27.8 in the producing regions Zona da Mata and Sul de Minas, respectively (Anonymous, 2007a). Another key factor that interferes with productivity is coffee's natural biannual yield oscillation. This is clearly seen in Brazil's total production, which was 30.9, 33.1, 31.3, 48.5, 28.8, 39.2, 32.9 and 42.5 million bags in the harvests between 1999/2000 and 2006/2007 (Anonymous, 2007b).

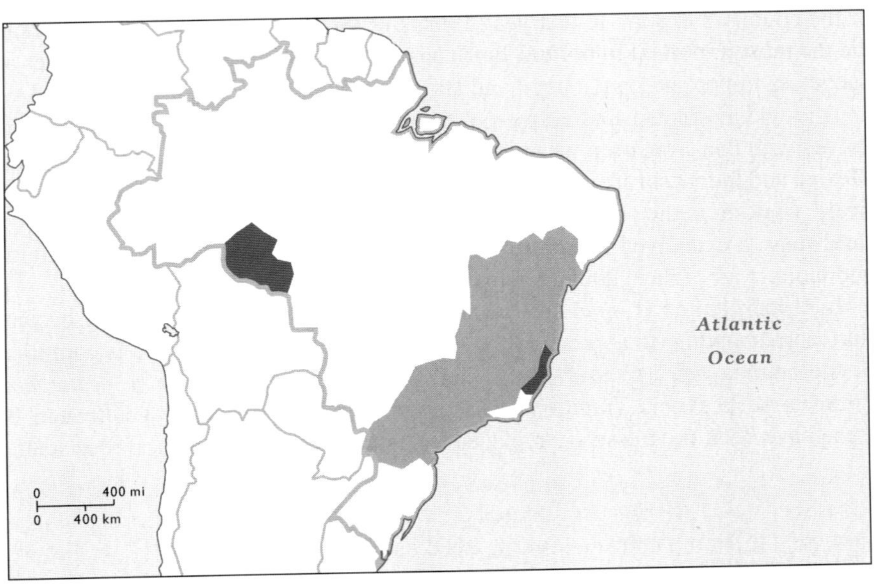

Fig. 12.1 Brazil's robusta and arabica coffees growing regions (*dark and light grey*, respectively). Map by UENF/GRC

The economic status of Brazilian coffee growers also determines productivity. About two thirds are smallholders (less than 10 ha) who often keep the use of high-technology practices to a minimum due to restricted access to subsidies. A small proportion of growers with large properties and strong financial support, mostly in Espirito Santo, Minas Gerais and São Paulo, practice high input production systems, which include seedling preparation, chemical control of pests and diseases, fertilizer application, irrigation and automated harvest.

As regards coffee policy, during the mid- and late nineteenth century the regulations were dictated almost entirely by the coffee 'barons', great producers in the then Provinces of Rio de Janeiro and São Paulo who had their production based on slave labor. The abolition of slavery in 1888 and the proclamation of the Republic the next year reduced their influence. Large coffee-producing areas soon failed, and they were occupied by other crops such as sugarcane (Fernandes, 2003). The production remained concentrated in São Paulo, based on European immigrants (mainly from Italy) until 1929, when the global economic crisis swept away many 'barons' and their plantations. In 1952, the IBC (Portuguese acronym for Brazilian Coffee Institute) was created, which coordinated policies for nearly four decades. Nowadays, the PNP&D/Café (National Program for Coffee Research and Development) is responsible for establishing policies, defining marketing strategies and supporting basic and applied research and technology transfer (Anonymous, 2004a).

12.2 Nematological Problems

12.2.1 Incidence and Economic Importance

12.2.1.1 Root-Knot Nematodes (*Meloidogyne* spp.)

The first report of problems in coffee plantations due to nematode parasitism was presented by Jobert (1878), who did not provide a precise identification of the organism involved. Göldi (1892) published a landmark work dealing with the same subject – the incidence of nematodes causing heavy damage to plantations in what is now the State of Rio de Janeiro. This article includes the description of *Meloidogyne exigua*, the causal agent, and recommendation of a variety of control measures. Apparently, this article was made available by the author for the first time in 1887, as an advanced reprint (Chitwood, 1949). In addition, a brief technical report which summarizes the most relevant aspects about the incidence of *M. exigua* in Rio de Janeiro was published in Germany (Göldi, 1888).

In 1929, Rahm reported *M. exigua* in São Paulo. Since then, it has been found in all major coffee-producing States (Campos and Villain, 2005), and it is the most widespread *Meloidogyne* species in Minas Gerais (Campos and Melles, 1987; Santos et al., 1998). The pathogenicity of *M. exigua* to coffee seedlings and trees was first confirmed through studies developed under greenhouse and field conditions by Arruda (1960) and Arruda and Reis (1962). One year after inoculating seedlings with *M. exigua*, their growth had fallen by 30% in comparison to non-inoculated ones. The yield of trees cultivated in infested soil had fallen by 50% compared to those grown in chemically disinfested soil.

Since then, slight to strong adverse effects of *M. exigua* on the growth of coffee plants have been reported in São Paulo (Macedo et al., 1974), Minas Gerais (Santos, 1978; Boneti et al., 1982; Guerra Neto et al., 1985; Souza, 1990) and Rio de Janeiro (Barbosa et al., 2004). Despite these results, since this nematode induces typical root galls but rarely causes disorganization of the root's cortical tissue, it is feasible for a grower to sustain a profitable production through a combination of nematicide and fertilizer applications, especially in Minas Gerais and São Paulo. A comprehensive investigation of the *M. exigua*-coffee interactions, mainly on aspects related to epidemiology, is currently underway in Rio de Janeiro (Souza et al., 2008a,b).

A second root-knot nematode (RKN), *M. coffeicola* Lordello and Zamith, was found parasitizing coffee in Paraná, São Paulo and Minas Gerais (Lordello and Zamith, 1960; Lordello, 1967; Guerra Neto et al., 1983), but it has not been reported from other countries. Apparently, coffee is the only economically important host of this nematode, which also parasitizes the weeds *Eupatorium pauciflorum* Kunth and *Psychotria nitidula* Cham. and Schltdl. (Jaehn et al., 1980). On coffee, *M. coffeicola* reproduces well on eight to 10 year-old plants only, but no root galls are induced (Figs. 12.2 and 12.3); however, the nematode induces a severe disorganization of the cortical tissue, which often leads the plants to show symptoms of defoliation and chlorosis (Fig. 12.4), and a marked yield reduction. Plant death occurs within a variable period of time.

For many years *M. coffeicola* was considered the species with the highest damage potential among all coffee-parasitic nematodes in Brazil. Indeed, the recovery of parasitized plants was not possible, and their eradication often represented the sole alternative for growers just a few years after the nematode's incidence had been confirmed. Presently, this nematode is rarely found parasitizing coffee because in the infested areas this crop has been replaced by soybeans, wheat, corn and other annual crops. Hence, *M. coffeicola* is no longer of economic importance.

Another RKN, *M. incognita* (Kofoid and White) Chitwood, has caused the most devastating effects on coffee plantations in Brazil since it was first recorded in São Paulo (Lordello and Mello Filho, 1970). Subsequently, it was also reported from Espirito Santo, Paraná, Ceará and Minas Gerais (Lordello and Hashizume, 1971; Lordello and Lordello, 1972; Ponte and Castro, 1975; Guerra Neto and D'Antonio, 1984). It has been hypothesized that several records of a *M. exigua* variant population found affecting coffee in Paraná and São Paulo in the 1960s and early 1970s actually referred to *M. incognita* (Moraes and Lordello, 1977). Coffee plants parasitized by *M. incognita* are chlorotic and/or show strong defoliation, particularly during the dry season (Figs. 12.5, 12.6 and 12.7). Typical rounded root galls are not usually induced, but localized root swellings resembling galls may be seen; also, cortical tissues often appear detached (Fig. 12.8), resulting in a characteristic 'rough', heavily cracked root (Lordello, 1972). In the 1970s millions of infected coffee plants had to be eradicated in two large producing regions in São Paulo, Alta Paulista and Araraquarense, due to this nematode's aggressiveness and the

Fig. 12.2 Arabica coffee roots heavily damaged by *Meloidogyne coffeicola*, showing typical disorganization and detachment of the cortical tissue. (Photo by Luiz C.C.B. Ferraz) (see color Plate 13, p. 325)

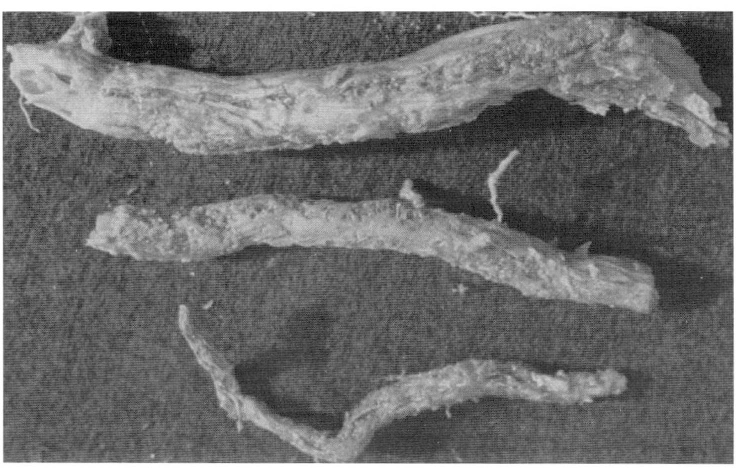

Fig. 12.3 Arabica coffee roots parasitized by *Meloidogyne coffeicola*, showing small rounded cavities in the cortical tissue from which nematode adult females have been removed. (Photo by Luiz C.C.B. Ferraz) (see color Plate 14, p. 326)

Fig. 12.4 Arabica coffee plants severely affected by *Meloidogyne coffeicola*, showing chlorosis and defoliation. (Photo by Luiz C.C.B. Ferraz) (see color Plate 15, p. 326)

low efficacy of the control measures recommended at that time. This forced many growers to replace their plantations with pasture (Curi et al., 1977) or rubber trees.

M. paranaensis Carneiro, Carneiro, Abrantes, Santos and Almeida was described from coffee in 1996, adding to the group of the most important parasitic nematodes in Brazil (Carneiro et al., 1996). Before its description, this nematode had been reported as a new *M. incognita* pathotype named 'biotype IAPAR' (Carneiro, 1993). It had also been referred to as an 'unidentified *Meloidogyne* population from coffee' (Esbenshade and Triantaphyllou, 1985), often found in São Paulo and Paraná (Santos and Triantaphyllou, 1992; Carneiro, 1993). Its incidence in Minas Gerais seems to be limited (Castro et al., 2005).

The symptoms shown by *M. paranaensis*-parasitized coffee plants resemble those induced by *M. incognita*: chlorosis, defoliation, reduced growth and often death. These symptoms are related to the splitting and cracking of cortical root tissue, especially on the tap-root. Typical root galls are not induced (Carneiro et al., 1996). Although comprehensive assessments of the damage caused by *M. paranaensis* on different coffee varieties and cultivars have not been undertaken in Brazil, it is suspected this species may have a high economic impact on production (Gonçalves, 2000).

Fig. 12.5 Young arabica coffee plants heavily affected by *Meloidogyne incognita*, showing chlorosis and partial defoliation. (Photo by Luiz C.C.B. Ferraz) (see color Plate 16, p. 327)

Other *Meloidogyne* species that have occasionally been found parasitizing coffee in Brazil are *M. hapla* Chitwood and *M. javanica* (Treub) Chitwood (Lordello and Monteiro, 1974; Ponte, 1977). However, in Brazil, Central America and Africa these species are reported as causing little damage to coffee plantations, and so they are considered of minor importance to this crop.

Fig. 12.6 Leaves collected from a *Meloidogyne incognita*-affected arabica coffee plant showing typical symptoms of nutritional deficiency. (Photo by Luiz C.C.B. Ferraz) (see color Plate 17, p. 327)

Fig. 12.7 Arabica coffee replanting in a sandy soil heavily infested by *Meloidogyne incognita* in the State of São Paulo, Brazil. (Photo by Luiz C.C.B. Ferraz) (see color Plate 18, p. 328)

12.2.1.2 Root-Lesion Nematodes (*Pratylenchus* spp.)

Two root-lesion nematodes, *Pratylenchus brachyurus* (Godfrey) Filipjev and S. Stekhoven and *P. coffeae* (Zimmerman) Filipjev and S. Stekhoven, have been found associated with coffee plants in Brazil. Both species have a large host range and are widely distributed in the country, in particular the former (Kubo et al., 2004). A single account exists of arabica coffee-parasitism by *P. vulnus* Allen and Jensen,

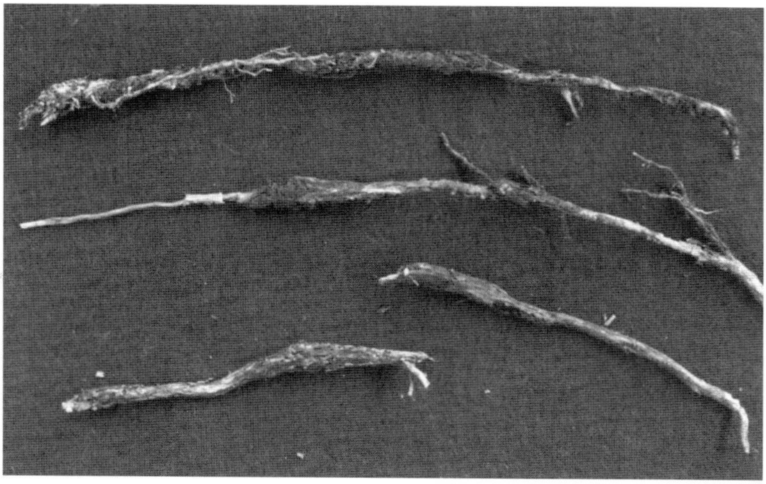

Fig. 12.8 Arabica coffee roots heavily parasitized by *Meloidogyne incognita* showing disorganized, detached cortical tissue and atypical swellings. (Photo by Luiz C.C.B. Ferraz) (see color Plate 19, p. 329)

but the plant had been cultivated for ornamental purposes in a park in the city of São Paulo, Brazil (Monteiro et al., 2001). This species has never been found in coffee plantations.

P. brachyurus has more often been reported from plantations in São Paulo (Lordello et al., 1968; Gonçalves et al., 1978; Kubo et al., 2004) and Minas Gerais (D'Antonio et al., 1980; Castro et al., 2005). Although this species' reproductive rate on different coffee cultivars is usually very low, it may cause poor development, especially in young plants (Inomoto et al., 1998; Oliveira et al., 1999b). Indeed, this species' adverse effect on the development of seedlings of arabica coffee 'Mundo Novo' and robusta coffee 'Apoatã' has been clearly demonstrated in a greenhouse study (Oliveira et al., 1999a). In this study, the inoculum used was two, six, 18 or 54 nematodes/cm^3 soil. Plant height, fresh root and shoot dry weights and nematode reproduction factor were assessed 90 days after soil infestation. The seedlings showed no tolerance to *P. brachyurus*, as indicated by the reduction of all variables evaluated in the inoculated seedlings, in comparison to the health controls. Plant height was reduced even at the lowest inoculum level. Nonetheless, the nematode reproduction factor was below one, indicating that those cultivars are not suitable hosts for *P. brachyurus*. In the field, such a delay in the plants' development occurs frequently when the coffee plantation is established in an area previously cultivated for a long time with pasture or other suitable hosts of *P. brachyurus* (Fig. 12.9), which allows the nematode soil population to reach high level (Lordello, 1984).

In Brazil, *P. coffeae* was first found in coffee roots in São Paulo (Monteiro and Lordello, 1974), where it is less disseminated than *P. brachyurus* (Gonçalves et al., 1978). The former species has also been reported causing high yield losses in other coffee-producing regions, such as the State of Pernambuco (Moura et al., 2002).

Fig. 12.9 Arabica coffee plants affected by *Pratylenchus brachyurus*. This field had been cultivated with pastures for many years before being cultivated with coffee. (Photo by Luiz C.C.B. Ferraz) (see color Plate 20, p. 329)

It has been demonstrated that morphological, biological and molecular differences exist among *P. coffeae* isolates from around the world (see Chapters 3 and 5) and that populations can vary with respect to host preference. Indeed, coffee has not been listed among the most suitable hosts of *P. coffeae* in a greenhouse trial that assessed the host preferences of two isolates from Brazil, K_5 and M_2 (Silva and Inomoto, 2002). The K_5 isolate, originally collected around coffee roots, has had its pathogenicity to seedlings of 'Mundo Novo' demonstrated under greenhouse conditions (Kubo et al., 2003). The inocula applied were 333, 1,000, 3,000 or 9,000 nematodes/seedling, with twelve replicates for each. Nine months after inoculation, all the plants that had been inoculated with 9,000 nematodes and most of those inoculated with 3,000 were dead. The seedlings' growth and photosynthesis were reduced at inoculum levels of as few as 333 and 1,000 nematodes, respectively, in comparison to healthy controls. In the infected plants, root necrosis was very common. The seedlings had no tolerance to *P. coffeae* in the variables height and shoot dry weight, which were reduced significantly at the lowest inoculum level. In a second experiment the *P. coffeae* isolate M_2, originally collected around *Aglaonema* sp. roots, was inoculated at the rate of 8,000 nematodes/coffee seedling. This isolate was also pathogenic to coffee, but to a much lesser extent than K_5. Since in both trials the nematode reproductive rate was very low, 'Mundo Novo' was considered a poor host of those isolates.

12.2.1.3 Other Nematodes

In addition to *Meloidogyne* spp. and *Pratylenchus* spp., certainly the main nematode problems for coffee production in Brazil, several other genera and species have occasionally been recorded during surveys in plantations. However, these reports just briefly mention their findings, giving no details on the symptoms or damage that these nematodes may cause to coffee plants. Some of the genera found associated to arabica coffee in Brazil are *Aorolaimus* sp., *Discocriconemella* sp., *Dolichodorus* sp., *Helicotylenchus* sp., *Hemicycliophora* sp., *Mesocriconema* sp. (= *Macroposthonia* sp.), *Trichodorus* sp. and *Xiphinema* sp. (Manso et al., 1994). A complete list, with comments, is given in Chapter 11.

Radopholus similis (Cobb) Thorne, the burrowing nematode, is considered a major problem in banana production around the world. It has also been considered a threat to coffee in Java (Zimmerman, 1898). In Brazil, this nematode has not been found parasitizing coffee. However, while assessing the host range of a *R. similis* population from Brazil, Zem and Lordello (1983) grew five seedlings of 'Mundo Novo' for 90 days in a heavily infested field; three seedlings died, one was severely affected by the nematode and one remained healthy. Despite these results, the burrowing nematode has not been considered an important problem for coffee production in Brazil.

Rotylenchulus reniformis Linford and Oliveira, also known as the reniform nematode, is considered a major threat to cotton, pineapple and soybean production in Brazil and many other countries. In India and the Philippines, this species has been reported on coffee. In Brazil, *R. reniformis* has sporadically been reported associated with coffee plants (Lordello, 1980; Castro et al., 2005); thus, it is not regarded as a problem for its cultivation.

12.2.2 Control of Coffee-Parasitic Nematodes in Brazil

Because of the impact of parasitic nematodes, particularly *Meloidogyne* spp. and *Pratylenchus* spp., on national coffee production, the control of these nematodes represents a permanent challenge to Brazilian researchers. It should be emphasized however, that some of the most efficient nematode control measures known today were actually taught to coffee growers in the nineteenth century. Indeed, as early as 1887, Göldi proposed a number of essential actions in his report on the decline of plantations parasitized by *M. exigua* in Rio de Janeiro. These actions were designed to recover nematode-infested areas and to prevent nematode dispersal into new, nematode-free ones. Göldi's wise recommendations regarding the control of coffee-parasitic nematodes are thus considered milestones.

The control of coffee-parasitic *Pratylenchus* spp. and *Meloidogyne* spp. is discussed in detail in Chapters 5 and 8. This section discusses additional issues and recommendations drawn from decades of experience of Brazilian coffee growers and researchers dealing with these nematodes.

12.2.2.1 The Origin and Sanitation of Coffee Seedlings

Göldi stated that any grower who intended to start a coffee plantation in a nematode-free area should necessarily (i) produce his own seedlings using soil collected from places situated quite apart from coffee-growing areas or (ii) acquire healthy seedlings only, refusing any plants of unknown or suspected origin. A careful examination of the seedlings' above and underground parts prior to definitive transplantation in the field should become routine among growers.

It is regrettable that although these lessons are of undisputed merit, they were ignored almost completely for many decades by the majority of Brazilian growers and government authorities. Consequently, both *Meloidogyne* spp. and *Pratylenchus* spp. became widespread in coffee plantations in Brazil. Growers and nursery owners became aware of the issue of seedling sanitation in the early 1950s, when Dr. Luiz Gonzaga Engelberg Lordello, considered the father of Nematology in Brazil, started publishing a series of technical notes in newspapers and magazines dealing with the importance of seedling sanitation.

Furthermore, until the late 1960s nearly no legislation existed to prevent the commercialization of nematode-infected coffee seedlings in the main producing regions. As a result, millions of *M. incognita*-infected seedlings produced mostly in private nurseries in Paraná were introduced into non-infested areas of São Paulo during an extensive program coordinated by the IBC during the 1970s to stimulate the renewal of coffee plantations (Jaehn, 1984). This phytossanitary disaster was only no worse because dedicated professionals from the extension service network inspected and destroyed many infected seedlings that were about be commercialized. In 1976/1977 around 3.3 million seedlings were destroyed in São Paulo alone (Gonçalves and Martins, 1993).

The high yield losses caused by *M. incognita* made clear to coffee growers, in particular to smallholders, that more attention should be paid to the preparation and acquisition of seedlings. Therefore, from the 1980s on some nematode-exclusion techniques assessed by researchers were promptly adopted in many nurseries. For example, for many years the production of nematode-free seedlings was possible through soil disinfestation with the application of methyl bromide at the rate $100\,cm^3/m^3$ of soil (Moraes et al., 1977; Gonçalves, 2000). Attempts to control *M. incognita* in nurseries through the application of the granular nematicides aldicarb, carbofuran, phenamiphos and phensulphothion in the soil did not succeed (Jaehn et al., 1984). In the last decade, the production of seedlings has been increasingly carried out in small tubular plastic containers filled with disinfested substrates (Cunha et al., 2002), which are often enriched with different organic amendments (Gonçalves et al., 1998a). This technique has reduced the costs associated with seedling production. Some procedures have been improved to avoid nematode introduction into nurseries through soil sticking to machinery or in irrigation water (Krzyzanowski, 2000).

Today, legislation exists in most Brazilian regions to regulate the production and commercialization of coffee seedlings (Carneiro, 1993; Lima, 1993). In relation to plant-parasitic nematodes, RKNs are the main target; specific guidelines exist for the sampling of nurseries' seedlings and for destruction of any suspected material (Anonymous, 2006). Nonetheless, the enforcement of such regulations and their efficacy in halting nematode dispersal are variable in the different coffee-producing regions because not enough well-trained professionals are available to properly inspect the nurseries. Also, unregistered nurseries do exist that counterfeit regulations. Brazilian growers have, nonetheless, progressively learned Göldi's lessons on the crucial role played by seedlings in the dispersal of the most important nematodes for coffee production.

12.2.2.2 The Benefits of Controlling Nematodes Through Preventive, Not Curative, Measures

In Göldi's words, any attempt to recover coffee plants heavily parasitized by *M. exigua* should be compared to medical procedures designed to heal a man whose lungs are nearly destroyed. Instead, he strongly recommended growers to promptly eradicate their old, unproductive infested plantations and to cultivate such areas with annual crops for the eight to 10 following years, thus allowing a progressive, significant decrease in the nematode soil population. The immediate replanting of coffee in a highly infested area would be as ineffective as to 'one's efforts to fill with water a wicker basket'. Because this recommendation was not followed, during the second half of the twentieth century growers faced high yield losses in successive coffee replants in *M. incognita*-heavily infested areas in Paraná and São Paulo (Curi and Silveira, 1978; Lordello, 1984).

Göldi also advised growers who planned to expand their coffee cultivation into new areas to take into account the fact that nematode problems are much more frequent in sandy soils than in those with high clay content. This general rule on the relation between soil type and nematode damage was first established by Göldi for the interaction between *M. exigua* and coffee plants. The soundness of this rule was later confirmed throughout the world for many nematode-plant associations and also by coffee growers in other regions in Brazil. Indeed, most of the young plantations eradicated in São Paulo and Paraná due to severe parasitism by *M. incognita* occurred in areas of sandy soils (Jaehn, 1984; Gonçalves, 2000).

Once the association between nematodes and the severe damage observed in many coffee plantations became clear from the 1960s on, the studies dealing with the efficacy of control techniques under field conditions became more numerous in the 1970s. Apparently, the growers' demand for an urgent solution encouraged researchers to see the use of nematicides as the best choice among the available control approaches.

Soon a number of field experiments were conducted to assess the efficacy of fumigant (DBCP) and systemic (aldicarb, carbofuran, phenamiphos, phensulphothion, oxamyl) products, which were tested alone or in combination with different organic amendments. These experiments were conducted against *M. exigua* (Curi and Silveira, 1974) and *M. incognita* (Guidolin and Rebel, 1974; Curi et al., 1975; Rebel and Guidolin, 1975; Curi et al., 1977). These studies' results, sometimes contradictory or inconclusive, revealed a trend towards the inefficacy of nematicides to control RKNs on coffee, particularly *M. incognita*. Further studies in the 1980s and 1990s demonstrated that nematicides did not enable the formation of new plantations in areas heavily infested by *M. incognita*, nor did they recuperate severely affected coffee plants (Ferraz et al., 1983; Jaehn, 1984; Jaehn and Rebel, 1984; Jaehn et al., 1984).

In some instances however, i.e. in coffee plantations under slight infestation, the use of granular nematicides decreased *M. incognita* population for a few months after the product application, which allowed the plants to develop a good foliage cover and to survive the following years (Novaretti et al., 1993; Novaretti et al., 1997). As for *M. exigua*, nematicides have seldom been used under field conditions, not even in Minas Gerais where this is the prevailing species, although studies have shown a productivity increase in nematicide-treated plants (Huang et al., 1983).

Coffee plantations parasitized by *M. coffeicola* did not respond to nematicide applications; hence, their short-term eradication was usually the only alternative for the growers (Lordello, 1984). As for *M. paranaensis*, aldicarb and terbuphos were effective in reducing the soil population of second-stage juveniles and the total root population in comparison to non-treated plants (Lusvarghi and Santos, 1997). In Brazil, experimental data relative to chemical control of *Pratylenchus* spp. in coffee plantations are not available.

Because of the disadvantages associated with the use of nematicides, *a viz* toxicity to man, soil contamination with chemical residues and increase in production costs, other non-chemical approaches for controlling coffee-parasitic nematodes have been investigated. For example, studies dealing with coffee genotypes with nematode resistance, particularly against RKNs, were initiated in Brazil in the early 1970s.

In Brazil, the majority of the most cultivated arabica coffee cultivars resulted from the long-term, exceptional research program at the genetics section of the IAC (Agronomic Institute of Campinas), which was developed mostly under the leadership of Dr. Alcides Carvalho in the years 1935–1993. Unfortunately, despite their many agronomic attributes, the cultivars 'Mundo Novo', 'Catuai Vermelho', 'Catuai Amarelo', 'Bourbon Vermelho', 'Caturra Amarelo' and others are susceptible to several *Meloidogyne* species, particularly *M. exigua, M. incognita* and *M. paranaensis* (Gonçalves et al., 2004).

Due to their susceptibility to phytonematodes and to some important pests and diseases such as 'leaf miner' (*Perileucoptera coffeella* Guérin-Menèville) and 'leaf rust' caused by *Hemileia vastatrix* Berk and Br., it became a priority to search for sources of nematode resistance in the coffee germplasm available in Brazil. Again, the contribution of the IAC research team was crucial. From the 1970s on, many basic and advanced studies were carried out dealing with the host status of new coffee cultivars to *Meloidogyne* spp. and *Pratylenchus* spp. The genotypes assessed in these studies resulted from crosses between *C. arabica* and other *Coffea* species, especially *C. canephora, C. congensis* A. Froehner and *C. dewevrei* De Wild. and T. Durand (Fazuoli, 2004). Since most of these studies have been summarized by Gonçalves (1993), this chapter will discuss only the studies related to the coffee cultivars mostly grown in Brazil. A comprehensive discussion on several aspects of *Meloidogyne*-resistance is presented in Chapter 9.

'Apoatã IAC 2258', or simply 'Apoatã', is possibly the most relevant cultivar produced in Brazil in order to face the problem represented by nematode parasites. It is resistant to *M. exigua, M. incognita, M. paranaensis* and *P. coffeae* (Fazuoli, 2004), as well as to *H. vastatrix*, although slight infections may occasionally be observed

Fig. 12.10 Plants of arabica coffee 'Mundo Novo' grown in a *M. incognita*-infested field. Dead, self-rooted, nematode-susceptible plants are in the foreground. Healthy plants grafted onto nematode-resistant *C. canephora* 'Apoatã' are in the background. (Photo by Luiz C.C.B. Ferraz) (see color Plate 21, p. 330)

in a number of plants under field conditions. This robusta cultivar is often used as a rootstock for the most productive arabica cultivars, and it is highly recommended for planting in the extensive *M. incognita*-infested areas in São Paulo and Paraná (Fig. 12.10) (Gonçalves et al., 2004). Interestingly, when planted in nematode-free areas, 'Mundo Novo' grafted onto 'Apoatã' yielded equally or better than self-rooted 'Mundo Novo', thus confirming the high compatibility between these genotypes (Costa et al., 1989).

Timor Hybrid, which is phenotypically an arabica coffee, is possibly a natural hybrid between *C. arabica* and *C. canephora* that has frequently been used in the Brazilian genetic breeding program as a source of resistance to some *Meloidogyne* spp. and to *H. vastatrix*. Among its derivatives are 'Obatã' (IAC 1669-20), 'Tupi' (IAC 1669-33) and 'IAPAR-59', which are resistant to *M. exigua* and *H. vastatrix* (Salgado et al., 2002; Fazuoli, 2004). Since these cultivars' plants present short stature, they are highly recommended for planting in high density/ha; their cultivation has progressively increased in some regions of São Paulo, Paraná and Minas Gerais (Mattielo, 2004a).

Progenies of 'Icatu Vermelho IAC 4160' resulted from crosses between *C. arabica* and *C. canephora* have been rated as resistant to *M. paranaensis* under greenhouse and field conditions in the Alta Paulista region, in São Paulo (Gonçalves et al., 1998b). Progenies of the arabica coffee 'IPR-100' have also recently been considered resistant to *M. paranaensis* (Sera et al., 2007). The tetraploid form of *C. congensis* has also been used in the Brazilian genetic breeding program as a source of resistance to *M. exigua, M. incognita* and *H. vastatrix* (Fazuoli et al., 1983).

In relation to *Pratylenchus* spp., the coffee cultivars most cultivated in Brazil are susceptible or intolerant to *P. brachyurus* and/or *P. coffeae* (Kubo et al., 2004).

Possible sources of resistance to these nematodes have been found in the interspecific hybrids 'Icatu' and 'Sarchimor' and in the robusta coffees 'IAC 4764' and 'IAC 4765' (Oliveira et al., 1999b; Tomazini et al., 2005). Therefore, the search for new *Pratylenchus*-resistant genotypes clearly merits further investigation.

According to the principles of 'nematode integrated management', which is currently in use by many Brazilian coffee growers, some other control methods are employed in addition or alternatively to nematicide application and use of resistant cultivars. In most cases, such methods play an indirect, positive effect on the plants' development rather than a direct, negative effect on the nematode population. This is the case of application of chemical and/or organic fertilizers and weed control. As the plants become more vigorous, with expanded and more efficient root systems, they are often able to tolerate nematode parasitism and yield better. These procedures are recommended by technical personnel and usually practiced by growers.

In some coffee-producing areas, such as Noroeste and Alta Paulista in São Paulo, intercropping or crop rotation using antagonistic plants is employed to enhance the control of *M. incognita, P. coffeae* and *P. brachyurus*. *Crotalaria* spp. (Fig. 12.11) and velvetbean (*Mucuna* sp.) used as green manure are among the most preferred plants (Gonçalves et al., 1998a).

Biological control *strictu sensu*, that is, the use of bacteria or fungi that parasitize nematode eggs, juveniles or adults, has not been applied in coffee plantations in Brazil, and no marketable bioproducts have been routinely used against nematodes. However, nematophagous organisms, mostly fungi, have been collected from soil of coffee plantations (Silva and Campos, 1990; Naves and Campos, 1991) and on at least one occasion the low incidence of *M. exigua* has been correlated with a high soil population of these beneficial organisms (Campos, 1992).

Fig. 12.11 Nematode-antagonistic *Crotalaria* sp. intercropped with coffee to reduce the soil nematode population. (Photo by Luiz C.C.B. Ferraz) (see color Plate 22, p. 330)

As in many other countries around the world, basic and applied studies have been conducted in Brazil to evaluate the potential of two well-known, promising agents of nematode biocontrol: the fungus *Paecilomyces lilacinus* (Thom) Samson and the bacterium *Pasteuria penetrans* (Thorne) Sayre and Starr. However, just a few investigations have been carried out on coffee-parasitic nematodes, and mostly under laboratory and greenhouse conditions. The results are stimulating, but yet inconclusive (Santiago et al., 2006; Cirotto et al., 2006).

12.3 Research and Extension on Coffee-Parasitic Nematodes

As outlined above, the first milestone in nematology research in Brazil was Göldi's classic report, which was made available in 1887; this work was officially published in 1892. However, this was an isolated event in nematology research in this country because political events started a decline in the coffee industry in Rio de Janeiro; therefore, the studies on *M. exigua* were discontinued.

During the first half of the twentieth century the Brazilian government kept a tight rein on the coffee industry (Anonymous, 2001). During this period, the policies were defined mostly by governmental organisms, such as the CNC (National Coffee Council) in the years 1931–1933 and the DNC (National Coffee Department) in 1933–1946. Due to the nearly complete absence of plant nematologists working actively in this country during this period, contributions to research on coffee-parasitic nematodes apparently do not exist.

In the early 1950s, two relevant events took place simultaneously: (i) in 1951, Dr. Luiz G.E. Lordello, Brazil's pioneer in nematology research, started his long and productive career at Esalq/Universidade de São Paulo, during which he published an extensive series of articles dealing with nematode problems in a variety of important crops, including coffee and (ii) in 1952, the IBC was created, which played a much more significant role in the coffee industry than the two previous regulatory bodies.

Alerted by Lordello's publications or stimulated during short training courses taught by him, many IBC researchers initiated national cooperative research programs that systematically included 'nematological implications' among their most relevant topics. During the 1970s and 1980s the technological improvements resulting from such research programs, including those related to nematodes, were periodically transferred to extension service professionals and to coffee growers through the CBPCs (Brazilian Congress on Coffee Research), which were organized and supported by the IBC. Until it was abolished in 1989, the IBC decisively supported comprehensive nematology research and technology transfer to growers.

Notwithstanding the relevance of the IBC, from the 1970s through the 1990s the personnel from State research institutions and public universities in São Paulo, Paraná, Minas Gerais, Espirito Santo, Bahia and Rio de Janeiro also contributed with studies dealing with the identification, biology, pathogenicity and especially control of coffee-parasitic nematodes. During this period, many greenhouse and field trials were conducted regarding the efficacy of nematicides, alone or in combination with the application of organic matter, and regarding the host status and agronomic performance of new nematode-resistant cultivars. Basic studies under laboratory conditions, mostly designed to improve RKN taxonomic identification through electrophoresis and other techniques apart from the examination of perineal patterns, were also conducted. On the other hand, the transference of technology and specific field activities, such as the regular inspection of nurseries, were mainly performed by extension service professionals.

With the opening of the Brazilian economy in the early 1990s coupled with the end of the IBC, the coffee industry was restructured on 'free market' principles. Created in 1997, the PNP&D/Café is a consortium comprised of representatives of coffee growers, companies that operate in the domestic market, exporters and researchers associated with key governmental agencies, public universities and extension services (Anonymous, 2001; 2004a). During the period 1998–2003, the PNP&D/Café received approximately US$ 30 million, of which 73% was committed to supporting initiatives on research and transference of technology and 27% to acquisition of equipment and improvement of facilities. These resources supported many advanced studies on coffee, the spread of new technologies, the publishing of high-quality specialized publications and a number of graduate scholarships. Unfortunately, the support from PNP&D/Café for research activities dwindled in the following years, down to US$ 6 million in 2005, resulting in the temporary interruption or cessation of many research programs (Anonymous, 2004a; Carvalho, 2006).

12.4 Concluding Remarks

During the last two decades, the Brazilian coffee industry has become significantly stronger due to (i) employment of new cultivation technologies, such as high density planting, (ii) renewal of old plantations, (iii) development of mechanized harvest systems, (iv) expansion of the crop to areas not prone to frost, such as Espirito Santo and Bahia and (v) high investments into new coffee bean processing techniques, such as pulped-natural systems. These improvements have increased yields and provided the market with a wide array of coffee types (Anonymous, 2001). For example, 'organic' coffee production has increased at an annual rate of 100% (Caixeta and Pedini, 2002).

Concurrently with these advances, in Brazil the research on coffee-parasitic nematodes has also addressed some important issues. A good example is the more precise taxonomic identification of *Meloidogyne* sp. and *Pratylenchus* sp. populations based on a combination of classical (morphological and morphometrical) and modern (biomolecular) methods. Also, traditional techniques for quantitative sampling of RKNs in soil and coffee roots are under reevaluation, the efficacy of selected soil fungi and rhizobacteria for the biocontrol of nematodes has been assessed, the potential of some nematode-resistant cultivars for use in new producing areas and/or under different cultivation systems has been assessed, and the influence of different coffee management techniques on the structure of soil nematode communities has been demonstrated.

Furthermore, an ambitious national research program, the Coffee Genome Project, was initiated in 2002. It is supported by the PNP&D/Café, FAPESP (State of São Paulo Research Foundation) and EMBRAPA/CENARGEN (Brazilian Agricultural Research Corporation/Genetic Resources and Biotechnology Center). Other research institutes, such as the IAC, IAPAR, EPAMIG and INCAPER, as well as public universities such as USP, UNESP, UNICAMP, UFLA and UFV participate in this program. In 2004, when the first phase of the genome sequencing was completed, a data bank containing more than 200,000 DNA sequences was made available to the program's associate members. Since approximately 30,000 genes have already been identified, the following years should bring a better understanding of coffee's different development mechanisms, as well as speeding up its genetic breeding program, with the development of new insect-, pathogen- and nematode-resistant cultivars (Anonymous, 2007c,d).

It should be emphasized that the long-term development of all these research efforts is strongly dependent on the existence of full-time job positions for professionals working with plant-parasitic nematodes, in particular those associated with coffee, both in private and governmental research institutions. The number of phytonematologists in Brazil is very limited, especially considering the continental size of this country. Also, several experts on coffee-parasitic nematodes have retired and their positions have not always been filled with professionals with the same profile. Therefore, forming new, talented human resources is absolutely essential, and to accomplish it graduate students should be stimulated to get involved with nematology research on coffee and receive proper financial support. Special attention must also be paid to the subsequent incorporation of these well-trained personnel into the professional market, thus avoiding their leaving from Nematology and the waste of high investments made in their training.

All these positive initiatives have significantly contributed to reinforcing Brazil's top position among the world's coffee producers and exporters, as well as to supporting the expectation of an even more favorable scenario in the coming years. However, at least two issues must be urgently addressed: (i) the provision of special subsidies to indebted growers, in particular smallholders, so as to minimize their financial problems and improve their plantation management and (ii) the need for a gradual but consistent expansion of the national export of roasted and ground coffees, which have a higher market value. This would enable Brazil to compete with countries that are true 'non-producing' coffee exporters, such as Italy and Germany.

Unlike other major producing countries, prospects for the coffee industry are quite positive in Brazil. No critical, restrictive factor exists nowadays for the production of both arabica and robusta coffees and some commercial barriers that have been raised can be overcome. In the next years, the Brazilian coffee industry should evolve, incorporating all technological, environmental and social requirements, and not only short-minded economic drives.

References

Anonymous (2001) Brazil coffee to 2010: Implications for global coffee players: management summary. Promar International, Alexandria.

Anonymous (2004a) Embrapa Café – A unidade: Histórico. http://www22.sede.embrapa.br/café/unidade/historico.htm. Visited on April 15, 2007

Anonymous (2004b) Embrapa Café – Gestão da pesquisa do café no Brasil. http://www22.sede.embrapa.br/cafe/outros/Arq_Relat_Gestao/Gest%E3o_Pesquisa.pdf. Visited on April 16, 2007

Anonymous (2006) Portaria disciplina a produção, a entrada, o comércio e o trânsito de mudas de café e de eucalipto no estado de Minas Gerais. Instituto Mineiro de Agropecuária, Belo Horizonte.

Anonymous (2007a) Associação Brasileira da Indústria de Café. Estatisticas – Produção Agricola. http://www.abic.com.br/estat_pagricola.html. Visited on August 30, 2007

Anonymous (2007b) Avaliação da safra agricola cafeeira 2007/2008. Companhia Nacional de Abastecimento, Ministério da Agricultura, Pecuária e Abastecimento, Brasilia.

Anonymous (2007c) Embrapa Cenargen – Projeto Genoma Café: Impactos potenciais. http://www.cenargen.embrapa.br/biotec/genomacafe/impactos.html. Visited on April 23, 2007

Anonymous (2007d) Embrapa Cenargen – Projeto Genoma Café: O projeto. http://www.cenargen.embrapa.br/biotec/genomacafe/projeto.html. Visited on April 23, 2007

Arruda HV (1960) Efeito depressivo de nematóides sobre mudas de cafeeiro formadas em laminados. Bragantia 10: 15–17

Arruda HV, Reis AJ (1962) Redução nas duas primeiras colheitas de café devido ao parasitismo de nematóide. Biológico 28: 349

Barbosa DHSG, Vieira HD, Souza, RM et al (2004) Field estimates of coffee yield losses and damage threshold by *Meloidogyne exigua*. Nematol Bras 28: 49–54

Boneti JIS, Ferraz S, Braga, JM et al (1982) Influência do parasitismo de *Meloidogyne exigua* sobre a absorção de micronutrientes e sobre o vigor de mudas de cafeeiro. Fitopatol Bras 7: 197–207

Caixeta IF, Pedini, S (2002) Comercialização do café orgânico. Informe Agropecuário 23: 149–152

Campos VP (1992) Perspectivas do controle biológico de fitonematóides. Informe Agropecuário 16: 27–30

Campos VP, Melles CCA (1987) Ocorrência e distribuição de espécies de *Meloidogyne* em cafezais dos Campos das Vertentes e do Sul de Minas Gerais. Nematol Bras 11: 233–241

Campos VP, Villain L (2005) Nematode parasites of coffee and cocoa. In: Luc M, Sikora RA, Bridge J (eds) Plant-parasitic nematodes of tropical and subtropical agriculture, 2nd ed CABI, Wallingford.

Carneiro RG (1993) Fitonematóides na cafeicultura paranaense: situação atual. Proceedings XXVII Braz Cong Nematol: 42–44

Carneiro RMDG, Carneiro RG, Abrantes, IMO et al (1996) *Meloidogyne paranaensis* n.sp., a root-knot nematode parasitizing coffee in Brazil. J Nematol 28: 177–189

Carvalho, G.R. (2006) Conjuntura agropecuária: o mercado de café em perspectiva. Embrapa Monitoramento por Satélite, Campinas.

Castro JMC, Campos VP, Pozza EA et al (2005) Levantamento de fitonematóides em cafezais do Sul de Minas. Nematol Bras 29: 110

Chitwood, BG (1949) Root-knot nematodes – Part I. A revision of the genus *Meloidogyne* Göldi, 1887. Proc Helm Soc Wash 16: 90–104

Cirotto, PAS, Mesquita LFG, Mota FC et al (2006) Controle biológico de *Meloidogyne incognita* em cafeeiros com *Pasteuria penetrans* isolado P10. Nematol Bras 30: 99

Costa WM, Gonçalves W, Fazuoli LC (1989) Efeito de vários porta-enxertos de *Coffea canephora* na produção de café 'Mundo Novo' em áreas com *Meloidogyne incognita*. Proceedings XV Braz Cong Coffee Res: 130

Council LR, Hanrahan CE (2006) CRS Report: Brazil's agricultural production and exports: selected data. Nat Counc Sci Environ, Washington.

Cunha RL, Souza CAS, Andrade Neto A et al (2002) Avaliação de substratos e tamanhos de recipientes na formação de mudas de cafeeiros em tubetes. Ciênc Agrotec 26: 7–12

Curi SM, Silveira SGP (1974) Estudos preliminares sobre o controle quimico do nematóide do cafeeiro, *Meloidogyne exigua*. Biológico 40: 337–345

Curi SM, Silveira SGP (1978) Distribuição geográfica, sintomatologia e significado dos nematóides *Meloidogyne incognita* e *M. exigua*, parasitos do cafeeiro no Estado de São Paulo. Biológico 44: 243–251

Curi SM, Silveira SGP, Elias JG Jr (1975) Resultados preliminares do controle quimico do nematóide *Meloidogyne incognita*, parasito do cafeeiro, em condições de campo. Biológico 41: 67–72

Curi SM, Silveira SGP, Elias JG Jr (1977) Resultados de produção e da proteção do sistema radicular de cafeeiros sob controle quimico do nematóide *Meloidogyne incognita* em condições de campo. Nematol Bras 2: 93–99

D'Antonio AM, Libeck PR, Coelho AJE et al (1980) Levantamento de nematóides parasitas do cafeeiro que ocorrem no Sul de Minas Gerais. Proceedings VIII Braz Cong Coffee Res: 440–443

Esbenshade PR, Triantaphyllou AC (1985) Use of enzyme phenotypes for identification of *Meloidogyne* species. J Nematol 17: 6–20

Fazuoli LC (2004) Melhoramento genético do cafeeiro. Proceedings V Reunião Itinerante Fitos do Inst Biol: 2–28

Fazuoli LC, Costa WM, Gonçalves W et al (1983) Identificação de resistência em *Coffea canephora* e *C. congensis* ao nematóide *Meloidogyne incognita*, em condições de campo. Cienc Cult 35: 20

Fernandes AA (2003) Vassouras, e a região fluminense, na produção cafeeira do Brasil Império. http://www.jbcultura.com.br/Anibal/vassouras.htm. Visited on April 17, 2007

Ferraz LCCB, Rocha AD, Brancalion AM (1983) Considerações sobre a viabilidade do controle de *Meloidogyne incognita* visando à recuperação de cafezais infestados. Nematol Bras 7: 117–124

Göldi EA (1888) Der Kaffeenematode Brasiliens (*Meloidogyne exigua* G.). Zool Jahrb 4: 262–268

Göldi EA (1892) Relatório sobre a moléstia do cafeeiro na Provincia do Rio de Janeiro. Arch Mus Nac 8: 7–123

Gonçalves W (1993) Reações de cafeeiros a *Meloidogyne exigua* Göldi e a diferentes populações de *M. incognita* (Kofoid and White) Chitwood. Universidade Estadual Paulista, PhD thesis, Jaboticabal.

Gonçalves W (2000) Manejo de fitonematóides em cafeeiro no Estado de São Paulo. Proceedings XXII Braz Cong Nematol: 42–43

Gonçalves W, Fazuoli LC, Lima MMA et al (1998b) Café IAC 4160, promissora fonte de resistência a *Meloidogyne paranaensis*. Nematol Bras 22: 25

Gonçalves W, Martins ALM (1993) Fitonematóides da cafeicultura paulista: situação atual. Proceedings XVII Braz Cong Nematol: 51–58

Gonçalves W, Ramiro DA, Gallo PB et al (2004). Manejo de nematóides na cultura do cafeeiro. Proceedings Reunião Itinerante Fitos Inst Biol: 48–66

Gonçalves W, Silvarolla MB, Lima MMA (1998a) Estratégias visando a implementação de manejo integrado dos nematóides parasitos do cafeeiro. Informe Agropecuário 19: 36–47

Gonçalves W, Thomaziello R, Moraes MV et al (1978) Estimativas de danos ocasionados pelos nematóides do cafeeiro. Proceedings VI Braz Cong Coffee Res: 182–186

Guerra Netto EG, D'Antonio AM (1984) Nematóides parasitas em lavouras cafeeiras do sul de Minas Gerais. Proceedings XVI Braz Cong Coffee Res: 171

Guerra Netto EG, D'Antonio AM, Almeida SR et al (1983) Ocorrência do nematóide *Meloidogyne coffeicola* em lavouras de café no sul do estado de Minas Gerais. Rev Agric 58: 45–48

Guerra Netto EG, D'Antonio AM, Freire ACF (1985) Influência de *Meloidogyne exigua* no desenvolvimento de lavoura de *Coffea arabica* variedade Mundo Novo. Proceedings XVII Braz Cong Coffee Res: 36–37

Guidolin JA, Rebel EK (1974) Estudo do efeito de nematicidas em covas de café visando controle do nematóide *Meloidogyne incognita*. Proceedings II Braz Cong Coffee Res: 23–24
Huang SP, Resende IC, Souza PE et al (1983) Effect of aldicarb, ethoprop and carbofuran on control of coffee root-knot nematode, *Meloidogyne exigua*. J Nematol 15: 510–514
Inomoto MM, Oliveira CMG, Mazzafera P et al (1998) Effects of *Pratylenchus brachyurus* and *P. coffeae* on seedlings of *Coffea arabica*. J Nematol 30: 362–367
Jaehn A (1984) Viabilidade do uso de nematicidas em cafezal novo instalado em solo infestado por *Meloidogyne incognita*. Nematol Bras 8: 275–284
Jaehn A, Rebel EK (1984) Uso de palha de café, leguminosas e nematicida em mudas de cafeeiro plantadas em área infestada por *Meloidogyne incognita*. Nematol Bras 8: 309–317
Jaehn A, Rebel EK, Matiello, JB (1984) Viabilidade de recuperação de mudas de cafeeiro infestadas por *Meloidogyne incognita* através de nematicidas. Nematol Bras 8: 295–300
Jaehn A, Rebel, EK, Lordello LGE (1980) A origem do nematóide *Meloidogyne coffeicola*. Nematol Bras 4: 159–161
Jobert C (1878) Sur une maladie du caféier observée au Brésil. C R Acad Sci Paris 87: 941–943
Krzyzanowski, AA (2000) Nematóides do cafeeiro. Proceedings XXII Braz Cong Nematol: 40–41
Kubo RK, Oliveira CMG, Inomoto MM (2004) Manejo dos nematóides das lesões, *Pratylenchus* spp, em cafeeiros. Proceedings V Reunião Itinerante Fitos Inst Biol: 149–168
Kubo RK, Silva RA, Tomazini MD et al (2003) Patogenicidade de *Pratylenchus coffeae* em plântulas de cafeeiro cv Mundo Novo. Fitopatol Bras 28: 41–48
Lima RD (1993) Nematóides na cafeicultura mineira: situação atual. Proceedings XVII Braz Cong Nematol: 45–50
Lordello LGE (1967) O nematóide cofeicola invade São Paulo. Rev Agric 42: 162
Lordello LGE (1972) Nematode pests of coffee. In: Webster JM (ed) Economic Nematology. Academic Press, New York.
Lordello LGE (1980) Estado atual do nematóide reniforme como parasita do cafeeiro. Rev Agricult 55: 62
Lordello LGE (1984) Nematóides das plantas cultivadas. Nobel, São Paulo.
Lordello LGE, Hashizume H (1971) Suscetibilidade da variedade Konillon de *Coffea canephora* a um nematóide. Rev Agric 46: 157–158
Lordello LGE, Lordello RRA (1972) *Meloidogyne incognita* ataca cafeeiro no Paraná. Solo 64: 27
Lordello LGE, Mello Filho AT (1970) Mais um nematóide ataca o cafeeiro. Rev Agric 45: 102
Lordello LGE, Monteiro AR (1974) Informação preliminar sobre um nematóide nocivo ao cafeeiro. Nematol Bras 1: 13–15
Lordello LGE, Monteiro AR, D'Arce, RD (1968) Distribuição geográfica dos nematóides nocivos ao cafeeiro. Rev Agric 43: 79–81
Lordello LGE, Zamith APL (1960) *Meloidogyne coffeicola* sp.n., a pest of coffee trees in the state of Paraná, Brazil. Rev Bras Biol 20: 375–379
Lusvarghi HN, Santos JM (1997) Eficácia de terbufós e de aldicarb, em mistura com cyproconazole, no manejo de *Meloidogyne paranaensis* e da ferrugem do cafeeiro. Nematol Bras 21: 12
Macedo MCM, Haag HP, Lordello LGE (1974) Influência de *Meloidogyne exigua* na absorção de nutrientes em plantas jovens de cafeeiro. Ann Esc Super Agric Luiz de Queiroz 31: 91–104
Manso ESBGC, Tenente RCV, Ferraz LCCB et al (1994) Catálogo de nematóides fitoparasitos encontrados associados a diferentes tipos de plantas no Brasil. Serviço de Produção de Informação-Embrapa, Brasilia.
Mattielo, JB (2004a) A cafeicultura no Brasil: diversidade é a sua principal caracteristica. http://www.coffeebreak.com.br/ocafezal.asp?SE=6&ID=34. Visited on April 16, 2007
Mattielo, JB (2004b) Outras regiões: a produção de café é menor. http://www.coffeebreak.com.br/ocafezal.asp?SE=6&ID=45. Visited on April 16, 2007
Monteiro AR, Lordello LGE (1974) Encontro do nematóide *Pratylenchus coffeae* atacando cafeeiro em São Paulo. Rev Agric 49: 164
Monteiro AR, Antedomenico SR, Ferraz LCCB et al (2001) Primeira ocorrência de *Pratylenchus vulnus* em cafeeiro. Nematol Bras 25: 116–117

Moraes MV, Lordello LGE (1977) Estudo de três populações de nematóides nocivos ao cafeeiro. Nematol Bras 2: 249–253

Moraes MV, Thomaziello RA, Lordello LGE et al (1977) Ensaios de tratamento de solo por produtos quimicos para uso em viveiros de café. Nematol Bras 2: 231–244

Moura RM, Pedrosa EMR, Prado MDC (2002) Incidência de *Pratylenchus coffeae* causando severa nematose em cafeeiro no Nordeste. Fitopatol Bras 27: 649

Naves RL, Campos VP (1991) Ocorrência de fungos predadores de nematóides no Sul de Minas Gerais e estudo da capacidade predatória e crescimento *in vitro* de alguns de seus isolados. Nematol Bras 15: 152–162

Novaretti WRT, Papa G, Coelho JVG (1997) Controle quimico de *Meloidogyne incognita* em cafeeiro com o nematicida terbufós. Nematol Bras 21: 12

Novaretti WRT, Tihohod D, Lusvarghi HN (1993) Controle quimico de *Meloidogyne incognita* em cafeeiro com o nematicida terbufós. Nematol Bras 17: 22–23

Oliveira CMG, Inomoto MM, Vieira AMC et al (1999a) Efeito de densidades populacionais de *Pratylenchus brachyurus* no crescimento de plântulas de *Coffea arabica* cv Mundo Novo e *C. canephora* cv Apoatã. Nematropica 29: 215–221

Oliveira CMG, Monteiro AR, Antedomênico SR et al (1999b) Host reaction of *Coffea* spp. to *Pratylenchus brachyurus*. Nematropica 29: 241–244

Ponte JJ (1977) Nematóides das galhas: espécies ocorrentes no Brasil e seus hospedeiros. Brascan Nordeste, Mossoró.

Ponte JJ, Castro FE (1975) Lista adicional de plantas hospedeiras de nematóides de galhas no Estado do Ceará, referente a 1969–1974. Fitossanidade 1: 29–30

Rahm G (1929) Nematódeos parasitas e semiparasitas de diversas plantas cultivadas no Brasil. Arch Inst Biol 2: 67–137

Rebel EK, Guidolin JA (1975) Efeito de nematicidas sobre cafeeiros jovens infestados por *Meloidogyne incognita*. Proceedings III Braz Cong Coffee Res: 119–120

Rice R (2003) Coffee production in a time of crisis: social and environmental connections. SAIS Rev 23: 221–245

Salgado SML, Campos VP, Resende MLV et al (2002) Reprodução de *Meloidogyne exigua* em cafeeiros 'Iapar 59' e 'Catuai'. Nematol Bras 26: 205–208

Santiago DC, Cadioli MC, Hoshino AT et al (2006) Crescimento micelial e patogenicidade de isolados de *Paecilomyces lilacinus* sobre ovos de *Meloidogyne paranaensis* em diferentes temperaturas *in vitro*. Nematol Bras 30: 100–101

Santos JM (1978) Efeitos de fertilizantes sobre *Meloidogyne exigua* e influência de seu parasitismo sobre a absorção e translocação de nutrientes em mudas de *Coffea arabica*. Universidade Federal de Viçosa, MS dissertation, Viçosa.

Santos JM, Triantaphyllou HH (1992) Determinação dos fenótipos enzimáticos e estudos comparativos da morfologia de 88 populações de *Meloidogyne* spp. parasitas do cafeeiro. Nematol Bras 16: 88

Santos MA, Pinheiro JB, Santos CM et al (1998) Ocorrência de nematóides em cafeeiros do Triângulo Mineiro e Alto Paranaiba. Nematol Bras 22: 3–4

Sera GH, Sera T, Ito DS et al (2007) Progênies de *Coffea arabica* cv IPR-100 resistentes ao nematóide *Meloidogyne paranaensis*. Bragantia 66: 43–49

Silva JF, Campos VP (1990) Fungos endoparasitas de nematóides associados a diferentes ecossistemas no Sul de Minas Gerais. Nematol Bras 14: 14

Silva RA, Inomoto MM (2002) Host-range characterization of two *Pratylenchus coffeae* isolates from Brazil. J Nematol 34: 135–139

Souza ES (1990) Dinâmica populacional de *Meloidogyne exigua* em cafeeiros novos. Universidade Federal de Lavras, MS dissertation, Lavras.

Souza RM, Volpato AR, Viana AP (2008a) Epidemiology of *Meloidogyne exigua* in an upland coffee plantation in Brazil. Nematol Mediterr (*in press*)

Souza RM, Volpato AR, Viana AP (2008b) Field assessment of different sampling strategies for coffee plantations parasitized by *Meloidogyne exigua*. Nematropica 37: 345–355

Tomazini MD, Silva RA, Oliveira CMG et al (2005) Resistência de genótipos de cafeeiro a *Pratylenchus coffeae* e *Meloidogyne incognita*. Nematol Bras 29: 193–198
Zem AC, Lordello LGE (1983) Estudos sobre hospedeiros de *Radopholus similis* e *Helicotylenchus multicinctus*. Nematol Bras 7: 175–187
Zimmerman AWP (1898) De Nematoden der Koffiewortels. Deel I. Meded Lds Pltuin 27: 1–64

Chapter 13
Colombia

Alvaro Gaitán, Carlos Alberto Rivillas and Hernando Cortina

13.1 Brief Outline of the Crop

Colombia is currently the world's third coffee producer, with an annual yield of around 11 million 60 Kg-bags, which are worth US$ 1.6 billion on the international commodity market. Coffee represented only 8% of Colombian exports in 2000 (Anonymous, 2002); however, its production has a tremendous social impact. The coffee industry generates 800 thousand direct jobs (37% of national agricultural positions), and 1 million indirect jobs (8% of the national work force), which represents economic support for over one tenth of the population (Anonymous, 1997). Arabica coffee (*Coffea arabica* L.) plantations spread over 3.6 million hectares (ha) (Fig. 13.1).

The coffee plantations are restricted to the Andean mountains, at an altitude ranging from 1,000 to 2,000 masl, and with rainfall of between 1,200 and 4,000 mm/year. The great diversity of this region in terms of ethnicity, geography, microclimate, and edaphics determines the agricultural practices, disease and pest incidence, and ultimately, coffee growth and productivity. This diversity was sorted by Gómez et al. (1991) into 86 agroecological regions (ecotopos).

For over 200 years, coffee cultivation was carried out with traditional cultivars, at densities below 2.5 thousand trees/ha, as a shaded monoculture or as the main species in an intercropping system. In this latter system, coffee farms are patchy, composed by a mosaic of pastures, vegetable production, secondary woods, and coffee plots (Ramirez et al., 2002; Guhl, 2004). Nonetheless, two thirds of the coffee hectarage is currently cultivated in an intensive system, under full sun, and with plant density ranging from 5 to 10 thousand plants/ha.

About 64% of the coffee growers are smallholders (less than 5 ha), producing 15% of the national coffee yield, while only 5% of the growers have large properties, being responsible for 45% of the national production (Ramirez et al., 2002). The farmers either produce their own seeds or buy them of the cultivars 'Colombia' and 'Castillo'. Many growers prepare their own seedlings, although it is common

A. Gaitán
Centro Nacional de Investigaciones del Café, Colombia
e-mail: alvaro.gaitan@cafedecolombia.com

R.M. Souza (ed.), *Plant-Parasitic Nematodes of Coffee*,

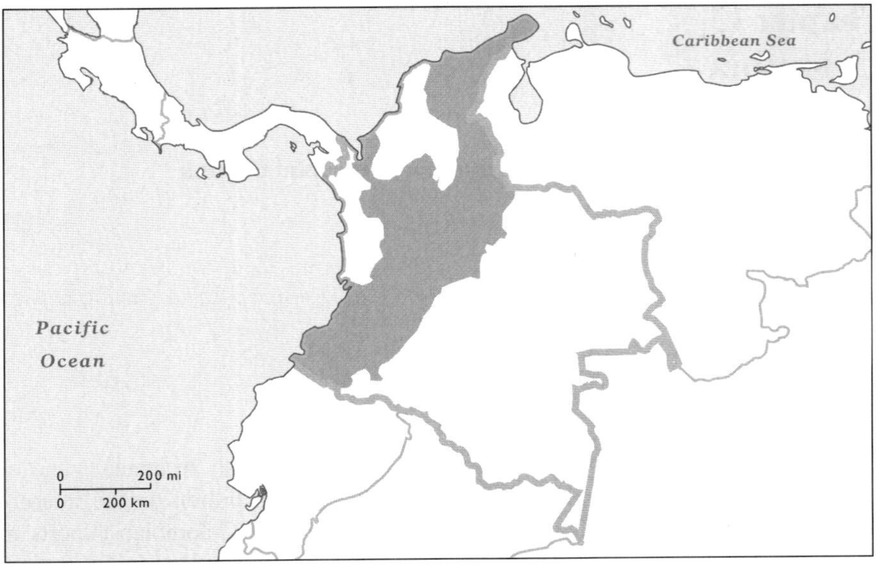

Fig. 13.1 Colombia's arabica coffee growing region. Map by UENF/GRC, adapted from a map by Flor Pulido (Cenicafé) and Agustin Codazzi (Geographical Institute, Colombia), with permission

practice to buy them from private nurseries, which are not subject to any certification or sanitary inspection. The production system is considered of low input, except for the minimal use of fungicides for the control of the fungus *Hemileia vastatrix* Berk and Br., which causes 'leaf rust' disease, and fertilizer applications twice a year. The national average productivity is 1,250 kg of parchment coffee/ha/year, with some farms producing up to 5,500 kg/ha (Anonymous, 1997).

An important aspect of the Colombian coffee industry is the role played by the FNC (the Spanish acronym for the National Federation of Coffee Growers of Colombia), to which 560 thousand families devoted to coffee production are affiliated. It coordinates the official coffee policies with the government, controls prices in the country, and promotes the marketing strategy for the brand Café de Colombia®. In addition, the FNC has continuously supported Cenicafé (the Spanish abbreviation for the National Coffee Research Center), which is responsible for developing and transferring technology to coffee growers.

13.2 Coffee-Parasitic Nematodes

13.2.1 Surveys

In 1929, Toro reported in the department of Cundinamarca a yellowing of coffee shrubs associated with root protuberances, similar to nodules on legumes. The disease was attributed to nematodes, at the time already known in Brazil, Guadeloupe and Martinique. The recommendation given then was to sterilize with boiling water the soil used in seed beds, and to increase the fertilization of the coffee plants, although the debilitated root system rendered useless the products applied.

In 1936, Obregon reported *Tylenchus* sp., *Cephalobus brevicaudatus* Zimmerman, and the abundant *Caconema radicicola* (Greef) Cobb (now a synonym for *Meloidogyne* sp.) in coffee plantation soils, as well as the same disease reported by Toro, now in the department of Caldas. The observation that *C. liberica* W. Bull ex Hiern was immune to the disease led to the proposal of using this species as a rootstock or in coffee breeding.

In 1972, Leguizamón and López reported that coffee plants parasitized by root-knot nematodes (RKNs) (*Meloidogyne* sp.) in Valle, Quindio and Risaralda presented poor growth, defoliation, increased susceptibility to foliar pathogens such as *Cercospora coffeicola* Berk and Cooke, and a root system characterized by a corky (suberous) primary root, and an odd abundance of secondary ones. Samples sent to the Commonwealth Institute of Helminthology (England) and to the Nematology Department at Wageningen, Netherlands, resulted in the identification of *M. javanica* (Treub) Chitwood, *M. incognita* (Kofoid and White) Chitwood, and *M. exigua* Göldi.

In a survey carried out in Quindio, Risaralda and Caldas, Baeza (1974) reported that RKNs and *Helicotylenchus* sp. were the most frequent nematodes in Colombian coffee plantations. *M. javanica* and *M. incognita* were associated with symptoms such as swelling of the main root, roots with a 'corklike' appearance and with longitudinal fissures, and atypical emission of roots at the plant's collar region. In addition to the nutritional deficiencies in the shoot, the enhanced susceptibility of the coffee plants to *C. coffeicola* was clear to Baeza. Although less important, *Pratylenchus coffeae* (Zimmerman) Filipjev and Schuurmans Stekhoven and *H. erythrinae* (Zimmermann) Golden, have been associated with lesions in the coffee's secondary and tertiary roots, leading to invasion and destruction of the root system by *Fusarium* sp. and *Rosellinia* sp. (Anonymous, 1975).

As part of the International Meloidogyne Project, in 1978 Navarro published the results of a survey in the Colombian crops located at altitudes ranging from 0 to 2,600 masl, reporting the same *Meloidogyne* spp. observed by Leguizamon and López (1972). While studying the RKN populations found in Colombia, Cano and Gil (1980) proposed a race 5 of *M. incognita* for some populations that combined the results of *M. arenaria* (Neal) Chitwood race 2 and *M. javanica* in the differential host test created by Taylor and Sasser (1978). Villalba et al. (1983) studied the *M. incognita* race 5 life cycle.

Blancos et al. (1982) surveyed the Sierra Nevada, the northernmost coffee region in Colombia, indicating that RKNs were present in 94% of the coffee plants sampled, followed by *Pratylenchus* sp. (1.15%), *Tylenchus* sp. (0.21%), and *Aphelenchoides* sp. (0.19%). *M. javanica* was the most common species (64% of the positive samples), followed by *M. incognita* (21%), and *M. exigua* (15%). A comparison of these frequencies with the soil fertility and altitude of the sampling sites did not suggest any correlation.

More recently, *M. arenaria* was reported for the first time in samples from Risaralda (Vergel et al., 2000). RAPD analysis and mitochondrial intergenic spacer marker analysis, with DNA extracted from individual egg masses, confirmed the species identification (Quintana et al., 2002).

13.2.2 Estimated Yield Losses

In Colombia, nematode parasitism has been considered of minor importance for coffee production. Nonetheless, Leguizamón (1976) observed a direct relationship between the levels of root infection, shoot symptoms, and yield loss. More recently, Leguizamón (1997) calculated a net yield loss of 78 grams of coffee berry and 4 grams of foliar dry weight for every 1% of root infection during the nursery period and later planting in the field. Additional, uncalculated losses could be added from the plant's increased susceptibility to *C. coffeicola*, and the ineffective application of fertilizers by the growers, in an effort to increase the plant's productivity. More studies are necessary to compare these losses to the costs and benefits of managing infested areas with practices such as chemical or biological control, removing the plantation, recovering the area through crop rotation, and replanting it.

13.2.3 Nematodes in Coffee-Associated Plants

In Colombia, weeds and coffee-associated crops, such as plantain and guamo (*Inga* sp.), are often found parasitized by RKNs, although no secondary effects are observed in the aerial part of the plants. In a survey carried out in 11 localities distributed in the departments of Quindio, Caldas, Valle, Risaralda and Tolima, Baeza et al. (1978) reported 23 hosts for *Meloidogyne* spp. (Table 13.1). *Physalis nicandroides* Schltdl. and *Talinum paniculatum* (Jacq.) Gaertn. were considered the best hosts for *M. javanica*, while *Spananthe paniculata* Jacq. was best for *M. incognita*, and *Solanum nigrum* L., *Hydrocotyle* sp. and *Galinsoga caracasana* (DC.) Sch.Bip. for *M. exigua*. All *Meloidogyne* isolates but *M. exigua* obtained from *S. nigrum* were pathogenic to coffee seedlings. Table 13.1 also includes results by Mayorga (1996) and Giraldo and Leguizamón (1997).

13.2.4 Chemical Control

In Colombia, the principle that has always guided nematode control is the production of nematode-free seedlings with the use of nematicides. When the coffee seedlings are transplanted to the fields, usually at the age of six months, the RKN second-stage juveniles do not get established in the roots in high numbers. This is due to a combination of the maturity of root tissues (Baeza, 1977), nematode-antagonistic soil microbiota (Angarita, 2000), and minimal disturbance of the plantations during the next few years of coffee cultivation.

Table 13.1 Host status of coffee-associated plant species for *Meloidogyne* spp., in the departments of Quindio, Caldas, Valle, Risaralda and Tolima, Colombia

Plant species	Family	*M. exigua*	*M. javanica*	*M. incognita*	*M. hapla*	*Meloidogyne* sp.	Reference
Amaranthus dubius Mart. ex Thell.	Amaranthaceae	–[a]	–	+	–	–	1
Anethum graveolens L.	Apiaceae	–	–	–	–	+	1
Antirrhinum majus L.	Plantaginaceae	+	–	–	–	–	1
Bidens pilosa L.	Asteraceae	–	+	–	–	–	1
Borreria laevis (Lam.) Griseb.	Rubiaceae	–	+	+	–	–	1
Brassica oleracea L.	Brassicaceae	–	–	–	–	–	1
Commelina difusa Burn.	Commelinaceae	+	–	–	–	–	1
Cucurbita maxima Duchesne	Cucurbitaceae	–	–	+	–	–	1
Cuphea racemosa (L.f.) Spreng.	Lythraceae	+	+	–	+	–	1
Cyperus rotundus L.	Cyperaceae	+	–	+	–	–	1
Galinsoga caracasana	Asteraceae	+	–	+	–	–	1
Heliopsis buphthalmoides (Jacq.) Dunal	Asteraceae	–	+	+	–	–	1
Hydrocotyle sp.	Araliaceae	+	–	–	–	–	1

Table 13.1 (continued)

Plant species	Family	*M. exigua*	*M. javanica*	*M. incognita*	*M. hapla*	*Meloidogyne* sp.	Reference
Impatiens balsamina L.	Fabaceae	nd[b]	nd	nd	nd	+	3
Inga sp.	Fabaceae	+	−	+	−	−	1
Matricaria chamomilla L.	Asteraceae	−	+	+	−	−	2
Melissa officinalis L.	Lamiaceae	−	+	+	−	−	2
Musa paradisiaca L.	Musaceae	−	−	+	−	−	1
Oxalis latifolia Kunth	Oxalidaceae	+	−	+	−	−	1
Phaseolus vulgaris L.	Fabaceae	+	−	+	−	−	1
Physalis nicandroides	Solanaceae	+	+	+	−	−	1
Setaria scandens Schrad. ex Schult.	Poaceae	−	−	−	−	−	1
Sida acuta (Burm.) Bors. Waalk.	Malvaceae	+	−	+	−	−	1
Solanum nigrum	Solanaceae	+	−	−	−	−	1
Spananthe paniculata	Apiaceae	−	−	+	−	−	1
Talinum paniculatum	Portulacaceae	+	+	−	−	−	1

[a](−) denotes resistance, (+) denotes susceptibility; [b] (nd) denotes not determined.
References: 1: Baeza et al. (1978), 2: Giraldo and Leguizamón (1997), 3: Mayorga (1996).

The first recommendation to maintain nurseries free of nematodes was to disinfect the soil in the germination beds with carbon bisulfide (Obregón, 1936). Later, successful results were obtained with foliar applications of oxamyl on the seedlings, which also induced a significant increase in the seedling's height, leaf surface area and weight, when compared to non-treated plants (Leguizamón and Baeza, 1972). By 1975, the suggested control at the nurseries consisted of application of 2 grams of Nemacur® or Dasanit® in the seedling's plastic bag (Anonymous, 1975). The currently recommended product, carbofuran, is required at 1 gram/bag before or during the first week of sowing. Doses higher than 2 grams/plant may cause plant toxicity, characterized by reddish-yellow spots on the leaves, with different sizes and shapes, that later necrotize (Baeza and Leguizamón, 1978). According to these authors, carbofuran application after the nematode has infected the seedlings resulted in significant increase in the fresh weight and reduced number of root-knots, when compared with untreated seedlings.

On the other hand, coffee plantations with severe symptoms caused by RKNs did not respond to any nematicide dose in field trials (López, 1978). Similar results were obtained in infested fields when chemical treatments were compared to planting healthy plants (Baeza, 1975).

13.2.5 Biological Control

The high human toxicity of nematicides has prompted the search for biological alternatives. The use of bacteria, fungi, predatory nematodes, and 'trapping' plants against coffee-parasitic nematodes has been suggested since the 1950s (González, 1950; Machado, 1951; Baeza, 1977). However, it was not until the 1990s that the interest in biological control measures resulted in intensive bioprospection of soils. *Paecylomyces lilacinus* (Thom.) Samson was the most frequently found fungus in RKN eggs and adult females, while a hyphomycete fungus (isolate Cenicafé 9501) was the most common in eggs (Cardona and Leguizamón, 1998). These authors also isolated the bacterium *Pasteuria penetrans* (Thorne) Sayre and Starr from all nematode stages. Although these organisms were as efficient as the recommended chemical products (Giraldo et al., 1998), the difficulty with the substrates needed to produce the fungi inocula, and the high dosages needed to obtain the LD_{50} values, made biological control economically unviable. Similar experiments carried out with native isolates of *Verticillium chlamydosporium* Goddard (Hincapié and Leguizamón, 1999), *Beauveria bassiana* (Bals.) Vuill., and *Metarhizium anisopliae* (Metsch.) Sorokin (Padilla and Leguizamón, 2001), resulted in lower doses necessary to obtain nematode control, but more studies are necessary on formulation technology. Also, the development of devices for proper application of biological products in coffee nurseries was addressed by Ibarra (2001).

The protective effect of mycorrhizae against nematodes is currently being evaluated in coffee and associated crops. Bioprospection in the Colombian coffee region resulted in the identification of the genera *Glomus* sp., *Scutellospora* sp.,

Acaulospora sp., *Gigaspora* sp., and *Entrospora* sp. associated with banana and plantain. A positive effect on coffee shoot and root growth, as well as root protection in nematode-infested seedbeds, was observed when *G. manihotis* Howeler, Sieverding and Schenck and *G. fistulosum* Skou and I. Jakobsen were applied in nurseries (C. Rivillas, unpublished results). The current recommendation to recover nematode-infested areas combines the production of healthy seedlings through early inoculation with mycorrhizae and *P. lilacinus*, in addition to crop rotation with maize to promote the beneficial soil microflora.

13.2.6 Genetic Control

Colombia has always produced arabica coffee, first with the introduction of the var *Typica*, followed by *Bourbon* and 'Caturra' (Krug et al., 1949). Although Obregón (1936) reported that *C. excelsa* A. Chev., today considered a variety of *C. liberica*, was resistant to *Meloidogyne* sp., no screening or breeding effort for nematode resistance was carried out for decades.

The susceptibility of 'Caturra' to 'leaf rust' led Cenicafé's researchers to cross it with the Timor Hybrid, later resulting in the release of 'Colombia'. This cultivar combines the agronomic characteristics of the 'Caturra', resistance to nematode, 'leaf rust', and *Colletotrichum kahawae* Waller and Bridge, and the largest bean in the market (Bettencourt, 1973; Rodrigues et al., 1975; Castillo, 1988; Alvarado, 2002). Currently, one third of the coffee areas are planted with the var *Typica* and *Bourbon*, 40% with 'Caturra', and 30% with 'Colombia'.

In 1977, Arango studied the histology of the interactions between *Coffea* spp., *M. javanica* and *M. incognita. C. liberica, C. canephora* Pierre ex A. Froehner and, in particular, *C. congensis* Froehner exhibited smaller root-knots and poorly developed giant cells, resulting in restricted nematode growth and prolificity. López (1978) reported that seedlings of *C. dewevrei* de Wild. and T. Durand were the most resistant to *M. javanica*, followed by *C. liberica, C. canephora*, and *C. congensis*, in comparison to *C. arabica* 'Caturra' and *C. eugenioides* S.Moore.

While evaluating Ethiopian accessions of *C. arabica*, Vergel (1999) found a wide range of responses to RKNs among the genotypes, from resistance to high susceptibility. On going studies show that Timor Hybrid derivatives are less susceptible to *M. incognita* and *M. javanica* than 'Caturra', under the same soil conditions and pathogen pressure (C. Rivillas, unpublished results).

13.3 Concluding Remarks

Colombia has a long tradition in research on coffee. Created in 1938, Cenicafé is responsible for developing and transferring technology to coffee growers, in order to increase productivity, reduce production costs, and preserve the environment and the quality of Colombian coffee. Based on five-year plans, the 60 scientists at Cenicafé (18 PhDs, 13 Masters, and 29 BSs) address key topics on the biology, chemistry, agronomy, engineering, and economics of coffee production. These research projects are funded mostly by the FNC, but also by government and international agencies. Scientific and technological findings are passed on through publications, courses, and a web site (www.cenicafe.org), which provide technical support for over a thousand FNC extensionists throughout the country.

In contrast to Central America and Brazil, severe nematode damages to Colombian coffee plantations, particularly by RKNs, are limited to small areas and short periods of time, even when the same susceptible varieties are cultivated in the same locale for over 200 years. Several hypotheses can be put forward to explain this phenomenon: coffee seedlings are not shipped among regions, which limits the dispersal of nematodes from high to low incidence zones. Also, several cultural practices enhance an effective antagonistic soil biota, such as the limited use of pesticides, the use of biological control agents and growth enhancers, and the selective control of weeds. Finally, and most importantly, the production of healthy seedlings at the nurseries guarantees root protection against nematodes at the plant's most susceptible stage.

This favorable scenario could change, however, if the pathogen or its epidemiology change. Furthermore, nematode dispersal and establishment in new areas are expected to happen because of increasing commercialization and shipment of plant materials, the cultivation of highly dense plantations, and the increasing diversification of coffee farms. Also, new coffee plantations are being established at higher altitudes in order to reduce the problems with 'leaf rust' and the berry borer, *Hypothenemus hampei* (Ferrari), and in lower lands to increase the planted area or to facilitate harvesting technology.

As has always happened, the price reached by coffee on the international market will play a major role in defining the total area cultivated in Colombia, as well as the plantation densities, varieties, and the management of the crop and diseases. All these factors could be favorable or detrimental to nematode populations.

In the upcoming years, the Plant Pathology Department at Cenicafé will continue to stimulate growers to use nematode-free coffee seedlings with a well-developed root system as their planting materials. Furthermore, an integrated nematode management will be implemented based on chemical and biological control at the nurseries, use of nematode-resistant cultivars, reduction of nematode populations in the fields, and appropriate diagnostic tools.

Therefore, as predicted in the 2006–2010 strategic plan, one PhD and two MSc researchers at Cenicafé are associated with the Colombian Institute for Agriculture, and students at Colombian Universities are to develop research actions in the following areas:

(1) Evaluation of soil solarization, intercropping, and crop rotation as part of an integrated nematode management program. These control strategies are being evaluated in the field in partnership with FNC's extension service. Ongoing studies are also evaluating green cover plants and plant extracts with allelopathic properties for use in the nurseries and the field.

(2) Development of fast and reliable nematode identification systems, up to host race level, by DNA fingerprinting applied to soil samples, infected tissues or any nematode specimen. Such techniques will allow quantification of soil populations, detection of mixed populations, monitoring of control practices, and identification of alternative hosts. Real-time PCR and ribosomal gene sequencing are being explored to accelerate the characterization and quantification of coffee pathogens, and nematodes are expected to be tested as well.
(3) Improvement of the biological control of nematodes through studies that examine better product formulations and the efficiency of mixtures of species and/or isolates of different biocontrol agents. These agents are also being enhanced for environmental adaptability and virulence. Finally, Cenicafé is pursuing advances in molecular identification and characterization of fungi populations in order to evaluate species and isolates, or their mixtures.
Since close attention must be paid to quality certification of biological products, Cenicafé has collaborated with the Colombian government and other agricultural organizations in order to establish quality standards for microbial pesticides based on fungi and bacteria.
(4) Use of bioinformatics for comparisons between genomes of coffee and other plant species. Breeding for resistance is not being pursued at Cenicafé at the present time.

In conclusion, we believe that achievements in these areas should build up Cenicafé's capability to face present and future nematode threats in Colombia and elsewhere.

References

Alvarado G (2002) Mejoramiento de las caracteristicas agronómicas de la variedad Colombia mediante la variación de su composición. Av Téc Cenicafé, 304: 1–8

Angarita M (2000) Aislamiento, identificación y evaluación de endomicorrizas nativas y otros microorganismos en diferentes sustratos para almácigos de café. Pontificia Universidad Javeriana, MS dissertation, Bogotá.

Anonymous (1975) Nematodos en plantaciones de café en Colombia. Agro-Bayer 61: 1–4

Anonymous (1997) Estadisticas cafeteras. Informe final. Sistema de información cafetera. Encuesta Nacional Cafetera SICA. Federación Nacional de Cafeteros de Colombia, Bogotá.

Anonymous (2002) Exportaciones tradicionales [MS Spreadsheet]. Banco de la república. www.banrep.gov.co, visited on 15 January 2005.

Arango L (1977) Estudio del proceso infectivo y la histopatologia del complejo de nematodos *Meloidogyne incognita – M. javanica* sobre plantas del cafeto. Universidad Nacional de Colombia, MS dissertation, Bogotá.

Baeza C (1974) Nematodos fitoparásitos asociados con el cultivo del café en Colombia. Proceedings Cong de Fund Colomb Fitopatol y Cienc Afines: 11

Baeza C (1975) Nematodos fitoparásitos asociados con el cultivo del cafeto en Colombia. Noticias Fitopatol 4: 120

Baeza C (1977) Metodologia para la identificación de la resistencia en *Coffea spp* a *Meloidogyne spp*. In: Informe trimestral de labores en nematologia. Centro Nacional de Investigaciones de Café, Sección de Fitopatologia, Chinchina.

Baeza C (1977) Ciclo de vida de *Meloidogyne exigua* en *Coffea arabica* var. Caturra. In: Informe anual de labores de la sección de fitopatologia. Federación Nacional de Cafeteros de Colombia/Centro Nacional de Investigaciones de Café, Chinchina.

Baeza C, Leguizamón J (1978) Control de nematodos en almácigos. Av Téc Cenicafé, 74

Baeza C, Benavides G, Leguizamón J (1978) Plantas de la zona cafetera colombiana hospedantes de *Meloidogyne* Göldi. Cenicafe 29: 35–45

Bettencourt A (1973) Consideracões gerais sobre o 'Hibrido de Timor'. Instituto Agronômico de Campinas, circular 23, Campinas.

Blancos E, Fernandez J, Cabrales L (1982) Nematodos en raices de cafeto (*Coffea arabica* L.) en el distrito de Marinca (zona comprendida entre los 600 y los 1200 m.s.n.m). Rev Agron 5: 63–77

Cano A, Gil F (1980) Dinámica de la población de *Meloidogyne incognita* raza 5 a diferentes densidades en *Coffea arabica* variedad Caturra, en condiciones de vivero. Universidad de Caldas, BS Dissertation, Manizales.

Cardona N, Leguizamón J (1998) Aislamiento y patogenicidad de hongos y bacterias al nematodo del nudo radical del café *Meloidogyne spp.* Göldi. Fitopatol Colombiana 21: 39–52

Castillo J (1988) Breeding for rust resistance in Colombia. In: Kushalapa AC and Eskes AB (eds.) Coffee rust epidemiology, resistance and management. CRC Press, Boca Raton.

Giraldo M, Leguizamón J (1997) Identificación de especies de *Meloidogyne* asociadas con *Melissa officinalis* y *Matricaria chamomilla*. Cenicafé 48: 275–286

Giraldo M, Leguizamón J, Chavez B (1998) Evaluación de *Paecilomyces lilacinus* (Thom.) Samson para el control de *Meloidogyne spp.* en almácigos de café (*Coffea arabica* L.) variedad Caturra. Fitopatol Colombiana 21: 104–117

Gómez L, Caballero A, Baldión V (1991) Ecotopos cafeteros de Colombia. Federación Nacional de Cafeteros de Colombia, Bogotá.

González R (1950) Nematodos. Cenicafé 1: 24–27

Guhl A (2004) Café y cambio de paisaje en la zona cafetera colombiana entre 1970 y 1997. Cenicafé 55: 29–44

Hincapié D, Leguizamón J (1999) Efecto de *Verticillium chlamydosporium* en el control de *Meloidogyne spp.* en almácigos de café var. Caturra. Cenicafé 50: 286–298

Ibarra M (2001) Diseño, construcción y evaluación de dos prototipos para la aplicación de productos biológicos en almácigos de café. Universidad Autónoma de Manizales, BS dissertation, Manizales.

Krug C, Mendes J, Carvalho A (1949) Taxonomia de *Coffea arabica* L. II. *Coffea arabica* L. var. Caturra e sua forma xanthocarpa. Bragantia 9: 157–163

Leguizamón J (1976) Relación entre poblaciones de *Meloidogyne spp.* en el suelo y el daño causado en cafetales establecidos. Cenicafé 27: 174–179

Leguizamón J, Baeza C (1972) Acción del nematicida experimental Dpx 1410, en el control del nematodo nodulador del cafeto (*Meloidogyne exigua* Göldi 1887). Cenicafé 23: 98–103

Leguizamón J, López S (1972) Nematodos en plantaciones de café en Colombia. Av Téc Cenicafé 20: 1–4

Leguizamón J (1997) Efecto de *Meloidogyne spp.* en plantaciones establecidas de café variedad Caturra. In: Informe anual de actividades 1996–1997: Disciplina de Fitopatologia. Centro Nacional de Investigaciones de Café, Chinchiná.

López DS (1978) Control de nematodos para el establecimiento de cafetales sanos en suelos infestados con *Meloydogyne spp.* In: Informe Semestral de Labores 1977–1978. Centro Nacional de Investigaciones de Café, Chinchiná.

Machado A (1951) Los nematodos y la decadencia de muchos cafetales y cultivos. Rev Cafetera Colomb 10: 3572–3576

Mayorga L (1996) Conservación e incremento de inóculo de *Meloidogyne spp.* en *Impatiens sp.* Cenicafé, 47: 53–56

Navarro R (1978) Identificación y ubicación de cinco *Meloidogyne spp.* en diferentes pisos térmicos. In: Conferencia de trabajo sobre el proyecto internacional *Meloidogyne*, Regional II, Memorias. Instituto Colombiano Agropecuario, Bogotá.

Obregón R (1936) Un nematodo del café. Rev Cafetera Colomb 6: 2040–2042

Padilla B, Leguizamón J (2001) Efecto de *Beauveria bassiana* y *Metarhizium anisopliae* en el control del nematodo del nudo radical del café. Cenicafé 52: 29–41

Quintana JC, Gaitan A, Cristancho M et al (2002) Caracterización molecular de lineas puras del nematodo del nudo radical *Meloidogyne* spp. provenientes de café. Ascolfi Informa 28: 24–28

Ramirez LF, Silva G, Valenzuela LC et al (2002) El café, capital social estratégico. In: Informe final comisión de ajuste de la institucionalidad cafetera. Federación Nacional de Cafeteros de Colombia, Bogotá.

Rodrigues C, Bettencourt A, Rijo L (1975) Races of the pathogen and resistance to the coffee rust. Ann Rev Phytopathol 13: 49–70

Taylor A, Sasser J (1978) Biology, Identification and Control of Root-knot Nematodes (*Meloidogyne spp.*). International Meloidogyne Project. North Carolina State University Graphics, Raleigh.

Toro R (1929) Un nuevo parásito del café en Cundinamarca. Rev Cafetera Colomb 2: 437–438

Vergel D (1999) Metodologia de evaluación de la resistencia a nematodos del nudo radical *Meloidogyne spp.* en *Coffea spp* y evaluación de germoplasma de café. Universidad Nacional, Facultad de Agronomia, MS dissertation, Bogotá.

Vergel D, Leguizamón J, Cortina, H et al (2000) Reconocimiento y frecuencia de especies de *Meloidogyne* en una localidad de la zona cafetera. Cenicafé 51: 285–295

Villalba D, Fernandez O, Baeza C (1983) Ciclo de vida de *Meloidogyne incognita* raza 5 (Kofoid and White, 1919) Chitwood, 1949 en *Coffea arabica* var. Caturra. Cenicafé 34: 16–33

Chapter 14
Central America

Luc Villain, Adan Hernández and Francisco Anzueto

14.1 Brief Outline of the Crop

14.1.1 Socio-Economic Aspects

This chapter focuses on five Central American coffee-producing countries; from northwest to southeast these are Guatemala, El Salvador, Honduras, Nicaragua and Costa Rica. In these countries the coffee industry contributes relatively little to the national GDPs (from 1.3% in Costa Rica to 7.2% in Nicaragua), but it has an important social role since it employs around a quarter of their active population.

Central America contributes 13–15% of the world's coffee trading, despite its small geographic area; among the five countries focused in this chapter, the smallest is El Salvador (with just over 21 thousand km^2) and the largest is Nicaragua (129.5 thousand km^2). Central America's coffee production, hectarage and average yield are shown in Table 14.1. As seen in Table 14.2, smallholders predominate in this region, contributing 27% of its output. Technological status and inputs vary greatly among coffee growers, and productivity varies by a factor of two.

Although there has been an increase in domestic coffee consumption in the last 10 years, about 90% of Central America's production is exported. In comparison,

Table 14.1 Central America's coffee production, hectarage and productivity

	Countries				
	Guatemala	Honduras	Costa Rica	El Salvador	Nicaragua
Green coffee production (tons)	239,168[a]	182,876	132,606	85,814	70,099
Coffee hectarage	248,026	229,243	113,387	160,622	117,334
Average yield (ton/ha)	0.96	0.80	1.17	0.53	0.60

[a] Data are average of 2002 through 2006.
Adapted from Anonymous (2008)

L. Villain
Centre de Coopération Internationale en Recherche Agronomique pour le Développement TA A-98/IRD, Montpellier, France
e-mail: luc.villain@ird.fr

R.M. Souza (ed.), *Plant-Parasitic Nematodes of Coffee*,

Table 14.2 Typology of Central America's coffee farming

	Classes of farm size (in hectares)					
	< 3.5	3.5–14	14–35	35–70	> 70	General averages
Average size	0.8	3.6	3.8	18.2	103.2	3.1
Number of farms ($\times 1,000$)	200	47.9	33	7.3	2.9	(–)
Percentage of the production	11.6	14.7	15.9	21.3	36.5	(–)
Production ($\times$1,000 tons)	85	108	117	156	268	(–)
Total area ($\times$1,000)	162	170	126	133	301	(–)
Average yield (ton/ha)	0.53	0.63	0.93	1.17	0.89	0.82

Adapted from Anonymous (2002).

arabica coffee (*C. arabica* L.) is grown much more often than robusta (*C. canephora* Pierre ex Froehner). In Guatemala, the region's largest robusta producer, it represents only 0.7% of the total exports.

14.1.2 Agro-Ecological Aspects

The five countries present a mountainous topography which, combined with their intermediate position between Northern and Southern hemispheres, makes them part of the Mesoamerican biological corridor. They present a great diversity of ecosystems and a huge biodiversity, which attract biologists from many areas, including nematologists.

As is the case in many other regions, coffee in these countries has a long history of monoculture, with large areas being cultivated since the end of the nineteenth century, often without any crop rotation. This may have favored the development and later dissemination of nematode populations well-adapted to coffee.

Coffee is primarily grown in highland areas with a climate characterized by heavy rainfall (mostly from May through October) coupled with a prolonged dry season, which results in water deficit for the plants. This agro-ecosystem is quite fragile, particularly in terms of its volcanic soils (mostly andosols), which present a slow rate of organic matter mineralization and a sandy texture that together make these soils highly prone to erosion (Bornemiza et al., 1999). Because these soils percolate rainfall well, and considering that many coffee plantations are established between 800 and 1,600 masl, it can be said that this crop influences water retention in basins, its surface runoff and underground infiltration. Therefore, the use of highly toxic water-soluble pesticides like nematicides may cause severe impact on the environment and on human health in those areas.

As regards nematodes, volcanic sandy soils are favorable to their development and dissemination; indeed, in Central America all major nematode problems have occurred along the volcanic cordillera. In this region, soil acidification has occurred wherever the increment of nitrogen fertilization required by intensive coffee cultivation has not been accompanied by pH management. In turn, soil acidification has resulted in plant nutrition imbalances (Bornemiza et al., 1999) and worsening of nematode problems.

In Central America, coffee is largely cultivated shaded by trees, with the exception of some regions with high nebulosity, such as Alta Verrapaz (Guatemala) and Costa Rica's central plateau, which are influenced by the humid Atlantic winds. Shading is managed through pruning the trees at the end of the dry season, just before coffee flowering. Since the 1970s many growers have switched from shaded to intensive, full sun coffee cultivation, using highly productive dwarf cultivars. This has led to the emergence of problems with soil erosion and premature decline in productivity; furthermore, full sun exposure has dramatically increased the impact of nematode parasitism, particularly by *M. paranaensis* Carneiro, Carneiro, Abrantes, Santos and Almeida and by root-lesion nematodes (RLNs), *Pratylenchus* sp.

Recently, shaded cultivation has received renewed interest, since it is a more sustainable production system, which is better adapted to today's coffee market crisis and responsive to the community's ecological concerns. The shade trees help to mitigate stressful climate factors, such as water deficit and high day temperatures, particularly in regions with a marked dry season; this is the case of coffee-growing areas that slope towards the Pacific coast. Hence, shade trees create a favourable microclimate for coffee plants (Wilson, 1985; Beer et al., 1998). Moreover, the organic matter supplied by shade trees through their fallen leaves and pruned branches helps to improve soil fertility, especially of poor volcanic or highly clay-based soils.

By decreasing abiotic stresses on coffee plants, shade trees indirectly enhance their tolerance to nematodes and favor their metabolism-based defence mechanisms, such as the phenolic pathway. This has been observed against RLNs (Toruan-Mathius et al., 1995; Villain et al., 2001c; 2004). For example, in a region with a marked dry season, the level of RLN root population in partially resistant robusta coffee rootstocks was negatively correlated to the degree of shading (Villain et al., 2000). Furthermore, the abundant litter produced in shaded plantations may favor the development of microfauna and microflora antagonistic to plant-parasitic nematodes, thus depressing their populations (Sayre, 1971; Norton, 1978; Stirling, 1991).

Another aspect of prime ecological and social importance in Central America is that the pruning of shade trees supplies large amounts of firewood, which is largely used by rural communities; this saves the natural forest. Recently, agroforestry has been stimulated to diversify coffee growers' income; timber tree species have thus been used for shading (Vaast et al., 2005; 2007).

14.2 The Importance of Nematodes to Coffee Production in Central America

In Honduras, relevant nematode problems have only been observed in a small area on the border with Nicaragua. In the remaining four countries, most of the coffee-producing regions have widespread infestation by nematodes (Villain et al., 1999; Campos and Villain, 2005). The exception observed in Honduras may be related to the fact that most of its plantations are located on calcareous and schistose highland soils, while in the other countries they are located on volcanic highland ones. The influence of soil properties on nematode incidence and severity has also been observed in Guatemala's western coffee-growing region (Villain et al., 1999).

Nematodes are not a recent problem for arabica coffee cultivation in Central America. In Guatemala, for example, severe damage caused by RLNs and RKNs was reported as early as 1935 by Alvarado. The coffee cultivars and varieties currently grown in Central America are susceptible to most pathogenic nematodes present in the region (Bertrand et al., 1999; Villain et al., 1999; 2002; Hernández et al., 2004b).

14.2.1 Coffee-Parasitic RKNs in Central America

Conventional taxonomic criteria, particularly the morphology of female perineal pattern, have proved to be deficient for reliable identification of *Meloidogyne* species (see Chapter 6). Since the 1990s, studies on isoenzyme systems (particularly esterase) have revealed a large and unexpected diversity of *Meloidogyne* species in Central America (Hernández 1997; Hernández et al., 2004a; Carneiro et al., 2004; Villain et al., 2007).

In Central America, the most widespread species on coffee is probably *M. exigua* Göldi; the same almost certainly applies to the whole of Latin America. Its presence has been confirmed in Honduras, Nicaragua and Costa Rica (Hernández et al., 2004a; Villain et al., 2007). *M. exigua* has not been found to cause a drastic impact on coffee plantations that are managed properly, including fertilization.

Two other *Meloidogyne* species that are very pathogenic to arabica coffee occur in Central America: *M. arabicida*, described in Costa Rica by López and Salazar (1989) and *M. paranaensis*, originally described in Brazil. Both species induce root 'corky' swellings, i.e., an extreme suberization of the root cortex; even the tap root is affected, which may result in complete destruction of the root system and plant death (Figs. 14.1 and 14.2). *M. arabicida* has been found to decimate entire plantations (López and Salazar, 1989).

Bertand et al. (2000a) have showed that a syndrome locally referred to as 'corchosis' seems to be caused by simultaneous parasitism by *M. arabicida* and *Fusarium oxysporum* (Schltdl.) W. C. Snyder et H. N. Hansen. *M. arabicida* was originally detected in Costa Rica's Turrilalba Valley. Although this nematode has been detected in restricted areas to which infected coffee seedlings had been shipped, it seems confined geographically; indeed, *M. arabicida* has not been reported from other countries or crops.

In Guatemala, *M. paranaensis* seems to be the predominant RKN. It is interesting to note that because of perineal pattern similarities this species was erroneously identified as *M. incognita*, which was considered the predominant species in this country (Chitwood and Berger, 1960). For decades the same misidentification also occurred in Brazil. Although *M. paranaensis* is present in Brazil and Hawaii (Carneiro et al., 2004; see Chapter 6), it has not been detected in other Central American countries.

Fig. 14.1 'Corky-root' symptom on *Coffea arabica* parasitized by *Meloidogyne paranaensis* in Guatemala (Photo by L. Villain) (see color Plate 23, p. 331)

Fig. 14.2 Adult *Coffea arabica* plants parasitized by *Meloidogyne paranaensis* in southwest Guatemala (Photo by L. Villain)

Guatemalan *M. paranaensis* populations have been thoroughly studied by Anzueto (1993) and Hernández (1997), who demonstrated that this species parasitizes own-rooted arabica coffee plants as well as those grafted onto susceptible robusta rootstocks. *M. paranaensis* is particularly damaging to seedlings infected at an early developmental stage. When healthy seedlings are planted in infested areas, the plants generally start to decline when they start production, and major mortality occurs after just two or three harvests. Plant mortality is more widespread in plantations under full sun.

Recent studies based on esterase diagnostics have found coffee-parasitic *M. incognita* (Kofoid and White) Chitwood in Central America; several populations have been collected in Costa Rica, one in Guatemala and another in El Salvador (Villain et al., 2007). However, this species' geographical distribution and economic impact on coffee production in these regions remain unknown.

M. izalcoensis has recently been described from El Salvador (Carneiro et al., 2005). In this country, it seems to be largely scattered in the southwestern region of the Izalco volcanic massif (Fig. 14.3), while it has not been found in any other region in El Salvador or Central America.

Some other *Meloidogyne* species have been reported on coffee in Central America; nonetheless, these species seem to be geographically restricted, and their

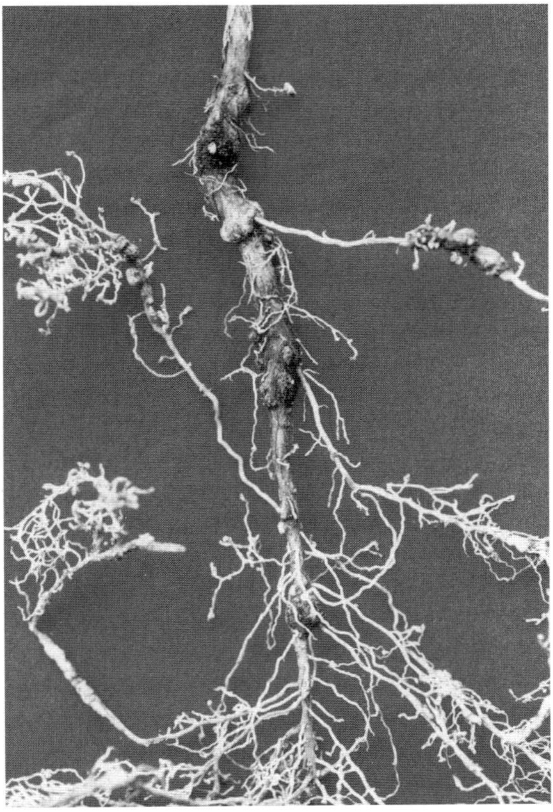

Fig. 14.3 Root symptoms on *Coffea arabica* parasitized by *Meloidogyne izalcoensis* in El Salvador (Photo by A. Hernández)(see color Plate 24, p. 331)

economic importance has not been established. Two populations of *M. mayaguensis* Rammah and Hirschmann have been reported from Costa Rica and Guatemala (Hernández et al., 2004a; Villain et al., 2007). *M. hapla* Chitwood, which is more adapted to temperate climate, has been observed in some highland coffee plantations in northern Guatemala, and in El Salvador on the Izalco volcano massif (Hernández et al., 2004a; Villain et al., 2007). Finally, *M. arenaria* (Neal) Chitwood has been found in El Salvador and Guatemala (Hernández et al., 2004a; Carneiro et al., 2004; Villain et al., 2007).

Several surveys in coffee-producing areas of Central America have detected *Meloidogyne* populations with atypical esterase phenotypes (Hernández, 1997; Hernández et al., 2004a; Carneiro et al., 2004; Villain et al., 2007). These populations warrant morphological and morphometric characterizations and taxonomic identification.

14.2.2 Coffee-Parasitic RLNs in Central America

Parasitism by RLNs frequently remains unnoticed by growers because they do not associate root necrosis with nematode parasitism (see Chapter 5). Nonetheless, it has been proved that RLNs are widely distributed in Central America's coffee-growing regions (Villain et al., 1999; 2004; Campos and Villain, 2005). From Guatemala, RLNs have been reported by Schieber and Sosa (1960), Chitwood and Berger (1960), Schieber (1966; 1971) and Villain (2000); from El Salvador by Abrego et al. (1961), Whitehead (1969) and Gutierrez and Jimenez (1970); from Nicaragua by Sequeira-Bustamente et al. (1979), and from Costa Rica by Salas and Echandi (1961), Tarjan (1971) and Figueroa and Perlazo (1982).

As is the case in other coffee-producing regions in the world, *P. coffeae* (Zimmerman) Filipjev and Schuurmans Stekhoven is the most reported RLN in Central America. Two species morphologically similar to *P. coffeae* have been described: *P. panamaensis* Siddiqi, Dadur and Barjas and *P. gutierrezi* Golden, López and Vilchez, from Panama and Costa Rica, respectively. Nonetheless, Siddiqi (2000) has synonymized the latter species to the former (see Chapter 3). The geographic distribution and pathogenicity of this species remain unknown.

Recent studies have been carried out to characterize RLN populations from coffee plantations in Central America. An integrated approach has been adopted, involving characterization of morphology through scanning electron microscopy, studies on mode of reproduction and mating tests, *in vitro* fitness as related to temperature, root penetration and reproduction patterning, pathogenicity on *Coffea* spp. and molecular aspects (Anzueto, 1993; Herve, 1997; Villain et al., 1998; 2000; 2001a; Villain, 2000). These studies have revealed a marked diversity among those RLN populations, whose mode of reproduction is always amphimitic. Through controlled inoculations, some RLN populations from Guatemala have shown a high degree of pathogenicity towards arabica coffee cultivars (Fig. 14.4), which confirmed the severe damage observed in field experiments (Fig. 14.5) and in commercial plantations (see Chapter 5).

Fig. 14.4 Six month-old *Coffea arabica* seedlings parasitized by *Pratylenchus* sp. from Guatemala, in comparison to a healthy control (left) (Photo by L. Villain)

Fig. 14.5 *Coffea arabica* plants parasitized by *Pratylenchus* sp. in southwest Guatemala. Own-rooted (foreground) and grafted onto a nematode-resistant *Coffea canephora* Pierre ex Froehner rootstock (background) (Photo by L. Villain) (see color Plate 25, p. 332)

Three RLN populations remain to be characterized and identified (or described) at the species level. Studies by Duncan et al. (1999) have confirmed the need for a thorough examination of *P. coffeae*-similar RLNs from Central America and other regions.

In conclusion, there seems to be a great diversity of coffee-parasitic RLNs (as well as RKNs) in Central America, which is a region that receives convergent influences from both North and South America (Dettman, 2006). The restricted or discontinued distribution of some *Pratylenchus* and *Meloidogyne* species could be related to the mountainous topography and/or with anthropogenic activities, particularly the traffic of coffee seedlings. Vegetative seedlings of intercropped plant species, such as *Musa* spp., probably also play an important role in disseminating nematodes.

14.3 Management of Coffee-Parasitic Nematodes in Central America

14.3.1 Biological Control, Crop Rotation and Intercropping

Just a few studies have been carried out in Central America to assess the effectiveness of biological control, crop rotation and intercropping to control coffee-parasitic nematodes; only a subset of these studies has been published. For example, some cover crops have been unable to suppress the variety of nematodes present in fields, or they have been difficult to employ in plantations, from an agronomic point of view (e.g., Herrera et al., 1999). In Nicaragua, *Desmodium ovalifolium* (Prain) Wall. ex Merr. and *Stizolobium* sp. have suppressed *M. incognita*; on the other hand, *D. ovalifolium* favored the development of *Rotylenchulus reniformis* Linford and Oliveira, and *Stizolobium* sp. inhibited coffee growth.

14.3.2 Chemical Control

Pesticides are currently used in coffee nurseries mostly as a prophylactic measure against plant-parasitic nematodes; the goal is to guarantee nematode-free seedlings. For disinfection of substrates, the most commonly used product is dazomet (granulated Basamid®), either in seed germination trays (at the concentration of 40 g/m^2) or in nursery bags (at 60 g/m^2). The application of granulated or liquid nematicides may continue during the whole six–month period prior to transplanting the seedlings to the field.

When a coffee field is found to be infested by pathogenic plant-parasitic nematodes, newly transplanted seedlings are sometimes treated with nematicides during the first two years of cultivation; the goal is to reduce nematode damage during the vegetative development of the plantation, prior to the first harvest. Chemical treatment is nonetheless not recommended as a curative approach for established plantations, because nematicide effectiveness is minimal when nematode-related symptoms are already being observed. In such cases, the recommended approach is to eradicate the declining plantation and replant it with a scion grafted onto resistant rootstock (see below).

14.3.3 Genetic Control

In Central America, genetic control has been the priority against coffee-parasitic nematodes; this has been pursued through screening of genotypes for nematode resistance. In 1966 Reyna developed the hypocotyledonar grafting technique to join arabica coffee scions with RLN-resistant robusta rootstocks (Fig. 14.6) (Reyna, 1968). This practice is now common among growers in areas of Guatemala and El Salvador where RLNs are widespread, especially on the volcanic cordillera. This approach is highly effective even when unscreened robusta genotypes are used (Villain et al., 2000; 2001b).

In 1976 the fungus *Hemileia vastatrix* Berk and Br. (causal agent of 'leaf rust') was introduced into Central America. Consequently, during the 1980s coffee breeding programs prioritized the search for resistance to that fungus, focusing on introgression of resistance genes from robusta coffees into arabica ones; such efforts led to the development of 'Catimor' and 'Sarchimor' (see Chapter 9). Although nematode-resistance was not the goal, some of the introgressed cultivars showed resistance to *M. exigua* (Bertrand et al., 1997; 1999; 2001a; Noir et al., 2003).

Since the 1990s, the description of nematode species which are very pathogenic to coffee, such as *M. arabicida* and *M. izalcoensis*, and the awareness of the

Fig. 14.6 Seedlings of *Coffea arabica* grafted onto *C. canephora* Pierre ex Froehner in Guatemala (Photo by L. Villain) (see color Plate 26, p. 332)

Fig. 14.7 *Meloidogyne paranaensis*-resistant (*left*) and -susceptible (*right*) seedlings of *Coffea canephora*. The seedlings on the left belong to one of the parent genotypes of the nematode-resistant rootstock cultivar 'Nemaya' (Photo by L. Villain)

widespread distribution, pathogenicity and economical importance of *M. paranaensis* in Guatemala (Anzueto, 1993; Hernández, 1997; Hernández et al., 2004b; Carneiro et al., 2004), have spurred on research into nematode resistance (see Chapter 9).

Resistance genes have been sought in *C. canephora* and *C. arabica* semi-wild Ethiopian accessions (Anzueto, 1993; Bertrand et al., 2000b; 2002; Anzueto et al., 2001; Anthony et al., 2003; Hernández et al., 2004b). Regarding *C. canephora*, this led to the identification of two coffee clones which are resistant to Central America's most important RKNs, such as *M. paranaensis* (Fig. 14.7); crosses between these clones have resulted in the creation of the new nematode-resistant rootstock cultivar 'Nemaya' (Anzueto et al., 1996; Bertrand et al., 2002). Moreover, 'Nemaya' is highly resistant to RLNs (Villain et al., 2004).

With regard to *C. arabica*, some Ethiopian semi-wild accessions have shown resistance to some of the major *Meloidogyne* species from Central America; this has created a new coffee breeding approach whereby *C. arabica* hybrid F1 cultivars have been created through crosses with resistant Ethiopian accessions (Bertrand et al., 2005). No source of resistance to RLNs has been found among a large group of semi-wild Ethiopian accessions (Anzueto, 1993; Villain et al., 2004).

14.4 Concluding Remarks

In Central America, research into and extension of all aspects of coffee cultivation are conducted by or in collaboration with national institutions: Icafe in Costa Rica, Anacafe in Guatemala, Procafe in El Salvador, Ihcafe in Honduras and Conacafe in Nicaragua. Some national universities also contribute to coffee research. To foster the development of a coffee research and development network, a cooperative program was started in 1979, under the aegis of the IICA (the Spanish acronym for Inter-American Institute for Cooperation on Agriculture). Today, IICA, CATIE (Tropical Agricultural Research and Higher Education Center), the five institutions cited above and those from Panama, Jamaica and the Dominican Republic form Promecafe (Regional Cooperative Program for Technological Development of Coffee in Central America, Panama, Dominican Republic and Jamaica).

Promecafe is supported by funds from the country members as well as international resources, primarily from USAID and the European Union. Since its creation, Promecafe has developed strong scientific cooperation with CIRAD (Agricultural Research Centre for International Development) and the IRD (Research Institute for Development), which are French institutions.

It is within the framework of this cooperative scientific program that nematode-related challenges should be addressed. For example, the diversity of coffee-parasitic RKNs and RLNs is only partially known. Particularly for the latter, studies have pointed out the need for better characterization of some populations from Central America; such examination might lead to a reconsideration of the importance of *P. coffeae* for coffee cultivation in this region (Herve, 1997; Villain et al., 1998; Duncan et al., 1999; Villain, 2000; Wayenberge and Moens, 2004).

For RKNs, some populations with unreported esterase phenotypes should be studied, while recently described *Meloidogyne* species should have their distribution accurately assessed through extensive or selective surveys throughout the different regions. Due to the complexity of *Meloidogyne* taxonomy, all RKN populations revealed by such surveys should be characterized on different, complementary aspects, such as enzymatic phenotyping, selective genome sequencing and pathogenicity towards key coffee genotypes.

Attention should also be paid to investigating whether other RKNs, in addition to *M. arabicida*, present interactions with soil-borne fungi, such as *Fusarium oxysporum* (Bertrand et al., 2000a). Indeed, such interaction may occur wherever 'corchosis' is observed, as in the case of *M. paranaensis*-parasitism in Guatemala.

As regards nematode control, new hybrid F1 cultivars are likely to represent an alternative for the control of RKNs, particularly in highland regions where grafting on *C. canephora* is more difficult because this species does not develop well in a mild climate (Bertrand et al., 2001b). Therefore, these new resistant genotypes should be assessed for resistance to all major RKNs present in Central America; such an assessment should also be carried out for RLNs.

As regards RLNs, their economic importance to coffee production warrants a thorough examination of *Pratylenchus*-resistance, as far as their genetic determinism and defence-mechanisms are involved. Such studies are crucial for the development of molecular-assisted breeding programs, and these have already been planned for RKNs (see Chapter 9).

Finally, alternative techniques should be developed to guarantee the sanitary status of nursery seedlings; indeed, no alternatives have been offered to growers since the ban on methyl bromide. Alternatives should also be developed to reduce nematode damage in infested fields, especially when resistant genotypes are employed, because this should increase the durability of nematode resistance.

References

Abrego L, Holdeman Q (1961) Informe de progresos en el estudio del problema de los nematodos del cafe en El Salvador. Instituto salvadoreño de investigaciones del cafe. Boletin Informativo, Santa Tecla

Alvarado JA (1935) Tratado de caficultura practica. Tipografia Nacional. Guatemala

Anonymous (2002) Centroamerica: el impacto de la caida de los precios del cafe em 2001. Comisión Económica para America Latina y El Caribe, Mexico

Anonymous (2008) Faostat. http://faostatclassic.fao.org, visited on March 28th 2008

Anthony F, Topart P, Anzueto F et al (2003) La resistencia genetica de *Coffea* spp. a *Meloidogyne paranaensis*: identificación y utilización para la caficultura latinoamericana. Manejo Integrado Plagas y Agroecol 67: 4–11

Anzueto F (1993) Etude de la resistance du cafeier (*Coffea* sp.) à *Meloidogyne* sp. et *Pratylenchus* sp. Ecole Nationale Supérieure Agronomique de Rennes, DS thesis, Rennes

Anzueto F, Bertrand B, Peña M et al (1996) Desarrollo de una variedad porta-injerto resistente a los principales nematodos de America Central. Proceedings Simp Caficultura Latino-americana: 7

Anzueto F, Bertrand B, Sarah JL et al (2001) Resistance to *Meloidogyne incognita* in Ethiopian *Coffea arabica* accessions. Euphytica 118: 1–8

Beer J, Muschler R, Kass D et al (1998) Shade management in coffee and cacao plantations. Agrofor Syst 38: 139–164

Bertrand B, Aguilar G, Bompard E et al (1997) Comportement agronomique et resistance aux principaux depredateurs des lignees de Sarchimor et Catimor au Costa Rica. Plante, Rech, Dev 4: 312–321

Bertrand B, Aguilar G, Santacreo R et al (1999) El mejoramiento genetico en America Central. In: Bertrand B, Rapidel B (eds) Desafios de la caficultura en Centroamerica. IICA-PROMECAFE, San José

Bertrand B, Anthony F, Lashermes P (2001a) Breeding for resistance to *Meloidogyne exigua* of *Coffea arabica* by introgression of resistance genes of *C. canephora*. Plant Pathol 50: 637–643

Bertrand B, Anzueto F, Moran MX et al (2002) Creation and distribution of a rootstock variety (*Coffea canephora*). Plante Rech Dev (special issue: research and coffee growing): 95–107

Bertrand B, Etienne H, Cilas C et al (2005) *Coffea arabica* hybrid performance for yield, fertility and bean weight. Euphytica 141: 255–262

Bertrand B, Etienne H, Eskes AB (2001b) Growth, production and bean quality of *Coffea arabica* as affected by interspecific grafting: consequences for rootstock breeding. Hort Sci 36: 269–273

Bertrand B, Nuñez C, Sarah JL (2000a) Disease complex in coffee involving *Meloidogyne arabicida* and *Fusarium oxysporum*. Plant Pathol 48: 383–388

Bertrand B, Peña-Durán MX, Anzueto F et al (2000b) Genetic study of *Coffea canephora* coffee tree resistance to *Meloidogyne incognita* nematodes in Guatemala and *Meloidogyne* sp. nematodes in El Salvador for selection of rootstock varieties in Central America. Euphytica 113: 79–86

Bertrand, B, Ramirez, G, Topart, P et al (2002) Resistance of cultivated coffee (*Coffea arabica* and *C. canephora*) trees to corky-root caused by *Meloidogyne arabicida* and *Fusarium oxysporum*, under controlled and field conditions. Crop Prot 21: 713–719

Bornemiza E, Collinet J, Segura A (1999) Los suelos cafetaleros de America Central y su fertilización. In: Bertrand B, Rapidel B (eds) Desafios de la caficultura en Centroamerica. IICA-PROMECAFE, San José

Campos VP, Villain L (2005) Nematode parasites of coffee and cocoa. In: Luc M, Sikora RA, Bridge J (eds) Plant parasitic nematodes in subtropical and tropical agriculture, 2nd edition. CABI, Wallingford

Carneiro RMDG, Almeida MRA, Gomes ACMM et al (2005) *Meloidogyne izalcoensis* n.sp. (Nematoda:Meloidogynidae), a root-knot nematode parasitizing coffee in El Salvador. Nematology 7: 819–832

Carneiro RMDG, Tigano MS, Randig O et al (2004) Identification and genetic diversity of *Meloidogyne* spp. on coffee from Brazil, Central America and Hawaii. Nematology 6: 287–298

Chitwood BG, Berger CA (1960) Preliminary report on nemic parasites of coffee in Guatemala with suggested and interim control measures. Plant Dis Rep 44: 841–847

Dettman S (2006) The mesoamerican biological corridor in Panama and Costa Rica: integrating bioregional planning and local initiatives. J Sustain For 22: 15–34

Duncan LW, Inserra RN, Thomas WK et al (1999) Molecular and morphological analysis of isolates of *Pratylenchus coffeae* and closely related species. Nematropica 29: 61–80

Figueroa A, Perlazo F (1982) Investigacion sobre *Meloidogyne* en Costa Rica. Proceedings third research and planning conference on root-knot nematodes, *Meloidogyne* spp., region I. North Carolina State University and USAID, Raleigh

Gutierrez G, Jimenez QMF (1970) Algumas observaciones sobre la injertación del cafe practicada en Guatemala y El Salvador como medio para el control de nematodos. Rev Cafetalera 98: 35–47

Hernández A (1997) Etude de la variabilite intra- et interspecifique des nematodes du genre *Meloidogyne* parasites des cafeiers en Amerique centrale. Université de Montpellier II, Sciences et Techniques du Languedoc, DS thesis, Montpellier

Hernández A, Fargette M, Sarah JL (2004a) Characterisation of *Meloidogyne* spp. (Tylenchida: Meloidogynidae) from coffee plantations in Central America and Brazil. Nematology 6: 193–204

Hernández A, Fargette M, Sarah JL (2004b) Pathogenicity of *Meloidogyne* spp. (Tylenchida: Meloidogynidae) isolates from Central America and Brazil on four genotypes. Nematology 6: 193–204

Herrera S, Isabel C, Marban-Mendoza N (1999) Efecto de coberturas vivas de leguminsosas en el control de algunos fitonematodos del café en Nicaragua. Nematropica 29: 223–232

Herve G (1997) Caracterisation biologique et moleculaire de populations de nematodes phytoparasites dans les plantations de cafeiers d'Amerique Centrale. Ecole Nationale Supérieure Agronomique de Rennes, D.E.A. dissertation, Rennes

Lopez R, Salazar L (1989). *Meloidogyne arabicida* n. sp. (Nematoda: Heteroderidae) nativo de Costa Rica: un nuevo y serio patógeno del cafeto. Turrialba 39: 313–323

Noir S, Anthony F, Bertrand B et al (2003) Identification of a major gene (*Mex*-1) from *Coffee canephora* conferring resistance to *Meloidogyne exigua* in *Coffea arabica*. Plant Pathol 52: 97–103

Norton DC (1978) Ecology of plant-parasitic nematodes. Wiley Interscience, New York

Reyna EH (1968) La tecnica de injerto hipocotiledonar del cafeto para le control de nematodos. Cafe 17: 5–11

Salas LA, Echandi E (1961) Parasitic nematodes in coffee plantations of Costa Rica. Coffee 3: 6–9

Sayre RM (1971) Biotic influences in soil environment. In: Zuckerman BM, Mai WF, Rohde RA (eds) Plant parasitic nematodes, vol 1: Morphology, anatomy, taxonomy and ecology. Academic Press, London

Schieber E (1966) Nematodos que atacan al cafe en Guatemala, su distribución, sintomatologia y control. Turrialba 16: 130–135

Schieber E (1971) The nematode problems of coffee in Guatemala. Nematologica 1: 17

Schieber E, Sosa ON (1960) Nematodes on coffee in Guatemala. Plant Dis Rep 44: 722–723

Sequeira-Bustamente F, Schuppener H, Cuarezma J et al (1979) Nematodos fitoparasitos asociados al cultivo del cafeto (*Coffea arabica* L.) en Nicaragua. Nematropica 9: 97

Siddiqi MR (2000) Tylenchida, parasites of plants and insects. CABI, Wallingford

Stirling GR (1991) Biological control of plant parasitic nematodes – progress, problems and prospects. CABI, Wallingford

Tarjan AC (1971) Some interesting associations of parasitic nematodes with cacao and coffee in Costa Rica. Nematropica 1: 5

Toruan-Mathius N, Pancoro A, Sudarmadji D et al (1995) Root characteristics and molecular polymorphisms associated with resistance to *Pratylenchus coffeae* in robusta coffee. Menara Perkebunan 63: 43–51

Vaast P, Salazar M, Martinez M et al (2007) Importance of tree revenues and incentives from the programme 'coffee-practices' of Starbucks for coffee farmers in Costa Rica and Guatemala. Proceedings XXI Int Conf Coffee Sci: 495–502

Vaast P, Van Kantent R, Siles P et al (2005) Shade: a key factor for coffee sustainability and quality. Proceedings XX Int Sci Colloq Coffee: 887–896
Villain L (2000) Caracterisation et bioecologie du complexe parasitaire du genre *Pratylenchus* (Nemata: Pratylenchidae) present sur cafeiers (*Coffea* spp.) au Guatemala. Ecole Nationale Supérieure Agronomique de Rennes, DS thesis, Rennes
Villain L, Anzueto F, Hernández A et al (1999) Los nematodos parásitos del cafeto. In: Bertrand B, Rapidel B (eds) Desafios de la caficultura en Centroamerica. IICA-PROMECAFE, San Jose
Villain L, Anzueto F, Sarah JL (2004) Resistance to root-lesion nematodes on *Coffea canephora*. Proceedings IV Int Congr Nematol: 289–302
Villain L, Baujard P, Anzueto F et al (2002) Integrated protection of coffee plantings in Central America against nematodes. Plante Rech Dev (special issue: research and coffee growing): 118–133
Villain L, Baujard P, Molina A et al (1998) Morphological and biological characterization of three *Pratylenchus* spp. populations parasiting coffee trees in Guatemala. Proceedings XXIV Int Symp Eur Soc Nematol: 600–601
Villain L, Figueroa P, Molina A et al (2001a) Reproductive fitness and pathogenicity of three *Pratylenchus* spp. isolates on *Coffea arabica*. Proceedings XXIII ONTA Reunion: 163
Villain L, Molina A, Sierra S et al (2000) Effect of grafting and nematicide treatments on damage by root lesion nematodes (*Pratylenchus* spp.) to *Coffea arabica* L. in Guatemala. Nematropica 30: 87–100
Villain L, Molina A, Sierra S et al (2001b) Evaluation of grafting and nematicide treatments for the management of a root lesion nematode, *Pratylenchus* sp., in *Coffea arabica* L. plantations in Guatemala. Proceedings XIX Int Sci Colloq Coffee, CD-ROM (no pagination)
Villain L, Pignolet L, Michau-Ferrière N et al (2001c) Evidence of resistance factors to *Pratylenchus* spp. on *Coffea canephora*. Nematropica 31:163
Villain L, Sarah JL, Hernández A et al (2007) Biodiversity of root knot nematodes, *Meloidogyne* spp., on coffee in Central America. Proceedings XXI Int Conf Coffee Sci: 1321–1324
Wayenberge L, Moens M (2004) Phylogenetic relationship of 24 *Pratylenchus* species based on the analysis of the D3 expansion region of the 28S rRNA gene. Proceedings XXVII Eur Soc Nematol Int Symp: 46
Whitehead AG (1969) Nematodes attacking coffee, tea and cocoa and their control. In: Peachey JE (ed) Nematodes of tropical crops. Tech commun n. 40. Commonwealth Bureaux of Helminthology. St Albans
Wilson KC (1985) Cultural methods. In: Clifford MN, Willson KC (eds) Coffee: botany, biochemistry and production of beans and beverage. Croom Helm, New York

Chapter 15
Indonesia and Vietnam

Soekadar Wiryadiputra and Loang K. Tran

15.1 Indonesia

15.1.1 Brief Outline of the Crop

Indonesia is the fourth coffee producer worldwide, with a total hectarage around 1.38 million hectares (ha), and an output of over 686 thousand metric tons in 2003 (Anonymous, 2004). Figure 15.1 shows the country's coffee-producing areas.

Ninety-three percent of the area is cultivated with shaded, robusta coffee (*Coffea canephora* Pierre ex A. Froehner), under a low input production system. On average, the 2.6 million smallholder families cultivate about 0.5 ha each, with a productivity of 0.5 metric ton/ha/year. The farms run by the government and by private companies cultivate about 54 thousand ha, with about the same productivity. Nonetheless, the potential productivity is believed to be around 2 tons/ha/year.

Several factors combine for the low productivity observed in Indonesia, amongst them the poor genetic potential of the coffees grown, poor crop maintenance (pruning, sucker removal, etc), poor socio-economic condition of the growers and their families, and unsatisfactory control of pests and diseases.

Pests and diseases cause significant yield losses in Indonesian plantations. The 'leaf rust' caused by *Hemileia vastatrix* Berk and Br. and the nematode *Radopholus similis* (Cobb) Thorne are of concern in arabica coffee (*C. arabica* L.) only, while the berry borer (*Hypothenemus hampei* Ferrari) and *Pratylenchus coffeae* (Zimmermann) Filipjev and Schuurmans Stekhoven are major concerns in robusta and arabica coffees. Altogether, most of the Indonesian plantations are affected by either one or both nematode species.

S. Wiryadiputra
Indonesian Coffee and Cocoa Research Institute, Jember, Indonesia
e-mail: soekadar@yahoo.com

R.M. Souza (ed.), *Plant-Parasitic Nematodes of Coffee*,

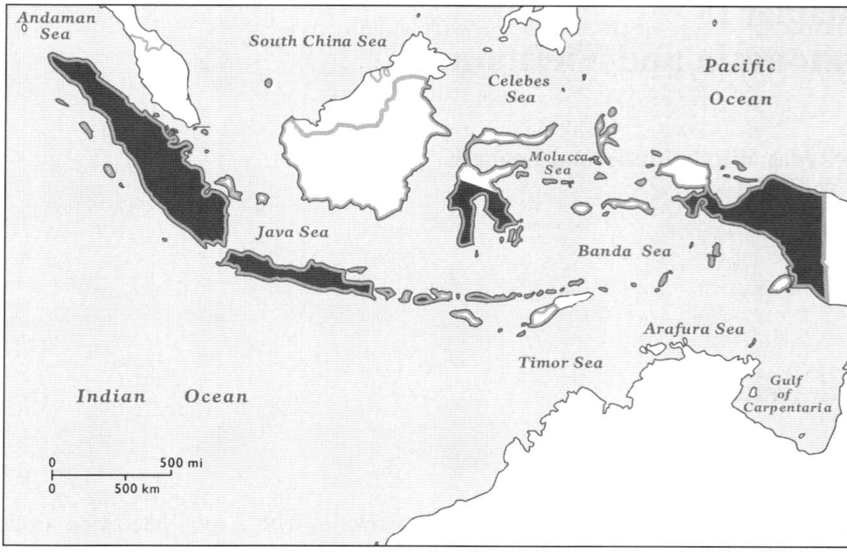

Fig. 15.1 Main robusta coffee-producing regions in Indonesia. Small arabica coffee-producing regions are not represented. Map by UENF/GRC

15.1.2 Coffee-Parasitic Nematodes

15.1.2.1 Main Species

Table 15.1 shows some data on the nematode species found associated with coffee (mostly robusta) in 1,341 samples collected in several provinces in Indonesia, during the period 1981–1991 (Wiryadiputra, 1991).

Some nematodes were found in just a small percentage of the soil and root samples, such as *Criconemoides morgensis* (Hofmanner in Hofmanner and Menzel) Taylor, *Hemicriconemoides chitwoodi* Esser, *Rotylenchus robustus* (de Man) Filipjev, and *Paratylenchus besoekianus* Bally and Reydon. Although found more frequently, *Helicotylenchus dihystera* (Cobb) Sher, *Rotylenchulus reniformis* Linford and Oliveira and *Meloidogyne* sp. (not identified at the species level) are not recognized as economically important in Indonesia, although no pathogenicity trials have been carried out for these species under Indonesian conditions.

As seen in Table 15.1, the incidence of *R. similis* was minimal until the early 1990s. Since then, the Indonesian government has aggressively distributed seedlings of the arabica coffee cultivars 'Kartika 1', 'Kartika 2' and 'S795', thus making *R. similis* a major concern, with plantations being heavily affected in the provinces of West Sumatera, Bengkulu, South Sumatera, Lampung, West Java, Central Java, East Java, Bali, and East Nusa Tenggara. No figures are yet available on the yield losses caused by *R. similis*, although it is widely accepted that it damages the

Table 15.1 Nematodes found associated with *Coffea* spp., the frequency of positive samples, population range in the samples, and distribution in Indonesia's provinces

Nematode	Coffee species		Frequency (%)	Population range	Provinces
	arabica	*canephora*			
Aphelenchus avenae Bastian	+	–	0.8	5–150[b]	1[c]
Criconemoides morgensis	+	–	10.6	5–1,590	4,5,6,8
Ditylenchus dipsaci (Kühn) Filipjev	–	+	0.08	30	5
Helicotylenchus dihystera	+	+	25.3[a]	5–865	1,2,4,5,6,7,8
Hemicriconemoides chitwoodi	+	+	4.6	5–390	1,2,5
Hemicycliophora arenaria Raski	–	+	0.08	5	5
Meloidogyne sp.	+	+	32.0	2–8,720	1,3,4,5,6,7,8
Paratylenchus besoekianus	+	+	1.7	2–120	1,5,8
Pratylenchus coffeae	+	+	44.5	2–22,508	2,3,4,5,6,7,8
Radopholus similis	+	–	0.3	10–367	5,8
Rotylenchulus reniformis	+	+	31.3	2–3,970	2,3,5,6,8
Rotylenchus robustus	+	+	0.3	5–15	2,5
Tylenchorhynchus dubius (Biitschli) Filipjev	+	–	0.08	30	6
Tylenchus davainei Bastian	+	+	0.6	5–30	5

[a] Combined frequency in *C. arabica* and *C. canephora*.
[b] Minimum and maximum population found in samples composed of 100 ml of soil and 10 g of roots.
[c] Provinces: 1, Aceh; 2, North Sumatera; 3, Lampung; 4, Central Java; 5, East Java; 6, Bali; 7, South Sulawesi; 8, East Nusa Tenggara.

plantations. Also, there have been no studies to identify the nematode race(s) present in these areas.

By far, *P. coffeae* is the most common and devastating nematode associated with coffee in Indonesia. It is present in almost all coffee-producing provinces, at altitudes ranging from zero to over 1,000 masl.

According to Wiryadiputra (1995), in robusta plantations the yield losses caused by *P. coffeae* may reach 78%, with an average around 57%. In arabica plantations, total loss has been observed, since the coffee plants may decline and die at the age of two.

15.1.2.2 Genetic Control

At present, the Indonesian coffee growers are advised to grow resistant genotypes in their properties, and to employ cultural methods to control plant-parasitic nematodes.

The efforts to employ genetic resources to control *P. coffeae* date back to the time of Indonesia's Dutch colonization. At that time, *C. excelsa* A. Chev. (= *C. liberica* Bull. ex Hiern) was considered more tolerant to *P. coffeae* than other coffee species (Bally and Reydon, 1931). Fluiter (1947) stated that in *P. coffeae*-infested fields, susceptible coffees could be successfully grown after grafting on the resistant hybrid 'Conuga' [*C. congensis* A. Froehner x *C. ugandae* Cramer (= *C. canephora*)] rootstock.

More recently, the coffee growers of the Malang district in East Java have grafted robusta coffee onto *P. coffeae*-resistant *C. liberica*. The grafting is not performed at the seedling phase, but rather in the fields, after the one year-old excelsa plants have been transplanted. In greenhouse, Wiryadiputra et al. (1994) confirmed the high resistance of the excelsa clone 'Bgn.121.09' towards *P. coffeae*. Nonetheless, since excelsa rootstocks present a relatively low compatibility with robusta and arabica scions, the research efforts on coffee resistance have been redirected towards robusta rootstocks.

In 1996, Wiryadiputra reported results from greenhouse and field showing that the robusta clone 'BP 961' was as resistant to *P. coffeae* as the excelsa 'Bgn.121.09'. Root histological sections revealed that 'BP 961' presents thick, lignified cell walls in the epidermis and in several layers of the cortical parenchyma. Also, near the epidermis, the parenchymal cortex presents dark stained idioblast cells, whose function in the storage of phenolic compounds is widely known. Analysis of the root concentration of phenolic compounds revealed that the clone 'BP 961' had the highest concentration amongst the six clones examined by Toruan-Mathius et al. (1995). Recently, Hulupi (2004) found the robusta clone 'BP 308' to be highly resistant to *P. coffeae* and *R. similis* (Fig. 15.2A; B), which led the Indonesian Ministry of Agriculture to release it to the growers.

In recent years, the Indonesian government has funded the Indonesian Coffee and Cocoa Research Institute (ICCRI) to produce and release to the growers millions of coffee seedlings (arabica scions 'S795' and 'Catimor', and robusta scions 'BP 42', 'BP 358' and others, grafted onto the robusta clone 'BP 308'). These seedlings have been planted in several nematode-infested areas. Hence, genetic resistance has a good prospect of solving nematode problems in Indonesia.

The research program on nematode coffee resistance will continue at the ICCRI. Recent results by Hulupi (2004) revealed the arabica coffee '542 A' as the most resistant to *R. similis* amongst all genotypes tested. Unfortunately, '542 A' is susceptible to 'leaf rust', and it has not been evaluated for resistance to *P. coffeae* yet.

15.1.2.3 Cultural Control

In Indonesia, several cultural practices have been recommended for controlling nematodes on coffee, but their effectiveness is unclear, and they are expensive to the growers.

For coffee plantations that are heavily infested, the growers are advised to uproot the plants and fallow the area for a minimum of one year. Alternatively, plant species known to be resistant or antagonistic to nematodes can be grown in the area, also

Fig. 15.2 Contrasting aspect of robusta coffee clones in a *Pratylenchus coffeae*-infested field. **A**: clone ‘BP 308’, resistant to the nematode. **B**: clone ‘BP 409’, susceptible (Photo by S. Wiryadiputra) (see color Plate 27, p. 333)

for a minimum of one year. These species include the French marigold (*Tagetes patula* L.), Guatemala grass (*Trypsacum laxum* Nash.), and *Crotalaria anagyroides* Kunth (Wiryadiputra, 1984; 1987).

Luki-Rosmahani et al. (2005) assessed the effectiveness of the African marigold (*T. erecta* L.) to control *P. coffeae* in robusta coffee plantations grown by smallholders in East Java. They reported effectively controlling the nematode after growing the marigold for two successive cycles, at a density of 25 plants/coffee tree.

The growers have adopted this practice despite the limited availability of African marigold seeds in Indonesia.

Another important aspect of nematode management in Indonesia is the susceptibility of the plant species used for coffee shading. The hoarypea (*Tephrosia* sp.) is often used for the temporary shading of arabica coffee because it grows fast, fixes nitrogen, and is resistant to pruning. Wiryadiputra et al. (1994) advised growers not to use this species, since it is a good alternate host for *P. coffeae*. Other plant species found to be suitable hosts to this nematode are cocoa, rubber tree, fish bean (*Tephrosia vogelii* Hook.f.), *Erythrina lithospermum* Miq., vegetable hummingbird (*Sesbania grandiflora* L.) and glory cedar (*Gliricidia maculata* (H.B.K.) Steud.).

On the other hand, Wiryadiputra (1994) found *Moghania macrophylla* (Willd.) Kuntze, *Crotalaria striata* DC., *C. usaramoensis* Baker f., *C. retusa* L., *C. anagyroides* Kunth, *C. juncea* L., sugarcane, leucaena (*Leucaena leucocephala* (Lam.) de Wit), *Adenanthera microsperma* Teijsm. and Binn., and pigeon pea (*Cajanus cajan* L.) to be resistant to *P. coffeae*.

The host status of banana to *P. coffeae* was also investigated since this plant is often intercropped with coffee in smallholding plantations. Under greenhouse conditions, the banana 'Giant Cavendish' presented a reproduction factor (the ratio between the final nematode population and the initial, inoculated one) of 3.44, while 'Barangan' presented a factor of 42.1. The bananas 'Mas', 'Kepok Kuning', 'Sukajaya', and 'Kayu' presented intermediate values (Wiryadiputra and Priyono, 1995).

The application of organic matter to the soil has also been practiced for the control of coffee-parasitic nematodes. Degraded coffee pulp has been routinely applied in large plantations, since it is known to significantly suppress *P. coffeae* population, in comparison to untreated plants (Wiryadiputra and dan Soenaryo, 1987). Cow manure has also been recommended for nematode-infested areas. A trial conducted on a two year-old plantation of the arabica coffee 'Kartika' showed that cow manure, applied at the dosage of 15 kg/plant, suppressed 91% of the *P. coffeae* population in the coffee roots, and 87.6% in the soil (Wiryadiputra, 1997).

15.1.2.4 Biological Control

In Indonesia, studies on the biological control of coffee-parasitic nematodes have not yet resulted in products or practices available to the growers. Several microorganisms have been assessed under greenhouse and field conditions, primarily against *P. coffeae*. For example, Baon et al. (1988) evaluated the effect of the vesicular-arbuscular mycorrhizal fungus *Gigaspora margarita* Becker Hall on *P. coffeae*-parasitized coffee seedlings. The inoculation of the seedlings with the fungus significantly increased the plant's vegetative growth (girth diameter, number of leaves, foliar area, and plant height), and reduced the nematode population in the root system. By reducing the nematode's negative effects on the plants, *G. margarita* may have increased the plant's tolerance to *P. coffeae*.

Baon and Wiryadiputra (2001) evaluated the effect of *G. margarita* and carbofuran, applied alone or combined, on new arabica and robusta plantations established in fields infested with *P. coffeae* and *R. reniformis*. At the age of three, both coffee plantations had had better growth and productivity when *G. margarita* had been combined with carbofuran. In the arabica plantation, carbofuran alone had a greater effect than the fungus alone, while in the robusta plantation these treatments had similar results. All treatments were statistically different from the control check. Both *G. margarita* and carbofuran reduced the *P. coffeae* population, but these treatments had no effect on *R. reniformis*.

The fungus *Paecilomyces lilacinus* (Thom.) Samson strain 251 (PL-251) and chitinolytic bacteria have also been assessed against coffee-parasitic nematodes. A field trial in a productive robusta plantation showed that PL-251, formulated as a bionematicide applied at the dosage of 4 g/coffee tree, suppressed the parasitism by *P. coffeae* and increased green coffee yield, in comparison with untreated plants (Wiryadiputra, 2002).

Under greenhouse conditions, Wiryadiputra et al. (2003) were able to suppress *P. coffeae* population on arabica coffee seedlings by treating them with chitinolytic bacteria and chitin powder. Nonetheless, chitin powder had a tendency to cause phytotoxicity in the higher doses. The best results in suppressing *P. coffeae* population and increasing the seedling's growth were obtained when bacteria isolated from shrimp waste was combined with the application of chitin powder at 10 g/pot.

15.1.3 Concluding Remarks

In Indonesia, research on coffee-parasitic nematodes is conducted at the ICCRI, the Bogor Agricultural and Gadjah Mada Universities, and the Biotechnology Research Institute for Estate Crops. Although these institutions have well equipped laboratories and facilities, their activities are not focused on coffee-parasitic nematodes only, since they must respond to problems with plant-parasitic nematodes in several other crops as well.

At the ICCRI, research on coffee emphasizes nematode ecology and control, with the latter being pursued primarily through genetic resistance and other environmentally-safe methods. More specifically, in the next five years research at the ICCRI will focus on the following areas:

(1) Molecular taxonomy of the main coffee-parasitic species, with special attention to *Radopholus* sp., since the identity of some coffee-associated populations is uncertain. In this area, the ICCRI will attempt to establish cooperation with research institutions abroad.
(2) Control of coffee-parasitic nematodes, primarily by non-chemical methods, such as biological control, plant resistance and botanical pesticides. For instance, the resistant robusta clone 'BP 308' still warrants studies on some of its agronomic and grafting aspects.

Most coffee growers are not aware of plant-parasitic nematodes and their harm to productivity. To educate them, extensionists and scientists must introduce new concepts, such as that microscopic organisms may cause symptoms similar to those caused by abiotic factors, such as nutrition and water inbalance, and by other pests and diseases. To educate growers and policy makers, the ICCRI has published brochures, booklets and leaflets on coffee-parasitic nematodes. Also, on several occasions basic nematological information and research results have been presented in nationwide newspapers and seminars, coffee symposia and field meetings. Also, the Directorate General of Estate Crops of the Department of Agriculture has established field laboratories with basic equipment for nematological work in most provinces of Indonesia. These facilities conduct basic activities, such as processing field samples and taxonomic identification of coffee-associated nematodes.

In Indonesia, just a handful of scientists are dedicated full-time to nematology, since most nematologists occupy administrative positions or act in other technical or scientific areas. A concerted effort is needed in the training of recently graduated plant pathologists or entomologists in the science of nematology. Several estate company staff members have been trained at the ICCRI to run nematology laboratories in their companies.

Despite the difficulties faced by coffee growers and nematologists in Indonesia, most obstacles will be overcome in the years to come.

15.2 Vietnam

15.2.1 Brief Outline of the Crop

Vietnam is situated in the centre of South-East Asia, stretching from 8°30′ to 23°22′ latitude north. Its climate is favorable for commercial cultivation of arabica and robusta coffees. The Hai Van mountain pass, with an altitude of over 1,000 masl, divides the country in two climatic regions: the tropical south, warm and humid, is suitable for robusta cultivation. Among its eight provinces, those in the Central Highlands concentrate 95% of the 500 thousand ha cultivated with coffee in Vietnam (Fig. 15.3). The Daklak province alone is responsible for half of Vietnam's production, which was 740 thousand metric tons of green beans in 2005. The north region presents a milder climate, with a cold and humid winter. It concentrates most of Vietnam's 25 thousand ha of arabica plantations, and it is a region of expanding coffee cultivation.

Coffee cultivation was introduced in northern Vietnam by French missionaries in 1857, and by the end of the nineteenth century plantations had been established in the northern midlands. Soon afterwards, the plantations had expanded to the Central Highlands, stretching through 10 thousand ha by 1945. In the 1970s a steady growth of the Vietnamese coffee industry began (Table 15.2).

Over the past 20 years, coffee has become a major industry in Vietnam, playing an important role in its economy. Indeed, coffee has become the most valuable agricultural product after rice, sustaining 600 thousand permanent and 1 million part-time jobs (Bau and Sung, 2005). Around 12% of the coffee hectarage is managed by the State, while more than 80% is owned by 300 thousand smallholders, which

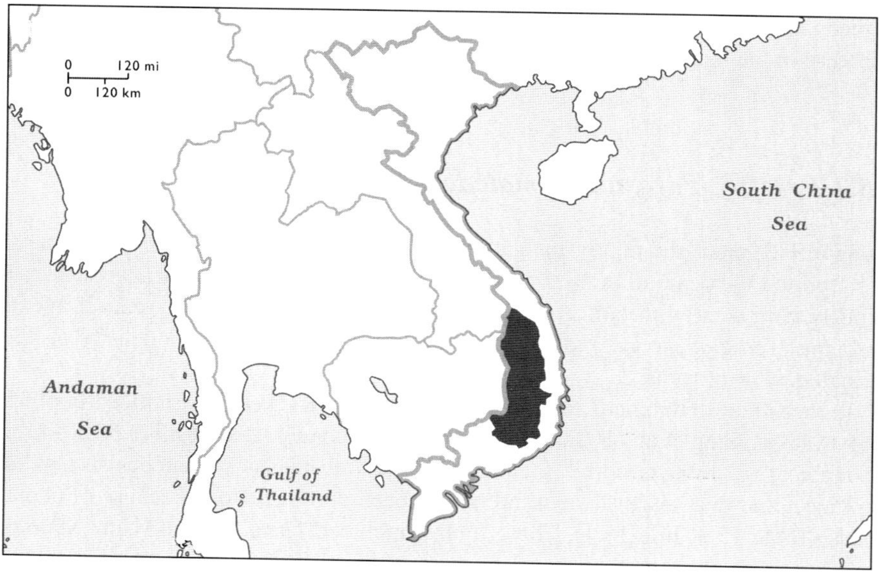

Fig. 15.3 Distribution of robusta coffee cultivation in Vietnam. Map by UENF/GRC

Table 15.2 Evolution of coffee hectarage, production and productivity in Vietnam

Year	Hectarage	Total production[a]	Productivity/ha[b]
1975	13,400	6,100	N/A[c]
1985	44,600	20,400	1.03
1995	205,000	245,000	1.81
2005	500,000	740,000	1.53

[a] production in metric tons of green beans.
[b] based on hectarage in production.
[c] data not available.
Source: Anonymous (2005).

cultivate between two and five ha each (Tiem and Minh, 2001). Vietnam outputs around 11% of world's coffee production, second only to Brazil, but it holds the top position in the robusta world market, with a 42% share worth between 400 and 600 million USD/year. This dramatic change in the Vietnamese coffee industry stemmed from changes in the official policies (allowing the farmers land property and profits), and from the favorable international market during the 1980s. Recently, a plan was put forward to expand also the cultivation of arabica coffee, from the present 25 thousand ha to 100 thousand by 2010 (Anonymous, 2005).

The fast growth of the Vietnamese coffee industry exacerbated rather than improved several constraints. Since most growers propagate coffee through unselected seeds, and the processing facilities are less than ideal, Vietnamese coffee beans achieve little quality and competitiveness in the world market. The water needs in the Central Highlands during its six month-dry season cannot be met by the irrigation infrastructure, while in certain areas the excessive irrigation has resulted in soil erosion, and reduction of the available underground water. Finally, there has been an increase in the incidence of coffee diseases and pests, with increasing yield losses.

15.2.2 Coffee-Parasitic Nematodes

Sung (1976) made the first investigation on coffee-parasitic nematodes in Vietnam. He reported the death of arabica coffee plants that had been planted in an area previously cultivated with coffee. In greenhouse, he demonstrated the causal agents as being *Meloidogyne* sp. and *P. coffeae*, with the latter reaching a density of 357 nematodes/5 g of roots.

In the Central Highlands, coffee-parasitic nematodes were first noticed in 1995. Surveys by Loang et al. (1997) and Loang (2002) concluded that more than 500 ha of arabica and robusta coffee plantations were infested by nematodes in Daklak province, and that nearly 1 thousand ha had been uprooted (lost) because of nematodes (Table 15.3; Fig. 15.4). These figures included farms owned by the State or private companies, which occupy 10 to 15% of the coffee hectarage only.

Specifically on robusta coffee, a survey by Cuc et al. (1990) in the provinces of Tiengiang, Bentre and Haugiang revealed *P. coffeae* in 45% of samples collected, with a relative abundance of 50%. Symptoms of *P. coffeae*-parasitism in young and mature plants include peeling and necrosis of secondary and feeder roots, resulting in impaired uptake of water and nutrients. The leaves become yellowish and fall, even during the rainy season, and the plant's growth is progressively reduced (Figs. 15.5 and 15.6). Replanting coffee in an area infested by *P. coffeae* results in the death of the seedling's tap root, so the plants may easily be pulled out of the soil by hand. During the rainy season, young plants typically put forth adventitious roots at the collar region (Fig. 15.7). These symptoms become apparent two to three years after the replanting (Loang, 2002).

Table 15.3 Survey sites and damage by nematodes to coffee plantations in Daklak province, Vietnam

State farm or company	Uprooted (lost)	Infestation			
		Minor[b]	Moderate	Serious	Total
Chuquynh State farm	424[a]	10	20	70	100
Eaktur company	400	200	00	00	200
Thangloi company	11	45	29	29	103
Easim company	14	100	00	16	116
Krongana company	105	00	00	25	25
Total	954	355	49	140	544

[a] areas in hectare.
[b] minor infestation means less than 20% of the coffee trees infected; moderate: 20–50%; serious: more than 50%.

Coffee surveys in Daklak, Gialai and Dongnai provinces during the years 1997–1999 revealed a widespread incidence of five nematode genera in robusta coffee roots, with *P. coffeae* and *Meloidogyne* sp. being considered the most important (Sung et al., 2001; Loang, 2002) (Table 15.4).

Fig. 15.4 Uprooted (foreground) and *P. coffeae*-parasitized coffee plants (background) in Krong Ana, Daklak province, Vietnam (Photo by Loam K. Tran)(see color Plate 28, p. 334)

Fig. 15.5 Mature robusta coffee tree presenting the *P. coffeae*-associated decline (Photo by Loam K. Tran) (see color Plate 29, p. 334)

Fig. 15.6 Young robusta plant planted into a *P. coffeae*-infested area (Photo by Loam K. Tran) (see color Plate 30, p. 335)

Fig. 15.7 *P. coffeae*-parasitized robusta coffee plant presenting rotten tap root and abundant adventitious roots at the collar region (Photo by Loam K. Tran) (see color Plate 31, p. 335)

Table 15.4 Nematode taxa found associated with roots of robusta coffee trees in three southern Vietnamese provinces

Taxa	Provinces			Maximum density found[b]	Percentage of positive samples
	DakLak	GiaLai	Dong Nai		
P. coffeae	+ + +[a]	++	++	4,784	85.6
Meloidogyne sp.	++	+	+	184	12.8
Tylenchus sp.	+	+	+	64	8.4
Rotylenchus sp.	+	+	+	40	7.3
Helicotylenchus sp.	+	+	+	24	1.8
Number of samples collected	212	60	20	(–)	(–)

[a]: + denotes fewer than 100 nematodes/5 g of roots; ++ denotes 100–500 nematodes; + + + denotes more than 500 nematodes.
[b]: maximum density found in all samples/5 g of roots.
Adapted from Sung et al. (2001) and Loang (2002).

A recent survey by Chau and Thanh (2001) in four provinces cultivated with arabica coffee revealed nearly 30 plant-parasitic nematode taxa associated with coffee plantations (Table 15.5).

Table 15.5 Abundance of nematode taxa associated with arabica coffee plantations in four Vietnamese provinces

Taxa	Provinces			
	Laichau	Nghean	Quangtri	Lamdong
Meloidogyne incognita (Kofoid and White) Chitwood	–[a]	–	–	++[b]
Pratylenchus brachyurus (Godfrey) Filipjev and S. Stekhoven	–	–	–	+
P. coffeae	–	–	+	+ + +
P. delattrei Luc	–	+	+	–
P. neglectus (Rensch) Filipjev and S. Stekhoven	–	+	–	–
P. penetrans (Cobb) Filipjev and S. Stekhoven	–	–	–	+
Radopholus sp.	–	–	++	–
Rotylenchulus reniformis	–	+ + ++	–	+
Tylenchorhynchus brassicae Siddiqi	–	–	–	+
Hoplolaimus seinhorsti (Luc) Shamsi	+	–	–	–
Helicotylenchus coffeae Eroshenko and Nguen Vu Thanh	+ + +	–	+	+ + +
H. concavus Roman	–	–	–	++
H. crassatus Anderson	–	–	–	++
H. crenacauda Sher	–	–	–	++

Table 15.5 (continued)

Taxa	Provinces			
	Laichau	Nghean	Quangtri	Lamdong
H. dihystera	++	+	+	+++
H. digonicus Perry in Perry, Darling and Thorne	–	–	+	+
H. dignus Eroshenko and Nguen Vu Thanh	+	+	+	–
H. erythrinae (Zimmermann) Golden	–	–	–	+
H. exallus Sher	–	–	–	++
H. paraconcavus Rashid and Khan	–	+	–	–
H. pseudorobustus (Steiner) Golden	–	–	–	++
Criconemella magnifica (Eroshenko and Tkhan) Raski and Luc[c]	++	++	–	–
C. goodeyi de Guiran[d]	–	–	–	+
C. onoensis (Luc) De Grisse and Loof[e]	+++	–	–	+
Crossonema fimbriatum (Cobb in Taylor) Mehta and Raski	–	–	–	+
Xiphinema insigne Loos	+	–	–	–

[a] '–' denotes not found;
[b] '+' denotes fewer than 50 nematodes/250 ml of soil; '++' denotes 50–100 nematodes; '+++' denotes 100–500 nematodes; '++++' denotes more than 500 nematodes;
[c] *Macroposthonia magnifica* Eroshenko and Tkhan, according to Siddiqi (2000).
[d] *Criconemoides goodeyi*;
[e] *M. onoensis*.
Adapted from Chau and Thanh (2001).

Recently, *Radopholus* sp. was found affecting young arabica coffee plants in some areas of Daklak and Gialai provinces (Sung et al., 2001). The nematodes damage the plant's collar region and rot the tap root, although the plant remains firmly attached to the ground. The leaves become yellowish and the shoot's growth stops. This nematode has been described as *R. arabocoffeae* n.sp. by Trinh et al. (2004), affecting arabica coffee 'Catimor'.

15.2.3 Concluding Remarks

In Vietnam, coffee nematology has just begun; therefore the results are somewhat limited. The major research focuses have been surveying the plant-parasitic nematode species associated with coffee, determining the causal agents of the coffee decline observed in some producing regions, and testing nematode control measures. The results show that unbalanced coffee cultivation, ie. excessive application of inorganic fertilizers and water irrigation, helps to weaken the coffee plantations, which creates favorable conditions for plant-parasitic nematodes.

Experiments have shown that no single control measure is effective. Management recommendations include application of manure and balanced soil and foliar fertilization, shading of the plants, and mulching during the dry season. Nematicides are to be applied in nurseries, on young plants or on those with a low level of parasitism (Sung et al., 2001). Although research efforts began in 1999, no nematode-resistant robusta rootstock is available yet.

In Vietnam, nematodes are highly damaging parasites of coffee. Although no exact figures are available, their incidence is nonetheless believed to be somewhat localized. Research efforts, focused primarily on *Meloidogyne* sp., *P. coffeae* and *Radopholus* sp., should include an effective breeding program for development of resistant rootstocks, and studies to understand the ecological conditions under which plant-parasitic nematodes become damaging to coffee production. This investigation should consider other organisms that could be involved in such a pathosystem, like soil-borne fungi.

References

Anonymous (2004) Statistical Estate Crops of Indonesia: Coffee, 2001–2003. Department of Agriculture Republic of Indonesia, Directorate General of Estate Crops Production, Jakarta

Anonymous (2005) Coffee Annual Report 2005. Vietnam Coffee and Cocoa Association, Hanoi

Bally B, Reydon GA (1931) De tegenwoordige stand van het vraagstuk van wortelaaltjes in the coffiecultuur. Archf. Koffiecult. Indonesie 5: 23–216

Baon JB, Wiryadiputra S (2001) Response of arabica and robusta coffee to *Gigaspora margarita* and carbofuran application in parasitic nematode infested land. Proceedings III Int Conf on Mycorrizas: 9

Baon JB, Wiryadiputra S, Sulistyowati dan E (1988) Infection of *Pratylenchus coffeae* on coffee plant as affected by mycorrhizal inoculation. Indones J Coffee and Cocoa Res 4: 22–30

Bau NL, Sung QP (2005) Scientific and technological achievements of Dak Lak coffee and the orientation for coffee sustainable development. Proceedings Semin 'Buonmathuot Coffee' Brand Dev. Daklak Peoples Committee, Daklak

Chau NN, Thanh VN (2001) Plant parasitic nematodes associated with coffee in some North and Tayngueyen highland pronvinces of Vietnam. Selected works on ecological and bio-resource studies 1996–2000. Agriculture Publishing House Hanoi

Cuc TTN, Phen VT, Hau VT et al (1990) Surveys on cofee nematode infection in Tiengiang, Bentre, Haugiang. Plant Prot News 3: 14–16

Fluiter HJ (1947) Het aaltjesprobleem in de koffiecultuur. Tijdschr PlZiekt 53: 101–109

Hulupi R (2004) Research on acceleration development of superior coffee varieties and clones. In: Final Report of Research Project. Fiscal year 2004. The Participatory Development of Agriculture Technology Project. Department of Agriculture, Jakarta

Loang KT (2002) Research on the causal agents of robusta coffee yellow and rot roots symptoms in Daklak province and assessment of the control potential. Hanoi Agriculture University, DS thesis, Hanoi

Loang KT, Hoa XN, Mao TH et al (1997) Results on surveys of coffee root diseases in Daklak province. Scientific Report in Annual Agronomy and Plant Protection Conference. Ministry of Agriculture and Rural development. Daklak

Luki-Rosmahani, Diding-Rachmawati, Sarwono et al (2005) Assessment of tagetes (*Tagetes erecta*) planting to control parasitic nematode, *Pratylenchus coffeae* on robusta coffee. Proceedings I Int Conf Crop Security: 18

Siddiqi MR (2000) Tylenchida Parasites of Plant and Insects, 2nd edn. CABI Publishing, Wallingford

Sung QP (1976) Initial investigation on coffee nematode infected arabica coffee in Phuquy (Nghean). Phuquy Tropical Plants Research Station, Nghean

Sung QP, Trung MH, Tiem TH et al (2001) Research on coffee yellow syndrome and control measures. Ministry of Science, Technology and Environment, Buonmathuot

Tiem TH, Minh DT (2001) Present status of coffee industry in Vietnam and opportunities for speciality/organic coffee production. Results of scientific works 2000–2001. Western Highlands Agro-forestry Scientific Technology Institute, Daklak

Toruan-Mathius N, Pancoro A, Sudarmadi D et al (1995) Root characteristics and molecular polymorphisms associated with resistance to *Pratylenchus coffeae* in robusta coffee. Menara Perkebunan 63: 43–51

Tring PQ, Nguyen CN, Waeyenberge L et al (2004) *Radopholus arabocoffeae* sp. n. (Nematoda:Pratylenchidae), a nematode pathogenic to *Coffea arabica* in Vietnam and additional data on *R. duriophilus*. Nematology 6: 681–693

Wiryadiputra S (1984) The prospects of Guatemala grass on the control of coffee parasitic nematodes. Menara Perkebunan 52: 12–16

Wiryadiputra S (1987) The use of antagonistic plants to control parasitic nematodes in the coffee plantation. Proceedings IX Cong and Nat Sem Indones Phytopathol Soc: 484–490

Wiryadiputra S (1991) Survey of plant parasitic nematodes on coffee in Indonesia. Proceedings XI Cong and Nat Sem Indones Phytopathol Soc: 10–12

Wiryadiputra S (1994) Host suitability of *Pratylenchus coffeae* on several estate crops and coffee shade trees. Pelita Perkebunan 10: 21–30

Wiryadiputra S (1995) Estimation of yield losses caused by *Pratylenchus coffeae* on robusta coffee. Proceedings XII Cong and Nat Sem Indones Phytopathol Soc: 980–985

Wiryadiputra S (1996) Resistance of robusta coffee to coffee root lesion nematode, *Pratylenchus coffeae*. Pelita Perkebunan 12: 137–148

Wiryadiputra S (1997) Effect of cow dung and drenching of oxamyl solution on the population of *Pratylenchus coffeae* and the growth of arabica coffee cv. 'Kartika'. Proceedings XIV Cong and Nat Sem Indones Phytopathol Soc: 186–189

Wiryadiputra S, dan Soenaryo ES (1987) The use of organic matters and rice hull ash to control parasitic nematodes in coffee nursery. Pelita Perkebunan 2: 146–151

Wiryadiputra S, Junianto YD, Indarti S et al (2003) Effect of chitinolytic bacteria and crude chitin on population of *Pratylenchus coffeae* and growth of coffee seedlings. Indones J Phytopathol 7: 45–53

Wiryadiputra S, Priyono (1995) Studies on the use of banana trees (*Musa* spp.) for coffee and cocoa shading: V. Development of *Pratylenchus coffeae* on some banana cultivars derived from tissue culture. Pelita Perkebunan 11: 132–139

Wiryadiputra S, Santoso AB, Mawardi dan S (1994) Resistance of some species in the genus of *Coffea* against *Pratylenchus coffeae* on coffee seedlings stage. Proceedings 3rd Plant Breed Symp Indones Coffee Cocoa Res Inst: 136–142

Wiryadiputra, S (2002) The effect of *Paecilomyces lilacinus* strain 251 as a bionematicide on the infestation of *Pratylenchus coffeae* on robusta coffee. Indones J Plant Prot 8: 18–26

Chapter 16
India

M. Dhanam and K. Sreedharan

16.1 Brief Outline of the Crop

Coffee (*Coffea* sp.) is an important crop in India. In plantations that spread over 350 thousand hectares (ha), mostly in the Southern States of Karnataka, Kerala and Tamil Nadu (Fig. 16.1), coffee is typically cultivated in an agroforestry system, in which the shaded coffee plants are intercropped with pepper, banana, orange, cardamom, areca nut and vegetables, among others. This integrated, low input production system is instrumental in preserving forest ecosystems, while sustaining economic development. As discussed below, the agroforestry system

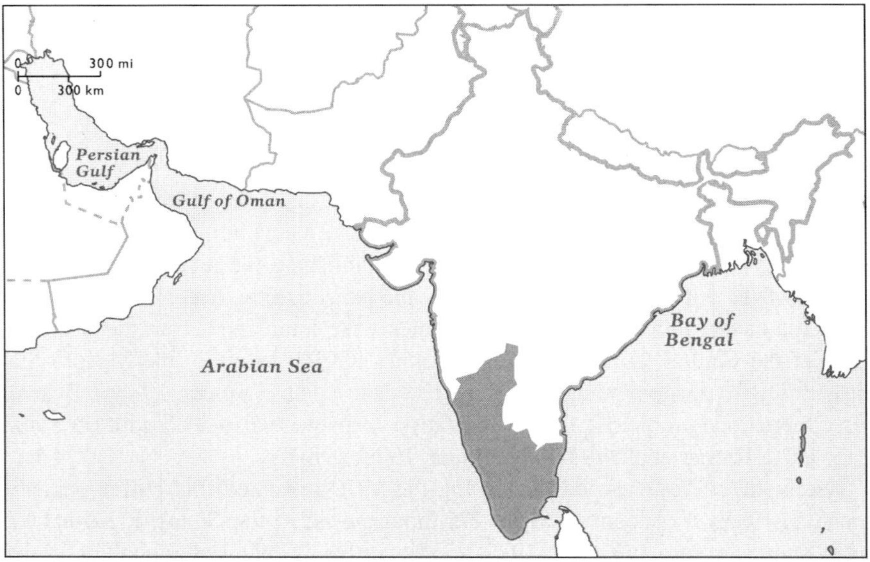

Fig. 16.1 India's main coffee growing region. Map by UENF/GRC

M. Dhanam
Central Coffee Research Institute, Coffee Research Station, Karnataka, India
e-mail: wahdhanjhs@rediffmail.com

R.M. Souza (ed.), *Plant-Parasitic Nematodes of Coffee*,

greatly facilitates the management of coffee pests and diseases, including nematodes (Anonymous, 2003; Jansen, 2005).

Such a sustainable system is essential for a crop typically cultivated by smallholders. Indeed, of the 178 thousand farms cultivated with coffee in India, about 77% have plantations that are less than 2 ha in area. Collectively, these smallholdings output just a little over 60% of India's production, of 100 and 183 tonnes of arabica (*C. arabica* L.) and robusta (*C. canephora* Pierre ex A. Froehner) coffees, respectively (Anonymous, 2006). Eighty percent of India's production is exported to the USA, European countries and Russia, among others.

16.2 Pests and Diseases of Coffee

A number of pests and diseases affect coffee in India, mostly the arabica group. Main pests include the white stem, coffee berry, and shot-hole borers [*Xylotrechus quadripes* Chevrolat, *Hypothenemus hampei* Ferrari and *Xylosandrus compactus* (Eichhoff), respectively], mealybugs (*Planococcus* spp.), and the green scale *Coccus viridis* (Green). The main pathogens are *Hemileia vastatrix* Berk and Br. (causing 'leaf rust'), *Koleroga noxia* Donk ('black rot'), and the root-lesion nematode, *Pratylenchus coffeae* (Zimmerman) Filipjev and Schuurmans Stekhoven.

16.3 Coffee-Parasitic Nematodes

16.3.1 Main Species

At the Mysore Coffee Experimental Station, Mayne and Subramanyan (1933) and Pattabiraman (1949) pioneered the studies on coffee-parasitic nematodes in India. In these early years, efforts were focused on establishing the relationship between poor growth of young arabica coffee plants and parasitism by *P. coffeae*.

Once a causal relation was proved, studies were intensified from the 1960 onwards at the Central Coffee Research Institute (CCRI, formerly the Mysore Station), aiming to develop nematode management strategies (Kumar, 1988a). Besides coffee, nematode problems of other plantation crops were also dealt with (D'Souza et al., 1970; Kumar et al., 1971a; b; Kumar, 1973; 1984b).

Systematic surveys carried out during the 1970 indicated that coffee-parasitic nematodes were a concern in about 3.5 thousand ha in the States of Karnataka, Tamil Nadu and Kerala (Kumar, 1979). Conservative estimates of annual losses due to nematodes reached 40 million Indian Rupees (1.2 million US dollars) (Kumar et al., 1995).

Among the several nematodes that parasitize coffee in India, *Meloidogyne* sp., *Radopholus* sp., *Rotylenchulus* sp., *Hemicriconemoides* sp., and *Pratylenchus* sp. have received the most attention (Kumar and Samuel, 1990).

In India, *M. hapla* Chitwood, *M. incognita* (Kofoid and White) Chitwood, *M. javanica* (Treub) Chitwood and *M. arenaria* (Neal) Chitwood are often retrieved from soil samples collected in coffee plantations. Although these nematode species occasionally invade coffee roots and induce root galls, they are not considered parasites of coffee. In an early greenhouse study, about one thousand second-stage juveniles (J2) of each of the above mentioned species were inoculated on young plants of *C. arabica* 'S.795' and *C. canephora* 'S.274'. Both cultivars were found to be resistant to all the *Meloidogyne* species tested. For all four species, the J2 were unable to penetrate the roots of *C. canephora*. As for *C. arabica*, the J2 of *M. hapla* were unable to penetrate the roots, while the J2 of *M. incognita, M. arenaria* and *M. javanica* penetrated the roots, but did not develop beyond the fourth stage (Anonymous, 1971). Kumar (1984e) confirmed the high resistance of both arabica and robusta coffees to *M. hapla*.

Although *R. similis* (Cobb) Thorne was considered pathogenic to coffee in Java, Brazil, Costa Rica and Natal (Zimmerman, 1898; Bally and Reydon, 1931; Tarjan, 1971; Sharma and Sher, 1973; Milne and Keetch, 1976), some South Indian populations isolated from black pepper and banana plants failed to penetrate and reproduce on coffee (D'Souza et al., 1969; Kumar, 1980b).

The reniform nematode, *R. reniformis* Linford and Oliveira, was reported as being parasitic to arabica coffee in Puerto Rico, Brazil, and the Philippines (Ayala, 1962; Curi, 1973; Sharma and Sher, 1973; Macedo, 1974). Sekhar (1963) and D'Souza and Srinivasan (1965) also reported this nematode as parasitic to arabica coffee in India, but later studies failed to confirm it. D'Souza and Kumar (1974) concluded that *R. reniformis*, although present in coffee plantations, were actually parasitizing weeds and shade trees.

In India, *Hemicriconemoides cassiae* Kumar, *H. mangiferae* Siddiqi, *H. chitwoodi* Esser, *H. coffeae* Kumar, and *H. gaddi* (Loof) Chitwood and Birchfield have often been found associated with arabica and robusta coffees (Kumar and D'Souza, 1969; Kumar, 1980a; 1982b; 1984b; 1985). The last two species, as well as *H. cocophillus* (Loof) Chitwood and Birchfield, cause coffee's 'crinkle leaf disorder': the nematode's feeding activity results in poorly developed feeder roots, with the above-ground symptoms becoming more obvious upon the beginning of the rainy season. The affected plants present shorter than normal stems with reduced internodes, which gives to the stems a 'bushy' or 'witch's broom' appearance. The leaves are small, crinkled, variously shaped, chlorotic and leathery. In severe cases, 'tip-burning' of shoots and plant death may occur. Most of the affected plants recover from the symptoms during the mid-monsoon period, when they put forth healthy leaves.

Indubitably, in India the most destructive nematode to arabica coffee is *P. coffeae*. *P. brachyurus* (Godfrey) Filipjev and S. Stekhoven parasitizes mostly robusta coffee, while *P. flakkensis* Seinhorst and *P. zeae* Graham have been reported as mild parasites by Kumar (1988b). When Southeast populations of *P. coffeae* were compared, South India's and Indonesia's were morphologically similar (Anonymous, 1973), while Kumar and Kasivisvanathan (1972) recognized the existence of the coffee and cardamom races. According to these authors, females of the coffee race were a little larger than those of cardamom (body length of 630–670 μm and 445–500 μm, respectively). Kumar (1988c) further identified two more races of *P. coffeae*, viz. pavetta and bamboo.

As in other countries, in India *P. coffeae* is a polyphagous nematode, parasitizing orange, cardamom, ginger, ornamental plants, fruit crops, pulses, cereals, spices, weeds, and shade trees associated with plantation crops, such as *Ficus* sp., *Glyricidia* sp., and *Bischofia javanica* Blume (Gadog) (Siddiqi 1964; Kumar et al., 1971a; b; Kumar, 1973; 1992).

In coffee, *P. coffeae* feeds on and destroys the cortical parenchyma cells of the tap, secondary and feeder roots. Consequently, the outer tissues of the tap and secondary roots peel off, and the feeder ones die (Kumar and Samuel, 1990). In young coffee plants, this condition was recognized as 'juvenile foot-rot' by several authors (Bally and Reydon, 1931; Pattabiraman, 1949; Abrego, 1960; Abrego and Holdeman, 1961; Salas and Echandi, 1961). Without a full root system, coffee plants cannot properly take in water and nutrients, failing to respond to fertilizer inputs and cultural practices, and they can easily be dislodged due to poor anchorage. During the rainy season, the plants may put forth adventitious roots at the collar region (Fig. 16.2).

Above ground, the *P. coffeae*-parasitized plants exhibit a myriad of symptoms. Young plants present lean stem, and the mature leaves become yellow and fall, leaving the lateral branches with few undersized, chlorotic and crinkled leaves at their tips, giving the lateral branches a 'tufted' appearance. The leaves produced during the pre-monsoon period (from April through June) are small, crinkled and chlorotic, while those produced during the monsoon (from July through October) are normal.

In contrast, mature bearing plants present poor foliage coverage and 'dieback' (Kumar and Samuel, 1990). This condition, known as 'Cannoncadoo dieback', in reference to the Indian State in which it was first observed, was studied in detail by Kumar (1984a; c; d). The lateral branches have shorter internodes, and the sparse flower buds produce many unfilled beans, which reduces the yield in amount and quality. Furthermore, a little delay in the 'blossom showers' induces the plants to produce vegetative buds instead of floral ones.

Should bearing plants be pruned at the collar region, they fail to put forth a vigorous shoot; new stems are reduced in length, with chlorotic and crinkled leaves. The parasitized plants decline progressively over a number of years, and finally die. Replanting of coffee in an infested field results in an early failure of the new plantation as it faces high nematode population in the soil.

16.3.2 Nematode Management

Since Zimmerman's report in 1898 on the pathogenicity of *P. coffeae* to arabica coffee plants, it has been a challenge for nematologists and growers to battle this nematode. Many years of studies have indicated an array of management practices, centered on cultural operations, for keeping the nematode population at a low level.

Fig. 16.2 Adventitious roots emitted at the coffee plant's collar region due to parasitism by *P. coffeae* (Photo by M. Dhanam)

Initially, it is recommended that growers do not use native (forest) soil for preparing seed beds, as *P. coffeae* does parasitize the native vegetation. Since coffee is cultivated among shade plants and other crops, some studies have focused on *P. coffeae*'s host range and population dynamics. Kumar (1991) reported that the nematode was persistent in the soil throughout the year, but at a higher population during the rainy season, from July through September, when an abundant root system was available to parasitism. Kumar (1988a) recommended uprooting the plants highly affected by *P. coffeae*, and fallowing the field for one summer season, thus reducing the nematode population. It has been demonstrated that day soil temperature around 29°C is favorable to *P. coffeae*'s reproduction, in comparison to day temperatures around 25°C (Anonymous, 1973). A light soil texture, with soil particles around 2 mm in size, was found to be favorable to the nematodes, in comparison to particle sizes of 0.05–1 mm.

In nematode-infested areas, a key practice is establishing new coffee plantations using grafted seedlings (Fernandez and Straube, 1966; Reina, 1966; D'Souza et al., 1969; Gutierraz and Jimenez, 1970; Loureiro and Cruz, 1970; D'Souza and Kumar, 1974; Kumar, 1974; 1979). This strategy, well accepted by the growers, uses the wedge graft method and seedlings at the 'soldier' stage. The grafting is performed from May through July, when the hot and humid weather favors the fusing of the plant tissues. Before establishing the new plantation, the *P. coffeae* population should be reduced (D'Souza et al., 1969; D'Souza and Kumar, 1974).

In a number of studies, more than 60 *Coffea* species and inter-specific hybrids were screened for resistance to *P. coffeae* (Anonymous, 1975; 1976; 1977; 1978; 1979). *C. robusta* (= *C. canephora*) genotypes were considered the most resistant to *P. coffeae*, followed by *C. excelsa* (= *C. liberica* W. Bull ex Hiern). All *C. arabica* genotypes were considered highly susceptible to this nematode. This pattern of results was also observed by Kumar (1979), while Kumar (1982c) observed that all phenological stages of the robusta coffees exhibit resistance to *P. coffeae*.

At the CCRI, efforts have been made to collect indigenous microorganisms that could be used for the biocontrol of *P. coffeae*. Screenings under laboratory conditions have shown that two species of blue green algae, *Microcoleus vaginatus* (Vauch.) Gom. and *M. lacustris* (Rabh.) Farlow, kill *P. coffeae* even within the root system (Kumar et al., 1993; Dhanam et al., 1993; 1994; Kumar et al., 1995). Also, the predatory nematode *Clarkus elongatus* Jairasjpuri and Khan was found to devour an average of 19 specimens of *P. coffeae* every 24 hours, under laboratory conditions (Dhanam, 1997). However, both organisms are obligatory to their prey, and mass culturing under laboratory conditions revealed to be a difficult task. More studies are needed for the utilization of these (and other) biocontrol agents in the integrated management of coffee-parasitic nematodes.

In the past, many chemical, granular nematicides, viz Hexanema® 5G, Temik® 10G, Thimet® 10G, Nemaphos® 10G, Rogor® 5G, and Nemacur® 5G, were assessed for their effectiveness against *P. coffeae* (D'Souza et al., 1971; Kumar, 1982a). These nematicides were not found effective; hence, nematicides are generally not recommended for nematode control. Recently, there have been several reports in India on the effectiveness of carbosulfan against *Heterderea zeae* Koshy, Swarup and Sethi, *M. incognita, Ditylenchus angustus* (Butler) Filipjev, *M. javanica, R. reniformis* and *M. graminicola* Golden and Birchfield, affecting crops such as maize, cowpea, rice, and chickpea (Panigrahi and Mishra, 1995; Kumar, 1996; Srivastava and Lal, 1997; Das, 1997; Verma and Gupta, 1997; Prasad et al., 1997). In most of these studies, carbosulfan was applied through root dipping, seed soaking or dressing, or foliar spray.

Carbosulfan was also effective against *P. thornei* Sher and Allen at the concentration of 100 ppm, and against *P. coffeae* at 0.06% a.i. (applied as the commercial product Marshal®), both under laboratory conditions (Sebastian and Gupta, 1997; Anonymous 2001; Dhanam et al., 2002).

In field studies herein reported, carbosulfan (0.06% a.i.) was used as soil drench in one, two or three applications, and compared to neem cake powder in its effectiveness to protect coffee seedlings against *P. coffeae*. The nematode population in the rhizosphere and roots was monitored monthly, and the development of the coffee plants (root dry and fresh mass, stem high and girth, and number of branches) was evaluated bimonthly.

Three applications of carbosulfan, in January, April and July, significantly reduced *P. coffeae* population in comparison to one (January) or two (January and April) applications. Three applications of carbosulfan resulted in lower nematode population until March of the following year, as well as increased plant growth and the emission of new roots. The application of neem cake powder did not affect *P. coffeae* population, in comparison to the blank (water) application.

In conclusion, nematode problems on arabica coffee in India are restricted to areas where *P. coffeae* or *Hemicriconemoides* sp. are endemic. In these areas, the causal relationship between these nematodes and coffee's 'Cannoncadoo dieback' and 'crinkle leaf disorder' is well documented (Kumar, 1984a; b; c; d). Although a survey conducted in the 1970s indicated that coffee-parasitic nematodes were restricted to about 3.5 thousand ha, routine sample processing at the CCRI indicates that those nematodes are being further spread. This suggests that another region-wide survey should be conducted in order to formulate appropriate management strategies.

As mentioned earlier, at present the management strategy against *P. coffeae* revolves around the adoption of phytosanitary measures and planting of grafted seedlings. The use of nematicides is not advised because of its high cost, inconsistent results, constraints for the application in the fields, and toxicology-related issues. Nonetheless, recent studies have shown that carbosulfan could be used to reduce nematode population and increase productivity, although not as a permanent solution. Hence, efforts should be directed towards developing sustainable management strategies, such as the use of biocontrol agents and biotechnology-derived nematode-resistant coffee cultivars.

To achieve these and other goals, intensive, coordinated research efforts are greatly needed. The formation of a worldwide network of nematologists working on coffee is urgently needed. Such international collaboration could greatly facilitate the exchange of expertise, and the development of coffee-specific research programs.

16.4 Concluding Remarks

In India, the studies on coffee-parasitic nematodes are being pursued at the CCRI, under the management of the Indian Coffee Board. The major areas of work are the survey and mapping of nematode-infested areas, and the development of biocontrol-based management strategies. The CCRI also processes soil samples sent by coffee growers at a nominal charge.

Regionally, the CCRI acts through five Regional Research Stations in different States of India, where local problems are addressed under the coordination of the Coffee Research Station in Karnataka. An extension network, also acting under the management of the Indian Coffee Board, is responsible for transferring research findings and new techniques to the coffee growers. The extensionists are stationed in local Junior Liaison Offices, each covering about five thousand ha, or a few villages. In a three-way interaction, the extension personnel provides feedback from the field, and scientists often participate in meetings with coffee growers, thus integrating research, extension and the growers. In the long term, the CCRI is committed to continuing the studies on coffee-parasitic nematodes, with the major goal of developing efficient, sustainable management strategies against these unseen enemies.

Acknowledgments The authors are grateful to Dr. B. K. Jayarama, Director of Research, CCRI/ Coffee Research Station, for encouragement during the preparation of this manuscript, and to Mr. B. M. Chulaki, Socio-economist, CCRI, for his help in preparing the manuscript.

References

Anonymous (1971) Annual Detailed Technical Report 1971–72. Indian Coffee Board, Research Department, Bangalore.

Anonymous (1973) Annual Detailed Technical Report 1972–73. Indian Coffee Board, Research Department, Bangalore.

Anonymous (1975) Annual Detailed Technical Report 1974–75. Indian Coffee Board, Research Department, Bangalore.

Anonymous (1976) Annual Detailed Technical Report 1975–76. Indian Coffee Board, Research Department, Bangalore.

Anonymous (1977) Annual Detailed Technical Report 1976–77. Indian Coffee Board, Research Department, Bangalore.

Anonymous (1978) Annual Detailed Technical Report 1977–78. Indian Coffee Board, Research Department, Bangalore.

Anonymous (1979) Annual Detailed Technical Report 1978–79. Indian Coffee Board, Research Department, Bangalore.

Anonymous (2001) Annual Report 2000–2001. Indian Coffee Board, Research Department, Bangalore.

Anonymous (2003) Annual Report 2002–2003. Indian Coffee Board, Research Department, Bangalore.

Anonymous (2006) Data Base on Coffee. Indian Coffee Board, Economic and Market Intelligence Unit, Bangalore.

Abrego L (1960) Los Nematodos, Seria Amenaza a el Café en Salvador. Inst Salvadoreno Investig Café, Bol Inf, Suplemento 1: 13–17

Abrego L, Holdeman QL (1961) Nematodos del café en El Salvador. Inst Salvadoreno Investig Café, Bol Inf, Suplemento 8: 1–16

Ayala A (1962) Pathogenicity of the reniform nematode on various hosts. J. Agric. Univ. Puerto Rico 46: 73–82

Bally W, Reydon GA (1931) De tegenwoordige stand van het vraag-stuk koffieculyuur. Archf. Koffiecult. Indonesie 5: 23–216

Curi SM (1973) New observation on coffee tree nematodes. Biológico 39: 206–207

Das P (1997) An integrated approach for management of rice stem nematode, *Ditylenchus angustus* in deep water rice in Assam. Indian J Nematol 26: 222–225
Dhanam M (1997) Predatory potentiality and host preference of *Clarkus elongatus* (Nematoda: Mononchida) under laboratory conditions. Proceedings First National Symp Pest Manag Hortic Crops: 315–317
Dhanam M, Kumar AC, Sowjanya (1993) *Microcoleus vaginatus (Oscillatoriaceae)* a new biocontrol agent of plant nematodes. Indian J Nematol 23: 1–2
Dhanam M, Kumar AC, Sowjanya (1994) *Microcoleus vaginatus* (Oscillatoriaceae) a blue-green alga (or cynobacterium) parasitising plant and soil nematodes. Indian J Nematol 24: 125–132
Dhanam M, Sreedharan K, Vinodkumar PK et al (2002) Field evaluation of Marshal (Carbosulfan) 25 EC against the coffee root-lesion nematode *Pratylenchus coffeae*. Proceedings Placrosym XV: 588–591
D'Souza GI, Shamanna HV, Kumar AC (1969) Control of the coffee nematode in South India – a progress report. Indian Coffee 33: 129
D'Souza GI, Kumar AC (1974) Studies on the root parasitic nematodes – *Pratylenchus* species, *Rotylenchulus* species and *Meloidogyne* species – their natural distribution, plant host range, ecology, host relations, pathogenicity, speciation and economic control in the coffee growing tracts of South Western India. Central Coffee Research Institute Final Technical Report Pl-480, Project No. A7-CR-260, Chikmagalur.
D'Souza GI, Srinivasan CS (1965) A note on the reniform nematode *Rotylenchulus reniformis* on arabica coffee in South India. Indian Coffee 29: 11–13
D'Souza GI, Kasiviswanathan PR, Kumar AC (1971) Granular treatment for the control of *Pratylenchus coffeae* in the coffee nursery. Indian Coffee 35: 306–307
D'Souza GI, Kasiviswanathan PR, Kumar AC et al (1970) Relative distribution and prevalence of plant parasitic nematodes in coffee tracts of South Western India. Indian Coffee 34: 330
D'Souza GI, Kumar AC (1969) Occurrence of *Radopholus similis* and *Pratylenchus coffeae* in south India. Proceedings All India Nematol Symp: 17
Fernandez CE, Straube E (1966) La injertacion como solucion a los problemas de las raices del cafeto. Proceedings XIV Meet Caribb Reg – Am Soc Hortic Sci 10: 1–6
Gutierraz ZG, Jimenez MFQ (1970) Some observations on grafting of coffee practiced in Guatemala an El salvador as a means of controlling nematodes. Rev Cafetera 98: 35–47
Jansen AE (2005) Plant Protection in Coffee. Recommendations for the Common Code for the Coffee Community. Deutshe Gesellschft Fur: Technische Zusammenarbeit (GTZ) GmbH.
Kumar AC (1973) New hosts of various plant parasitic nematodes. J. Coffee Res 3: 27
Kumar AC (1974) Studies on physiological races of *Pratylenchus coffeae*; Embryology of *P. coffeae*; Root-stock screening for resistance to *P. coffeae*; Standardisation of Grafting technique in coffee. In: Studies on the root parasitic nematodes – *Pratylenchus* species, *Radopholus* species and *Meloidogyne* species – Their natural distribution, plant host range, ecology, host relations, pathogenicity, speciation and economic control in the coffee growing tracts of South Western India. Central Coffee Research Institute, Final Technical Report PL-480, Project No. A7-CR-260, Chikmagalur.
Kumar AC (1979) Relative tolerance or susceptibility of arabica, robusta and excelsa coffees to *Pratylenchus coffeae*. Proceedings PLACROSYM II: 20–26
Kumar AC (1980a) Studies on nematodes in coffee soils in South India. 2. Occurrence of four species of *Hemicriconemoides* (Criconematidae) and studies on the life history of *H. coccophilus*. J Coffee Res 10: 4–11
Kumar AC (1980b) Studies on nematodes in coffee soils in South India. 3. A report on *Radopholus similis* and description of *R. colbrani* n. sp. J. Coffee Res 10: 43–46
Kumar AC (1982a) Evaluation of Nemacur 5G against the coffee root lesion nematode, *Pratylenchus coffea*e. J. Coffee Res 12: 1–7
Kumar AC (1982b) Studies on nematodes in soils of South India. 6. Occurrence of *Hemicriconemoides cassiae*. J. Coffee Res 12: 14–17

Kumar AC (1982c) Studies on nematodes in soils of South India. 7. Histopathology and host parasitic relationship of *Pratylenchus coffeae* and two species of coffee. J Coffee Res 12: 23–30
Kumar AC (1984a) The Nematode: casual organism of 'Cannoncadoo malady'. Indian Coffee 58: 5–6
Kumar AC (1984b) Pathogenicity of *Hemicriconemoides cocophillus* to coffee. J. Coffee Res 14: 36–37
Kumar AC (1984c) Investigation of 'Cannoncadoo die-back' in coffee. J. Coffee Res 14: 85–103
Kumar AC (1984d) The symptoms and diagnosis of the disorder, 'spreading decline' (Cannoncadoo die-back) with a note on spread and control of the casual agent, *Pratylenchus coffeae*. J. Coffee Res 14: 156–159
Kumar AC (1984e) Resistance in coffee to *Meloidogyne* spp. and occurrence of intersex in *M. thamesi*. Nematologica 30: 108–110
Kumar AC (1985) Investigations on the three species of *Hemicriconemoides* (Nematoda) associated with the 'Crinkle leaf' disorder in coffee. J. Coffee Res 15: 90–98
Kumar AC (1988a) A review of work done on nematology at Central Coffee Research Institute. Indian Coffee 62: 5–8
Kumar AC (1988b) Nematode problem of coffee and its management. Indian Coffee 52: 12–19
Kumar AC (1988c) Reaction of arabica and robusta coffees to five species of *Pratylenchus* (Nematoda) with a study on the biological races of *P. coffeae*. Proceedings PLACROSYM VIII: 291–294
Kumar AC (1991) Population fluctuation of *Pratylenchus coffeae* (Nematoda) in coffee and orange. J. Coffee Res 21: 99–102
Kumar AC (1992) Disorders of coffee and certain other plantation crops caused by the nematode, *Pratylenchus coffeae* in the South Indian coffee tracts. Indian Coffee 46: 3–5
Kumar AC (1995) Nematode Problems of Coffee. In: G. Swarup, DR, Das, G and Gill, JS (eds) Nematode Pest Management – An Appraisal of Eco-friendly Approaches. The Nematol Soc India, New Delhi.
Kumar S (1996) Efficacy of systemic insecticides as seed soaking for the control of *Meloidogyne incognita* on cowpea. Madras Agric J 83: 538–539
Kumar AC, D'Souza GI (1969) Pathogenicity of *Hemicriconemoides* species to coffee. Plant Dis Rep 53: 15–16
Kumar AC, Kasivisvanathan PR (1972) Studies on physiological races of *Pratylenchus coffeae*. J. Coffee Res 2: 10–15
Kumar AC, Samuel SD (1990) Nematode attacking coffee and its management – a review. J. Coffee Res 20: 1–27
Kumar AC, Dhanam M, Sowjanya (1993) Pathogenicity of *Microcoleus* spp. (Oscillatoriaceae) to plant and soil nematodes. Indian J Nemat 23: 8–9
Kumar AC, Dhanam M, Sowjanya (1995). Evaluation of *Microcoleus lacustris*, a blue -green alga (cyanobacterium) against *Pratylenchus coffeae* (Nematoda). Proceedings Nat Symp Nematol Probl India: 110
Kumar AC, Kasiviswanathan PR, D'Souza GI (1971a) New hosts of coffee root-lesion nematode, *Pratylenchus coffeae*. Indian Coffee 35: 59
Kumar AC, Kasiviswanathan PR, D'Souza GI (1971b) A study on plant parasitic nematodes of certain commercial crops in coffee tracts of South India. Indian Coffee 35: 222–224
Loureiro MC, Cruz J (1970) A survey of the occurrence of *Meloidogyne exigua* in *Coffea arabica* in the State of Minas Gerais. Seiva 70: 32–42
Macedo MCM (1974) Susceptibility of coffee to the reniform nematode. Solo 66: 15–16
Mayne WW, Subramanyan VK (1933) Nematode worms in relation to cockchafer and mealybug problems in Coorg. Bull Mysore Coffee Exp Stn 11: 1–13
Milne DL, Keetch DP (1976) Some observations on the host plant relationships of *Radopholous similis* in Natal. Nematropica 6: 13–17
Panigrahi D, Mishra C (1995) Effect of some pesticides on rice root-knot nematode, *Melidogyne graminicola*. Ann Plant Prot Sci 3: 74–75

Pattabiraman TV (1949) The coffee eelworm – *Anguillulina pratensis* (de Man, 1881). Indian Coffee Board Mon Bull 13: 112–115
Prasad D, Dubey KN, Mittal A (1997) An integrated approach for management of plant parasitic nematodes in groundnut (*Arachis hypogaea* L.). Ann Plant Prot Sci 7: 217–219
Reina EH (1966) La technical del injerte hipocotiledonar del cafeto para el control de nematodes. Turrialba 7: 5–11
Salas LA, Echandi E (1961) Parasitic nematodes in coffee plantation of Costa Rica. Coffee 3: 6–9
Sebastian S, Gupta P (1997) Laboratory screening of some nematicides and oil cake extracts against *Pratylenchus thornei*. Indian J Nematol 27: 133–134
Sekhar PS (1963) A note on nematodes in coffee plantations of South India. Turrialba 5: 1–4
Sharma R, Sher SA (1973) Nematodes associated with coffee in Bahia, Brazil. Arq Inst Biol 40: 131–135
Siddiqi MR (1964) Studies on nematode root rot of citrus in Uttar Pradesh, India. Proceedings Zool. Soc. Calcutta 17: 67–75
Srivastava AN, Lal J (1997) Effect of carbosulfan and sebuphos applied singly and in combination on reproduction of *Heterodera zeae* and growth of *Zea mays*. Ann Plant Prot Sci 5: 94–97
Tarjan AC (1971) Some interesting association of parasitic nematodes with cacao and coffee in Costa Rica. Nematropica 1: 5
Verma KK, Gupta DC (1997) Management of root-knot nematode, *Meloidogyne javanica* infecting chickpea (*Cicer arietinum*). Indian J Nematol 26: 261–263
Zimmerman A (1898) Die Nematoden de Koffieworrties. Meded. Pl Trin 27: 16–41

Chapter 17
The Ivory Coast and Uganda

Amoncho Adiko, Philippe G. Gnonhouri and Josephine M. Namaganda

17.1 The Ivory Coast

17.1.1 Brief Outline of the Crop

A country of 322 thousand square kilometers, and inhabited by 16 million people, the Ivory Coast is considered the 'economic lung' of West Africa, with a GDP of 16.3 billion dollars in 2005. Nonetheless, social unrest and the instability of the international commodity market made the country's annual growth rate fall from 7% in the 1990's to 2% (Anonymous, 2006).

The Ivory Coast's prosperity is based primarily on agriculture, which accounts for 35% of the GDP, 70% of the export earnings, and 66% of the employment positions (Anonymous, 1997a). Major agricultural products are coffee, cocoa (40% of the world's production), palm-kernel oil, cotton, rubber, banana, pineapple, and mango. The offshore reserves of oil and natural gas are also important assets for the national economy.

Although coffee (*Coffea* sp.) has been cultivated in the Ivory Coast since the 1880, it was only after the Second World War that the crop received a real impetus. Programs were undertaken nationwide to promote the establishment of new plantations, and the regeneration of old ones. Consequently, coffee farming spread into the entire southern forest zone (Fig. 17.1), reaching 1.2 million hectares (ha). The country's historical output of 250 thousand metric tonnes declined to around 100 thousand, due to consistently low prices on the international coffee market (Anonymous, 1997; 2005). Despite this decline, coffee still contributes 18 to 35% of the country's export earnings and 5% of the GDP, and employs 12% of the population (Anonymous, 1990).

In the Ivory Coast, coffee is produced by some 500 thousand growers, most of them smallholders. Twenty-five percent of the national output is produced in plantations that are up to 2 ha in area, and 70% is produced in 2 to 10 ha-plantations

A. Adiko
Laboratory of Nematology, Centre National de Recherche Agronomique, Abidjan, The Ivory Coast
e-mail: amoncho.adiko@yahoo.com

R.M. Souza (ed.), *Plant-Parasitic Nematodes of Coffee*,

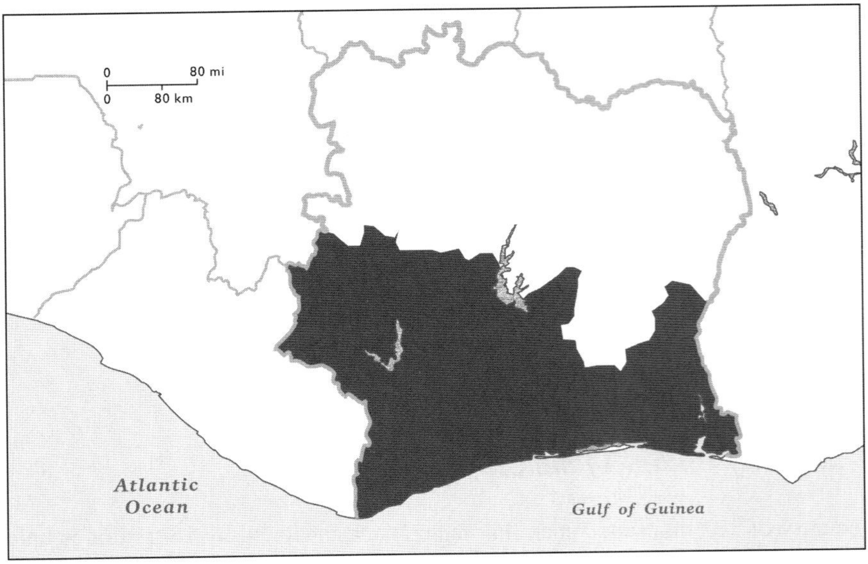

Fig. 17.1 The Ivory Coast's coffee-growing region. Map by UENF/GRC

(N'guessan, 2004). The production is essentially extensive: forests are cut down and replaced by fullsun coffee plantations intercropped with food crops, in a low input system. Fertilizers and pesticides are rarely used because of their high cost, and labor-intensive cultural practices, such as weeding, sucker removal and pruning, are insufficiently practiced. Ninety percent of the coffee plantations are established with the farmer's own seedlings, with 98% of the plantations being *C. canephora* Pierre ex A. Froehner variety (var) *Robusta* (Anonymous, 1988b; 2003; Montagnon et al., 2001).

The average productivity ranges from 200 to 250 kg/ha, although technology exists to produce ten times more (Montagnon et al., 2001; Anonymous, 2003). Since most growers believe that soil exhaustion is the primary cause of the plantations' low productivity, they simply abandon them, and move into new forest lands. In addition to *C. canephora*, some growers cultivate *C. liberica* W. Bull ex Hiern var *Indeniensis, C. liberica* var *Excelsa*, and *C. arabusta* Capot et Aké Assi, var *Arabusta* (Jacques-Félix, 1954; Meiffren, 1957).

17.1.2 Coffee-Parasitic Nematodes

Until recently, the only study involving coffee-parasitic nematodes in the Ivory Coast was a survey by Luc and de Guiran (1960). These authors gave an account of the plant-parasitic nematodes associated with the rhizosphere, roots or tubers of cultivated plants in West Africa. Luc and de Guiran reported 13 nematode species associated with coffee plants in the Ivory Coast. Of these species, three were found parasitizing the roots: *Helicotylenchus erythrinae* (Zimmermann) Golden, *Meloidogyne incognita* (Kofoid and White) Chitwood, and *Pratylenchus brachyurus* (Godfrey) Filipjev and S. Stekhoven.

In 2005, Adiko and Gnonhouri (herein reported) conducted a second survey in all but one of the Ivory Coast's coffee-producing areas. These authors identified five nematode genera or species parasitizing coffee plants (Table 17.1). The sedentary endoparasitic nematode *M. incognita* was the most frequent species, as it occurred in 31% of the plantations sampled. *Pratylenchus* sp. and *Paratylenchus* sp., which is reported for the first time on coffee in the Ivory Coast, were found in 13% and 15% of the plantations, respectively. The spiral nematode, *Helicotylenchus* sp., was the least frequent genus. Nematode density in coffee roots ranged from one to six specimens/g of root.

In addition to this survey, a test was conducted to assess the host status of *C. canephora* var *Robusta* and *C. arabusta* var *Arabusta* to *M. incognita, P. brachyurus* and *P. coffeae* (Zimmerman) Filipjev and Schuurmans Stekhoven, the most damaging nematode species in the southern region of the Ivory Coast (Kéhé et al., 1995; Adiko and N'guessan, 2002). Three months after separate inoculations with 20 thousand eggs/plant of *M. incognita*, five thousand nematodes/plant of *P. coffeae* and 1.3 thousand nematodes/plant of *P. brachyurus*, both coffee varieties exhibited a very poor host status, as shown by low (less than one) reproduction factors (Table 17.2).

These results show that plant-parasitic nematodes are not a constraint in the Ivory Coast's coffee agriculture. Indeed, no nematode damage has ever been reported in this country. During Adiko and Gnonhouri's survey in 2005, extension agents paid special attention to coffee plantations with possible nematode problems, such as those suspected of low productivity due to soil exhaustion. No nematode parasitism was observed in those plantations.

The low incidence of plant-parasitic nematodes on *C. canephora* could be ascribed to the high level of caffeine in the root system. In an early study, Rabéchault (1954) demonstrated that the high level of caffeine in *C. canephora* (1.5 to 2.5% of the dry matter, in the beans) is one of the key factors for resistance to *Fusarium xylarioides* Steyaert, the causal agent of 'coffee tracheomycosis'. He hypothesized that caffeine could be involved in a broad defense of *Robusta* plants to pathogens.

Table 17.1 Incidence and average root density of plant-parasitic nematodes, according to a survey in 85 coffee plantations in the Ivory Coast[a]

Nematode genera or species	Infested plantations[b]	Specimens[c]
Helicotylenchus sp.	7	2
Meloidogyne incognita	31	6
Paratylenchus sp.	15	1
Pratylenchus sp.	13	1
Scutellonema bradys (Steiner and LeHew) Andrassy	11	1

[a] The country's western war zone was not surveyed.
[b] in percentage of the total number of plantations sampled.
[c] number of specimens/g of roots.

Table 17.2 Initial and final populations, and reproduction factor of nematode species from the Ivory Coast, three months after inoculation on *Coffea canephora* var *Robusta* and *C. arabusta* var *Arabusta*

Nematode species and P_i^a	Robusta		Arabusta	
	P_f^b	R_f	P_f	R_f
Meloidogyne incognita, 20 thousand eggs/plant	440[c]	0.02	560	0.03
Pratylenchus coffeae, 5 thousand nematodes/plant	406	0.08	514	0.10
P. brachyurus, 1.3 thousand nematodes/plant	216	0.16	232	0.18

[a] P_i = Initial population inoculated;
[b] P_f = Final population in the plant;
[c] Values are means of five replicates.

17.1.3 Concluding Remarks

Considering the decrease in coffee prices on the international market, some strategic alternatives have been thought out for the Ivory Coast's coffee industry. These involve stimulating domestic consumption (from 12% to 30% of the national production by 2015), and a program to improve the 'cup quality' of the robusta coffee. For the latter aspect, coffee breeders are considering working towards a reduction of its caffeine level. On the basis of Rabéchault's hypothesis, new coffee hybrids or clones with less caffeine could be more prone to nematode parasitism and damage. Therefore, nematologists should work in association with breeders and agronomists in the assessment of new genotypes.

As mentioned above, apparently plant-parasitic nematodes are not a constraint on coffee production in the Ivory Coast. However, it is advisable that researchers and extensionists conduct surveys of and monitor plantation infestations, allowing for early action should new pathotypes emerge following changes in the agroecosystems. It is also advisable to alert coffee growers, extension agents and other agricultural services about the potential threat represented by nematodes. Such educational campaigns would contribute to an effective enforcement of the legislation regarding the introduction of planting materials into the Ivory Coast.

17.2 Uganda

17.2.1 Brief Outline of the Crop

17.2.1.1 Types of Coffee Grown in Uganda

The main types of coffee grown in Uganda are robusta (*Coffea canephora* Pierre ex A. Froehner) and arabica (*C. arabica* L.). *Coffea liberica* Bull Hiern and *C. abeokutae* Cramer have little commercial value, and they are found in coffee germplasm banks at the Kawanda Agricultural Research Institute and at the Coffee Research Institute in Kituuza, central Uganda. *C. liberica* var *dewevrei* occurs in the wild in the Semliki Valley and in the Zoka forest, near Gulu (Butt et al., 1970).

Two types of robusta coffee have been grown in Uganda, 'Erecta' and a spreading type ('Nganda'). According to Purseglove (1968), a field trial initiated by A.S. Thomas in 1935 showed that the 'Erecta' bushes gave higher yields, which peaked six years after planting, while 'Nganda' did not reach maximum production after 10 years.

17.2.1.2 The Development of the Coffee Industry

Robusta coffee is indigenous to Uganda. Long before coffee was developed as a commercial crop, a ritual meaning was given to this plant among the Baganda people. In the ceremony of 'blood brotherhood', two coffee beans taken from the same berry were moistened with each man's blood, and exchanged to be eaten (Thomas, 1940a). Coffee berries processed in a special way were also offered to gods and spirits, as well as to visitors to chew before a meal.

Arabica coffee var *arabica* was introduced into Uganda via Malawi (Nyasaland) in 1900 (Thomas, 1940b), and it soon attracted notice for its superior agronomic characteristics, in comparison to the indigenous robusta coffee. Seeds and seedlings were then distributed across the country. In the same year, a Catholic missionary introduced a seed stock of *C. arabica* var *Bourbon*, of which seeds harvested from two plants cultivated in Nandere were distributed to other Catholic mission stations and to farmers. Other introductions included var *Maragogipe*, probably from Kew (England) in 1901, and *Blue Mountain* from Guatemala in 1903.

Although commercial coffee production in Uganda started in the early 1920s, it was not until the 1950s that an extensive coffee production program was launched. By the late 1960s, coffee production had risen to 2.5 million 60 kg-bags of clean, export quality coffee (Loudon, 1970), and it reached 3.7 million bags in 1972 (Musoli et al., 2001). However, production declined substantially in the following years due to civil war, poor marketing system, and the low prices paid to growers as a result of the government's monopoly and over control. A recovery was observed in the 1980s and 1990s, with exports reaching 4.2 million bags in the period 1995–1997, of which 90% was robusta coffee (Anonymous, 1997b). This improvement in the coffee industry was associated with the government's actions towards liberalisation of the industry, including the abolition of the state's monopoly over coffee marketing. Consequently, farmers started to receive higher earnings, stimulating the rehabilitation of the coffee fields. A regulatory and development agency for the industry, the Uganda Coffee Development Authority, was established in 1991.

Despite fluctuating world market prices, and the diversification of Uganda's exports, coffee remains the major source of foreign exchange earnings, totaling US$ 204 million for the period 2006/2007, which represents 40% of the national export earnings (Nakkazi, 2007).

Consistent scientific efforts in the coffee industry began in 1956, when an effective robusta coffee breeding program was initiated, leading to the selection of six clonal varieties. In the early 1990s, these cultivars were used in the Coffee Rehabilitation Project, during which nurseries were established in all growing areas for mass propagation of those genotypes to replace old robusta coffee plantations. These nurseries were supplemented in 1993 by a tissue culture facility that was established at the Kawanda Agricultural Research Institute to carry on *in vitro* propagation. In the years 1998/1999, 10.3 million robusta clonal seedlings were produced under the Coffee Nursery Programme.

Due to the importance of coffee in the economy of Uganda, in 1996 the Coffee Research Programme of the National Agricultural Research Organisation at the Kawanda Agricultural Research Institute was upgraded to Coffee Research Centre. Shortly afterwards, it was further upgraded to the fully-fledged Coffee Research Institute. This institute is mandated to conduct research to improve coffee production, besides holding research mandate for cocoa, oil palm, tea and sugarcane.

17.2.1.3 Technological and Ecological Aspects of Coffee Production

Nowadays, nearly all robusta and most arabica coffee is grown by about 500 thousand smallholders on plots of less than 0.25 ha. Coffee is often cultivated lightly shaded, rarely mulched, and often intercropped with other crops, such as bananas.

The coffee-producing areas in Uganda meet the climate requirements for the crop, mainly for altitude and rainfall (Butt et al., 1970). Most of the arabica-producing areas lie between 1,500 and 2,300 masl, where 'leaf rust', caused by *Hemileia vastatrix* Berk et Br., is not a problem. Some areas below 1,500 masl, e.g. the lower slopes of Mount Elgon in eastern Uganda, and some areas in Western Uganda, produce-excellent arabica. The main robusta-producing area is the Lake Victoria crescent, at an altitude between 1,200 and 1,500 masl (Fig. 17.2).

Both regions receive heavy rainfall (1,140–1,520 mm/year), relatively well distributed throughout the year (Jameson and McCallum, 1970), and they present mean annual maximum and minimum temperatures around 28 and 16°C, respectively. Soils do influence the distribution of coffee cultivation in Uganda, but to a minor degree. The volcanic soils of Mount Elgon and some areas of Southwestern Uganda are excellent for growing arabica coffee. Acidic soils do limit coffee growing, unless fertilizers are applied.

17.2.2 Coffee-Parasitic Nematodes

In Uganda, nematodes have not been regarded as economically important parasites of coffee. This situation is not unique to coffee. With the exception of banana and cassava, very little research has been done on nematodes of important crops in Uganda. Certainly, plant nematology has not been given due attention, as attested by the reduced number of nematologists in this country. Just as in many other countries, the overall importance of nematodes in crop production is still not fully appreciated, mainly because these are unseen organisms. Furthermore, the damage caused by nematodes often resembles symptoms of abiotic plant stresses, such as moisture and nutrient deficiency. Therefore, research efforts in Uganda are concentrated on

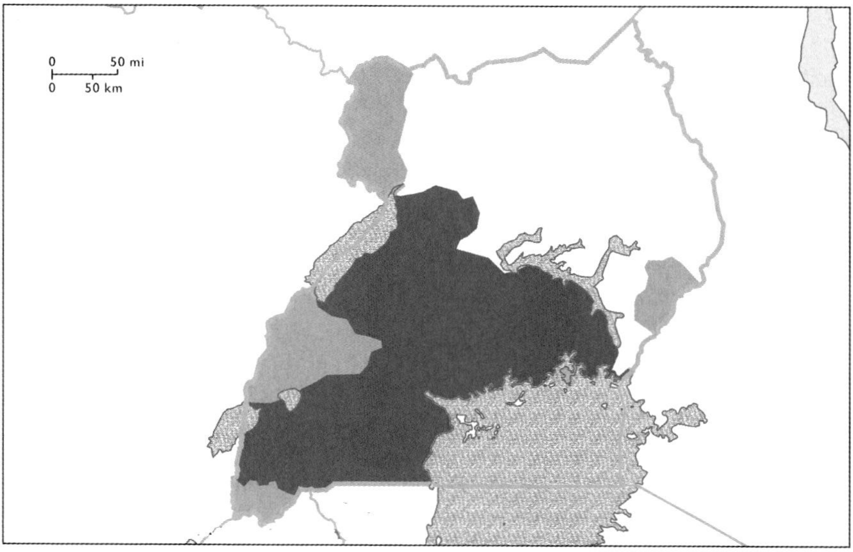

Fig. 17.2 Uganda's robusta and arabica coffees growing regions (*dark and light grey*, respectively). Areas with line pattern represent lakes. Map by UENF/GRC, adapted from the Coffee Farming Systems Development Project Draft Final Report (1988a). COWI Consult, Agriculture and Rural Development Division, with permission

the easily seen 'leaf rust', the 'coffee berry disease' caused by *Colletotrichum kahawae* Waller and Bridge, the 'coffee wilt disease' caused by *Fusarium xylarioides* Steyaert, antesia bugs (*Antesia lineaticollis* Stal. Brit), and the coffee berry borer *Hypothenemus hampei* Ferrari.

Whitehead (1969) reported *Meloidogyne megadora* Whitehead parasitizing coffee in Uganda. Bafokuzara and Bazirake (1993) also found *Meloidogyne* sp., with more conspicuous damage being caused in nurseries. Although their survey also revealed *Pratylenchus* sp. and *Radopholus similis* (Cobb) Thorne, the authors were uncertain whether these nematodes were parasitizing coffee or intercropped bananas.

In a recent survey in 2004/2005 to identify nematodes on economically important crops in Uganda, J. Namaganda isolated *Meloidogyne* sp., *Rotylenchulus reniformis* Linford and Oliveira, *Helicotylenchus dihystera* (Cobb) Sher and *Tylenchus* sp. from robusta coffee roots, while *Aphelenchus* sp., *Trichodorus* sp., *Xiphinema* sp. and *Paralongidorus* sp. were found in the coffee rhizosphere only. No parasitic nematodes were found associated with arabica coffee.

A considerable amount of work has been carried out on the effect of certain nematodes on wilt diseases of various crops. Nonetheless, no investigations have been carried out in Uganda to establish the role of nematodes in 'coffee wilt disease' (Adipala-Ekwamu et al., 2001).

17.2.3 Concluding Remarks

Considering how important coffee is to the economy of Uganda, there is an urgent need for a nationwide survey to identify the nematode species associated with this crop. Also, greenhouse and field experiments are necessary for assessment of the damage caused by the more prevalent nematode species to arabica and robusta coffees. Finally, attention should be paid to investigating whether nematodes, particularly *Meloidogyne* sp., are involved in 'coffee wilt disease'.

References

Adiko A, N'guessan AB (2002) Evolution of the nematofauna of plantain, *Musa* AAB, in Côte d'Ivoire. InfoMusa 10: 26–27

Adipala E, Opio F, Kyetere D et al (2001) The Coffee Wilt Disease: A literature review. In: Coffee Wilt Disease Research and Development in Uganda: Research Progress 1997–2001. National Agricultural Research Organisation, Entebbe.

Anonymous (1988a) Coffee Farming Systems Development Project. Draft Final Report. COW-IConsult, Agriculture and Rural Development Division, Entebbe.

Anonymous (1988b) La Côte d'Ivoire en Chiffres. Edition 1986–1987. Les tout derniers chiffres sur tous les secteurs d'activités. Edition Inter Afrique Presse. Paris.

Anonymous (1990) Annuaire des Statistiques Agricoles. Ministère de l'Agriculture et des Ressources Animales. Direction de la Programmation. Abidjan, Côte d'Ivoire.

Anonymous (1997a) La Côte d'Ivoire en Chiffres. Ministère de l'Economie et des Finances. Côte d'Ivoire. Edition 1996–1997. Dialogue Production. Abidjan.

Anonymous (1997b) Uganda Coffee Development Authority Annual Report, Vol.6: 1996/97, Kampala.

Anonymous (2003) Des semences hybrides de café aussi performantes que les meilleurs clones cultivés. In: Le CNRA en 2002. CNRA, Direction des Systèmes d'Information. Abidjan.

Anonymous (2005) Note de Conjoncture. Bimestriel d'Information sur les marchés du cacao et du café. Campagne 2005/06. Bourse Café Cacao. Abidjan.

Anonymous (2006) Coté D'Ivoire. www.eia.doe.gov/emeu/cabs/cdivoire.html, visited on July 3rd 2007

Bafokuzara ND, Bazirake CB (1993) Plant quarantine in Uganda: structure and functions. Proceedings of the TCP/RAF/2255(T): sub-regional workshop on establishment of a technical cooperation network for plant quarantine in East Africa, Dar-es-salaam.

Butt DJ, Buttlers B, Dance, DW et al (1970) Coffee: Local aspects. In: Jameson JD (ed). Agriculture in Uganda. Oxford University Press, London.

Jacques-Félix H (1954) Généralités sur la physiologie, la biologie, la génétique et l'écologie du caféier. In: Contributions à l'étude du caféier en Côte d'Ivoire. Travaux du Centre de Recherches Agronomiques de Bingerville. Abidjan.

Jameson JD, McCallum D (1970) Climate. In: Jameson JD (ed). Agriculture in Uganda. Oxford University Press, London.

Kéhé M, Gnonhouri GP, Adiko A (1995) Evolution des infestations du symphyle *Hanseniella ivorensis* et du nématode *Pratylenchus brachyurus* sur l'ananas en Côte d'Ivoire. Proceedings II Int Pineapple Symp Acte Hortic: 465–474

Loudon TM (1970) Coffee: Place in the world market. In: Jameson JD (ed). Agriculture in Uganda. Oxford University Press, London.

Luc M, de Guiran G (1960) Les nématodes associés aux plantes de l'Ouest Africain. L'Agronomie Tropicale 15: 434–449

Meiffren M (1957) Les maladies du caféier en Côte d'Ivoire. HAUT-COMMISSARIAT de l'AOF. Centre de Recherches Agronomiques de Bingerville. Abidjan.

Montagnon C, Leroy T, Charmetant P et al (2001) Outcome of two decades of reciprocal recurrent deletion applied to *Coffea canephora* in Côte d'Ivoire: new outstanding hybrids available for growers. Proceedings XIX Colloq Sci Int sur le café: 444–450

Musoli PC, Hakiza GJ, Birikunzira JB et al (2001) Coffee (*Coffea* spp.). In: Mukiibi JK (ed). Agriculture in Uganda. Vol. II: Crops. Fountain Publishers/CTA/NARO, Kampala.

N'guessan KE (2004) Gestion des Filières Café et Cacao en Côte d'Ivoire. Bilan et perspectives. Edition MUSE SARL. Abidjan.

Nakkazi E (2007) Uganda: coffee exports gain as Vietnam supply dwindles. http://allafrica.com/stories/200707030767.html, visited on July 3rd, 2007.

Purseglove JW (1968) Tropical Crops: Dicotyledons. Longman Scientific and Technical, Essex.

Rabéchault H (1954) Sur quelques facteurs de résistance du caféier à la Trachéomycose. In: Contributions à l'étude du caféier en Côte d'Ivoire. Travaux du Centre de Recherches Agronomiques de Bingerville. Abidjan.

Thomas AS (1940a) Robusta Coffee. In: Tothill JD (ed) Agriculture in Uganda. Oxford University Press, London.

Thomas AS (1940b) Arabica Coffee. In: Tothill JD (ed) Agriculture in Uganda. Oxford University Press, London.

Whitehead AG (1969) The distribution of root-knot nematodes, *Meloidogyne* spp., in tropical Africa. Nematologica 15: 315–333

Color Plates

A

B

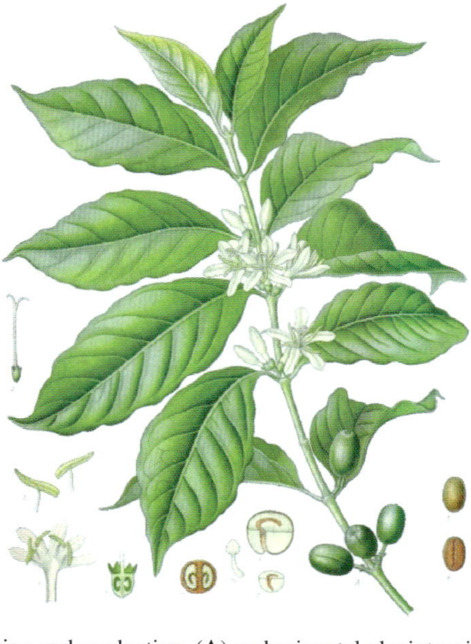

Plate 1 Coffee blooming and production. (**A**) on horizontal plagiotropic branches (Photo by H. Vieira). (**B**) anatomic details (from Köhler, 1887) (see Fig. 1.3, p. 6)

A

B

Plate 2 Coffee blooming. (**A**) inflorescence on the axiles of a plagiotropic branch (Photo by F. Partelli, with permission). (**B**) synchronous blooming (Photo by H. Vieira) (see Fig. 1.4, p. 7)

Plate 3 *Coffea* species. (**A, B**) *C. arabica*. (**C**) *C. dewevrei*. (**D**) *C. stenophylla* (Photos by H. Vieira) (see Fig. 1.5, p. 8)

Plate 4 Coffee seedling production and cultivation. (**A**) nursery. (**B**) seedlings vegetatively produced from orthotropic branches. (**C, D**) grafting of seedlings. (**E**) grafted seedling. (**F**) full sun cultivation (Photos by H. Vieira) (see Fig. 1.6, p. 10)

Plate 5 Coffee cultivation. (**A**) full sun plantation intercropped with beans (Photo by F. Partelli). (**B**) shaded plantation (Photo by K. Sreedharan, with permission) (see Fig. 1.7, p. 12)

A

B

Plate 6 Coffee cultivation and harvest. (**A**) plantation being irrigated (Photo by D. Barbosa, with permission). (**B**) harvesting of robusta coffee (Photo by K. Sreedharan, with permission) (see Fig. 1.8, p. 13)

Plate 7 Coffee harvest. (**A**) strip harvesting (Photo by F. Partelli, with permission). (**B, C**) harvested coffee in basket and fabric strip, respectively (from Anonymous, 1985, with permission) (see Fig. 1.9, p. 15)

Plate 8 Coffee harvesting and processing. (**A, B**) mechanical harvesting (from Anonymous, 1985, with permission). (**C**) coffee berries being sun dried. (**D, E, F**) damaged, high grade and roasted coffee beans, respectively (Photos by H. Vieira) (see Fig. 1.10, p. 16)

Plate 9 Root system of a *C. arabica* 'Caturra' seedling susceptible to *Meloidogyne exigua*, showing numerous galls of different sizes (Photo by F. Anthony) (see Fig. 9.1, p. 172)

Plate 10 Root system of a *C. arabica* 'Caturra' seedling susceptible to *Meloidogyne paranaensis*, showing symptoms of 'corchosis' on the main root (Photo by F. Anthony) (see Fig. 9.2, p. 172)

Plate 11 Coffee plantation affected by *Meloidogyne arabicida*, with several dead trees in the foreground (Photo by F. Anthony) (see Fig. 9.3, p. 173)

Plate 12 Nebulization room for extraction of infectious *Meloidogyne* sp. juveniles. The infected roots are cut in 5 mm long segments and placed on a sieve nested onto a funnel, to facilitate nematode descent to the bottom of the white flasks (Photo by P. Topard, with permission) (see Fig. 9.4, p. 175)

Plate 13 Arabica coffee roots heavily damaged by *Meloidogyne coffeicola*, showing typical disorganization and detachment of the cortical tissue. (Photo by Luiz C.C.B. Ferraz) (see Fig. 12.2, p. 229)

Plate 14 Arabica coffee roots parasitized by *Meloidogyne coffeicola*, showing small rounded cavities in the cortical tissue from which nematode adult females have been removed. (Photo by Luiz C.C.B. Ferraz) (see Fig. 12.3, p. 229)

Plate 15 Arabica coffee plants severely affected by *Meloidogyne coffeicola*, showing chlorosis and defoliation. (Photo by Luiz C.C.B. Ferraz) (see Fig. 12.4, p. 230)

Plate 16 Young arabica coffee plants heavily affected by *Meloidogyne incognita*, showing chlorosis and partial defoliation. (Photo by Luiz C.C.B. Ferraz) (see Fig. 12.5, p. 231)

Plate 17 Leaves collected from a *Meloidogyne incognita*-affected arabica coffee plant showing typical symptoms of nutritional deficiency. (Photo by Luiz C.C.B. Ferraz) (see Fig. 12.6, p. 231)

Plate 18 Arabica coffee replanting in a sandy soil heavily infested by *Meloidogyne incognita* in the State of São Paulo, Brazil. (Photo by Luiz C.C.B. Ferraz) (see Fig. 12.7, p. 232)

Plate 19 Arabica coffee roots heavily parasitized by *Meloidogyne incognita* showing disorganized, detached cortical tissue and atypical swellings. (Photo by Luiz C.C.B. Ferraz) (see Fig. 12.8, p. 233)

Plate 20 Arabica coffee plants affected by *Pratylenchus brachyurus*. This field had been cultivated with pastures for many years before being cultivated with coffee. (Photo by Luiz C.C.B. Ferraz) (see Fig. 12.9, p. 234)

Plate 21 Plants of arabica coffee 'Mundo Novo' grown in a *M. incognita*-infested field. Dead, self-rooted, nematode-susceptible plants are in the foreground. Healthy plants grafted onto nematode-resistant *C. canephora* 'Apoatã' are in the background. (Photo by Luiz C.C.B. Ferraz) (see Fig. 12.10, p. 239)

Plate 22 Nematode-antagonistic *Crotalaria* sp. intercropped with coffee to reduce the soil nematode population. (Photo by Luiz C.C.B. Ferraz) (see Fig. 12.11, p. 240)

Plate 23 'Corky-root' symptom on *Coffea arabica* parasitized by *Meloidogyne paranaensis* in Guatemala (Photo by L. Villain) (see Fig. 14.1, p. 265)

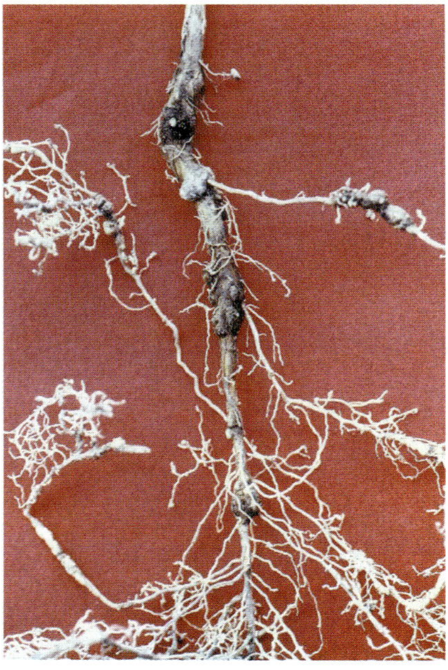

Plate 24 Root symptoms on *Coffea arabica* parasitized by *Meloidogyne izalcoensis* in El Salvador (Photo by A. Hernández) (see Fig. 14.3, p. 266)

Plate 25 *Coffea arabica* plants parasitized by *Pratylenchus* sp. in southwest Guatemala. Own-rooted (foreground) and grafted onto a nematode-resistant *Coffea canephora* Pierre ex Froehner rootstock (background) (Photo by L. Villain) (see Fig. 14.5, p. 268)

Plate 26 Seedlings of *Coffea arabica* grafted onto *C. canephora* Pierre ex Froehner in Guatemala (Photo by L. Villain) (see Fig. 14.6, p. 270)

A

B

Plate 27 Contrasting aspect of robusta coffee clones in a *Pratylenchus coffeae*-infested field. **A**: clone 'BP 308', resistant to the nematode. **B**: clone 'BP 409', susceptible (Photo by S. Wiryadiputra) (see Fig. 15.2, p. 281)

Plate 28 Uprooted (foreground) and *P. coffeae*-parasitized coffee plants (background) in Krong Ana, Daklak province, Vietnam (Photo by Loam K. Tran) (see Fig. 15.4, p. 287)

Plate 29 Mature robusta coffee tree presenting the *P. coffeae*-associated decline (Photo by Loam K. Tran) (see Fig. 15.5, p. 287)

Plate 30 Young robusta plant planted into a *P. coffeae*-infested area (Photo by Loam K. Tran) (see Fig. 15.6, p. 288)

Plate 31 *P. coffeae*-parasitized robusta coffee plant presenting rotten tap root and abundant adventitous roots at the collar region (Photo by Loam K. Tran) (see Fig. 15.7, p. 288)

Index